Mathematical Experiments on the Computer

This is a volume in
PURE AND APPLIED MATHEMATICS

A Series of Monographs and Textbooks

Editors: Samuel Eilenberg and Hyman Bass

A list of recent titles in this series appears at the end of this volume.

Mathematical Experiments on the Computer

ULF GRENANDER

DIVISION OF APPLIED MATHEMATICS
BROWN UNIVERSITY
PROVIDENCE, RHODE ISLAND

1982

ACADEMIC PRESS, INC.
Harcourt Brace Jovanovich, Publishers
Orlando   San Diego   New York
Austin   London   Montreal   Sydney
Tokyo   Toronto

ACADEMIC PRESS, INC.
Orlando, Florida 32887

United Kingdom Edition published by
ACADEMIC PRESS, INC. (LONDON) LTD.
24/28 Oval Road, London NW1 7DX

Library of Congress Cataloging in Publication Data

Grenander, Ulf.
   Mathematical experiments on the computer.

   (Pure and applied mathematics ;   )
   Includes index.
   1. Mathematics--Data processing. I. Title.
II. Series: Pure and applied mathematics (Academic
Press) ;
QA3.P8 [QA76.95]   510s [510'.724]   82-13718
ISBN 0-12-301750-5

Der stod kvinder bag ham
Emma-Stina, Angela, Charlotte

Contents

Preface

The working mathematician, pure or applied, uses a multitude of tools such as spectral decompositions, isomorphic representations, and functional transforms, to mention only a few. In this book we shall argue in favor of a less traditional but, we believe, highly useful tool, namely, the computer.

What we have in mind is using *the computer as the mathematician's laboratory,* in which he can perform experiments in order to augment his intuitive understanding of a problem, or to search for conjectures, or to produce counterexamples, or to suggest a strategy for proving a conjecture.

Hence the theme of this book is *experimental mathematics* on the computer. It will not deal with production computing, say, the solution of a large partial differential equation, the inversion of enormous matrices, or the numerical handling of combinatorial problems. Nor will it deal with topics like proving theorems on the computer or related questions belonging to the area of artificial intelligence.

The main thrust of the book is how to carry out mathematical experiments with the aims mentioned, how to do it quickly and easily, using the computer only as a tool, and with the emphasis on mathematics rather than on programming. For the purpose, interactive computing is almost indispensable, and we also claim that the programming language APL has no existing rival that can seriously compete with it for use in mathematical experiments on the computer.

We shall not argue the philosophical merits of mathematical experiments; we are more concerned with convincing mathematicians with different research orientations that it may be worth their while to try some experiments of their own on the computer. Obviously there are many areas in mathematics where experiments would be of little or no help. In others, however, experiments may be really useful, sometimes in situations where this does not appear likely at first.

Therefore the message is: try it! The programming side is not as formidable, or as boring, as some mathematicians tend to believe. The cost of experimentation is moderate or small, and will tend to decrease as computer technology continues its rapid development.

In Part I we shall examine a number of concrete cases where experiments helped to solve a research problem, or, in some cases, where they failed. We have selected the cases from different areas to show that the experimental technique is essentially area independent. The problems are therefore unrelated to each other. What they have in common, and what they are intended to illustrate, are how experiments can be designed, executed, and interpreted; also what are the scope and the limitations of computer-based mathematical experiments.

We hope that the reader can benefit from studying all the cases, including those from areas with which he may not be familiar. In the latter case it may be difficult to appreciate the analytical formulations, but the reader should at least try to follow the organization of the experiment, watching out for difficulties in the design and sometimes in the interpretation of experimental results. Although we claim that the effort that is needed for programming is usually small, the same is not true of the design and interpretation stages of the experiment. These require experience and are best learned by actually trying some small experiment oneself.

The programming language APL is presented in Part II. The past few years have witnessed a veritable APL epidemic, and the language is becoming widely available. The presentation is written from the point of view of the prospective mathematical experimenter. For this reason we have emphasized how to deal with those mathematical structures and operations that, in our experience, are likely to appear in experiments. On the other hand, we have given less attention to questions of APL programming that are less relevant for this particular use.

After discussing APL from the point of view of mathematical experiments in general, we present still another special experiment in Part III. This is of more ambitious scope and also requires a good deal more computing effort than the earlier ones. This is preceded by a general discussion of the techniques of mathematical experiments.

Part IV is of a different character. It contains software specially constructed for experiments, with the limitations that this implies. Since any two experiments are never quite the same, it is necessary that programs written earlier be easy to modify if we want to use them in a new experiment. We have tried to achieve this, often at the expense of computing efficiency, but with user efficiency in mind.

We do not intend Part IV to be read through systematically; that would be tough going! It is presented because of its utilitarian value, and a potential reader is encouraged to treat the programs only as suggestions that ought to be modified and improved in order to better fit the demands of a particular experiment. Although the programs have all been debugged and tested, they should be used with some caution. We do not claim that the program quality is on the level of industrially developed software, but it should be adequate for most situations.

Many dialects of APL are in use, but the differences between them are often negligible, so that this fact should not substantially restrict the usefulness of the software in Part IV. On the other hand, users of microcomputers with APL may run into difficulties because of limitations in speed, storage, and language facilities. When personal computers become more powerful, and perhaps cheaper, one will have to look more carefully into this: can experiments be of use for the mathematician who is limited to a personal computer? At this time we lack the experience needed for answering this question.

The material in the book has been used repeatedly in courses, both on the graduate and the undergraduate level. For the latter purpose the instructor should supplement the discussion of the cases in Part I that he selects for classroom discussion with background material to motivate the students. It may also be useful to include some problems from the instructor's own area of interest.

It is such a natural idea to use the computer as the mathematician's laboratory that many colleagues must be using it in this spirit. We have not been able to find much evidence of such activities in the literature (a few references are given in the bibliography), but they probably go on in many universities and industries. If so, the author would appreciate hearing about them in order to exchange experiences.

Acknowledgments

Many of my co-workers have contributed to the preparation of this volume. The technique of mathematical experiments described in the text was developed during the 1970s at Brown University in the Division of Applied Mathematics. I would like to thank its Chairman, Walter Freiberger, for his help and cooperation. It is very much due to his energy and foresight that we had access to efficient and well-run time sharing earlier than at most universities.

The software described in Part IV has been developed in the group, Mathematical Software for Experiments. Donald E. McClure has been one of the most active members in the group, contributing original ideas and expertise, both mathematical and computational, and stimulating the group with his enthusiasm. During the spring semester 1980, while I was away on a sabbatical, he led the group and further contributed to its success.

Another active member of the group was F. E. Bisshopp, who contributed much in discussions on the more sophisticated aspects of APL.

Many students have participated in the group, in particular during the years 1979–1981. They are B. Bennet, B. Bukiet, R. Cohen, P. DuPuis, A. Erdal, T. Ferguson, P. Forchioni, C. Huckaba, P. Kretzmer, A. Latto, D. McArthur, D. Peters, G. Podorowsky, D. Richman, J. Ross, E. Silberg, and R. Saltzman. Their enthusiastic cooperation has contributed much to our work in mathematical experiments.

Several colleagues have made helpful comments, and I would like to mention especially Y.-S. Chow, K. W. Smillie, and J. W. Tukey. The remarks made by D. G. Kendall were very encouraging and reinforced my intention to complete the manuscript.

I am grateful to E. Fonseca for her careful typing of the manuscript and to E. Addison for preparing the figures.

The proofreading has been demanding, especially for the APL code in Part IV. I would like to thank Y.-S. Chow, E. Fonseca, and W. Freiberger for their assistance with this. I hope that all errors introduced when transferring the code have been found and corrected.

The work reported in this book has been partially supported by the National Science Foundation, the Air Force Office of Scientific Research, and the Office of Naval Research. I thank Kent Curtis, Marvin Denicoff, Bob Grafton, and I. Shimi of these agencies for their sympathetic understanding.

I also thank Academic Press and its editors for their help in publishing this book.

PART I

EXPERIMENTS ON THE COMPUTER

CHAPTER 1

Definition of the Subject

Scientific reasoning has traditionally been classified as *inductive* or *deductive,* depending upon whether it involves arguing from special cases—say observations and measurements—or uses syllogistic procedures. In the natural sciences a mixture of both is used. Mathematics employs only deductive methods.

It has been argued by C. S. Peirce, however, that one should add a third mental activity to these two. Naming it *abduction,* he meant the conscious or subconscious act by which the scientist forms hypotheses, conjectures, and guesses.

How do mathematicians form conjectures? Poincaré, Hadamard, Pólya, and several others have written on this theme and tried to analyze their own thought processes, their abductions if you wish. In spite of this, little is known about what actually goes on in such processes.

Many, probably most, mathematicians would point to examples as a most powerful means by which abduction is carried out. One examines one special case after another, tries to avoid trivial, or (in some sense) singular cases that are thought to be atypical. In this way one hopes to strengthen the intuitive insight and arrive at conjectures that seem plausible and worthy of further explanation.

Let us try to imagine how the geometers in ancient Greece could have guessed some of their theorems in plane geometry, later to be collected by Euclid.

Proposition 5 in Book I—the *Pons Asinorum* feared by school-children—says that the angles at the base of isoceles triangles are equal. A simple drawing in the sand, repeated with different shapes, could have suggested the hypotheses and emphasized the symmetry of the problem.

Proposition 20—essentially the triangle inequality—could perhaps also have been guessed after looking at a few diagrams. It could have been supported by constructing the sum of the lengths of two sides using compasses and comparing it with the length of the third side.

Another example is related to the fact that the three bisectors in a triangle intersect in a single point. It could have been conjectured by actually drawing the bisectors in a few cases and observing that the lines seem to intersect, if not physically in a single point, at least close enough.

Of course, they did not accept mere induction by special cases as a proof; the essence of Hellenic genius was just their insight that *a proof was needed.*

Whether the Greek geometers actually proceeded like this is immaterial in the present context—we are not writing history. Our point is that this way of using small experiments (diagrams in the sand) for making guesses makes good sense.

This case has been better articulated by Euler, from whom we quote what could be chosen as the articles of faith for this project. He says, in "Specimen de usu observationum in mathesi pura":

> It will seem not a little paradoxical to ascribe a great importance to observations even in that part of the mathematical sciences which is usually called Pure Mathematics, since the current opinion is that observations are restricted to physical objects that make impression on the senses. As we must refer the numbers to the pure intellect alone, we can hardly understand how observations and quasi-experiments can be of use in investigating the nature of the numbers. Yet, in fact, as I shall show here with very good reasons, the properties of the numbers known today have been mostly discovered by observation, and discovered long before their truth has been confirmed by rigid demonstrations. There are even many properties of the numbers with which we are well acquainted, but which we are not yet able to prove; only observations have led us to their knowledge. Hence we see that in the theory of numbers, which is still very imperfect, we can place our highest hopes in observations; they will lead us continually to new properties which we shall endeavor to prove afterwards. The kind of knowledge which is supported only by observations and is not yet proved must be carefully distinguished from the truth; it is gained by induction, as we usually say. Yet we have seen cases in which mere induction led to error. Therefore, we should take great care not to accept as true such properties of the numbers which we have discovered by observation and which are supported by induction alone. Indeed, we should use such a discovery as an opportunity to investigate more exactly the properties discovered and to prove or disprove them; in both cases we may learn something useful.

As an example of this experimental approach—affectionately known as "eulerizing"—one can mention Euler's study of the sum of the divisors of a natural number; see Pólya (1954, Chap. VI). In it he starts by examining the first 99 numbers and calculates the sum of the divisors. He then experiments with the results, looking for regularities, and gradually strengthens the belief in a conjecture about the sums.

Quite apart from its mathematical content, Euler's study is of special interest in the present context since it describes the "abductive" process more openly than is usual in mathematical publications. The experimental aspect is clearly put forward with charm and enthusiasm.

Euler was, of course, far from alone in using inductive methods for generating hypothesis; in those days it may have been the rule rather than the exception. One of the most famous instances is Gauss's discovery of his "theorema aureum"—the law of quadratic reciprocity. Gauss was always computing, often to an incredible number of decimal places. We quote from Klein (1979, pp. 29–30):

> Here again was the untiring calculator who blazed the way into the unknown. Gauss set up huge tables: of prime numbers, of quadratic residues and non-residues, and of the fractions $1/p$ for $p = 1$ to $p = 1000$ with their decimal expansions carried out to a complete period, and therefore sometimes to several hundred places! With this last table Gauss tried to determine the dependence of the period on the denominator p. What researcher of today would be likely to enter upon this strange path in search of a new theorem?

Who, indeed! But for Gauss this path led to his goal.

For such experiments were needed only paper and pencil plus the unheard of energy: Gauss himself maintained that he differed from other men only in his diligence! We propose that computer technology—both as regards hardware and software—should be used in the same spirit, as a way of implementing mathematical experiments.

The hypotheses arrived at are likely to be less breathtaking—at least until another Gauss comes along and decides to play with the computer—but we shall show that it is possible to use such experiments as a routine tool, and sometimes with unexpected results.

In 1968 John Tukey and Frederick Mosteller stated about the influence of the computer (see Tukey and Mosteller, 1968):

> Ideally we should use the computer the way one uses paper and pencil: in short spurts, each use teaching us—numerically, algebraically, or graphically—a bit of what the next use should be. As we develop the easy back-and-forth interaction between man and computer today being promised by time-sharing systems and remote consoles, we shall move much closer to this ideal.

It has been technically feasible to operate like this for a long time; the crucial advance was not that computers became larger and faster but that access to interactive computing became widely available and that software became manageable in a drastically improved way. Mathematical experiments need not involve massive number crunching or data management, the requirements are different from other uses of the computer in several respects. Let us mention some of them.

P1.   Since one of the aims is to strengthen the experimenter's intuition, it is psychologically important that he be *close to the computer* with as few obstacles of communication as possible in between. Here "close" does not refer to physical distance but indicates ease of communication in both directions.

P2.   This requires *interactive computing*—fast batch processing is a poor second best.

P3.   Mathematical *analysis should be the main thing;* programming should be only a tool, distracting as little as possible from the analysis.

P4.   This requires a *programming language* directly related to the basic operations in mathematics.

P5.   *The results must be made intuitive,* which usually means graphical display.

These points express positive needs. On the negative side we can dispense with certain requirements for normal computing:

N1.   The CPU-time and memory requirements will often be small or moderate, so that we need not worry too much about *computing efficiency.* The user's time will be more valuable than CPU time.

N2.   Since programs will seldom be used for production computing, the *code need not be very streamlined.*

N3.   Graphical output will be used only as an intuitive guide: *high-quality graphics will usually not be required.*

N4.   The main idea is to search for conjectures, *not to prove theorems*. Machine proofs of theorems are certainly of great interest, and a few dramatic breakthroughs have occurred recently, but this does not concern us here. Therefore we shall usually not need large symbol manipulation systems.

These statements will be motivated and discussed in greater detail in Chapter 3 where several case studies are presented.

CHAPTER 2

History of the Project

The computer has been used for experimental mathematics since its advent in the 1950s. Number theory was probably the first area where this was done; see Lehmer (1954, 1968, 1974) for further references. This was directly in line with Euler's statement quoted above.

Other areas, especially algebra and, needless to say, numerical analysis, have also felt the impact of the computer, as will be discussed in Chapter 4. Most of these attempts should not be classified as experiments, however, since they have the character of production computing and therefore do not belong here.

An explicitly experimental activity was started at Brown University in 1967. Called the Computational Probability Project, it was carried out jointly with the IBM Cambridge Scientific Center and was later supported by the National Science Foundation (grant GJ-710). It was extended to cover statistics and, more recently, to certain areas of linear algebra, ordinary differential equations, geometry, and pattern theory.

The initial difficulties were considerable. At that time we had access only to a locally developed, interactive language BRUIN, similar to BASIC, in addition of course, to the usual batch processing systems. The interactive system was fairly fragile, with short average intercrash time, the debugging facilities were poor, and no interactive graphics output was available. Progress was slow and painful. If we had had to continue to be limited by this technology, the project would probably not have been carried to a successful end.

A quantum jump was experienced in 1968 when we were allowed to use APL as a programming language. This was done first via a telephone hookup to a dedicated 360/50 computer located at the Thomas J. Watson Research Center in Yorktown Heights, New York. Soon afterward we had APL implemented on the Brown University computer, then a 360/50, later a 360/67 and more recent machines.

It became obvious that APL was the ideal programming language for our purpose. Even today there exists no serious competitor well suited to our demands.

APL has several fundamental features that make it stand out when compared to other languages.

(a)   It is based on concepts that are natural for mathematics, with variables that can be Boolean, scalars, vectors, matrices, and higher-order arrays. Sets, functions, and operators are handled elegantly and concisely and correspond closely to mathematical usage.

(b)   The syntax is fairly close to traditional mathematical notation, although the symbols are often different.

(c)   The current implementations provide attractive debugging and editing facilities.

A second drastic improvement of our computing environment occurred when we acquired hardware for interactive computer graphics. A TSP plotter was connected via a control unit by direct line to the central computer and also to a small Tektronix scope with a bistable memory CRT.

This does not allow any fancy plotting; the TSP is too limited mechanically and the channel capacity of the line was quite low. In spite of this we experienced this somewhat primitive graphics capability as a breakthrough that enabled us to get geometric, often highly informative output.

A third step forward, less important than the other two, was when VS/APL became available. The "execute," "scan," and "format" functions add a great deal to the power of the language. They are not indispensable, however; they are more a convenience than a necessity.

Some programming was done in other languages, mainly in FORTRAN; for example, plotting on the CALCOMP. Another instance was when the IBM 2250 scope was used for displaying the fine details of sample functions from stochastic processes or for geometry. Also, some of the programs requiring much number crunching were written in FORTRAN.

During this period hundreds of programs were written and executed. This often led to a one-shot program: one that was used for a particular problem and then discarded or possibly saved in a personal work space. No effort was made to collect the programs systematically and make them generally available to our computing community.

In 1978 we began to save programs in order to pinpoint the need for new ones. In these cases new ones were found elsewhere and modified to our need, or completely new ones were developed. The main idea was that the resulting library should be uniformly written and so well documented that a user should be able to tailor the program as he saw fit. The external documentation should specify carefully the underlying mathematical conditions as well as the algorithms used. Accompanying examples should make this easier and references to the literature should be given.

The user should be encouraged to treat the programs only as a first attempt; he should go into the code himself, improve it, add features that he needs, and in general treat the given programs with little respect. In this way the library differs from most libraries, which are usually intended for production computing, not for an experimental activity.

This work is still in progress and will continue over the next several years.

CHAPTER 3

Case Studies

The only way of realizing the full power of the computer for experimental mathematics is to try it oneself. The next best thing is to study a few cases where this approach has been used, noticing both its potential and its limitations.

For this reason we present a list of case studies in this section. Each experiment is described briefly in terms of the area of research in which it originated, but a reader who happens not to be familiar with the particular area should still be able to understand the computational study that follows. More details are available in the references given at the end.

EXPERIMENT 1: From Statistics

Our first case is from statistics. Let x_t be a real-valued stationary process with finite variance and covariances

$$r_h = \text{Cov}(x_t, x_{t+h}) = E[(x_t - m)(x_{t+h} - m)] \tag{1}$$

where m is the mean. Of course, in the more sophisticated models the mean value is a more complicated linear combination of regression vectors, so that we have several unknown coefficients to estimate. For didactic reasons we have chosen the simplest case, where common sense tells us what is the best estimate. Common sense is sometimes bad sense; in the present case we shall see when and how this is so. The spectrum of the process will be assumed to be absolutely continuous with a spectral density f so that

$$r_h = \int_{-\pi}^{\pi} e^{ih\lambda} f(\lambda) \, d\lambda, \qquad f \in L_1[(-\pi, \pi)] \tag{2}$$

We observe x_t for $t = 1, 2, \ldots, n$ and want to find the best linear unbiased estimate, the BLUE m^*. This means that among all linear com-

10

binations $\hat{m}$ that are unbiased

$$\hat{m} = \sum_{t=1}^{n} c_t x_t, \qquad \sum_{t=1}^{n} c_t = 1 \tag{3}$$

we want to find the one solving the minimum problem

$$\text{Var}(\hat{m}) = E[(\hat{m} - m)^2] = \sum_{s,t=1}^{n} c_s c_t r_{s-t} = \text{minimum} \tag{4}$$

It is well known that the solution exists, is unique, and is given in vector form by

$$
\begin{aligned}
\hat{m} &= c^T x \\
x &= \text{col}[x_1, x_2, \ldots, x_n] \\
e &= \text{col}[1, 1, \ldots, 1] \\
c &= (e^T R^{-1} e)^{-1} R^{-1} e
\end{aligned}
\tag{5}
$$

where R is the $n \times n$ covariance matrix of the covariance in (1) for $t = 1, 2, \ldots, n$. The resulting minimum value in (4) is

$$\min \text{Var}(\hat{m}) = (e^T R^{-1} e)^{-1} \tag{6}$$

This optimal estimate $\hat{m}$ is of little practical interest, both because R^{-1} is difficult to compute for large n and because R is not always known to the observer. Therefore some mathematicians have tried to replace $\hat{m}$ by the common sense estimate

$$\bar{x} = \frac{1}{n} \sum_{t=1}^{n} x_t \tag{7}$$

the straight average. It is obvious that $\bar{x}$ is a good estimate, but how should one make this innocuous statement precise?

We do it in terms of asymptotic efficiency. (Finite efficiency simply does not hold!) It has been known for a long time that $\bar{x}$ is asymptotically efficient in the sense that

$$\lim_{n \to \infty} \frac{\text{Var}(\bar{x})}{\text{Var}(\hat{m})} = 1 \tag{8}$$

if (a) f is sufficiently well behaved, for example, continuous, and (b) $f(0) > 0$. As a matter of fact, condition (a) does not seem to play any important role and can be weakened a good deal. On the other hand, it was not clear whether the same holds for (b).

To get some insight into this the following computational experiment was carried out. A FORTRAN program was written that calculated both VAR($\hat{m}$) and Var($\bar{x}$) for given spectral density f. As a by-product the vector c given in (5) was obtained.

Another FORTRAN program plotted the spectral density as a function of the frequency and used the output of the first one to plot the vector $c = (c_t, t = 1, 2, \ldots n)$ as a function of the subscript t. This was done off-line on a CALCOMP plotter.

Once the software was written it was executed for a large number of fs and ns, but we can reproduce only a few here. For the spectral density shown in Fig. 1 [note that the coordinate axes shown do not intersect at (0, 0)] the ratio in (8), the relative efficiency, is seen in Fig. 2. It seems to converge fast to the value 1 in accordance with theory; note that $f(0)$ is not zero. The vector c is plotted in Fig. 3.

However, the spectral density in Fig. 4 leads to a completely different behavior. The relative efficiency, given in Fig. 5, is now a decreasing function of the sample size. It seems to converge to zero, or at least to a very small value. The vector c for the optimal estimate is plotted in Fig. 6 for $n = 10$ and in Fig. 7 for $n = 50$. There is a clear tendency to converge toward some vaguely parabolic shape.

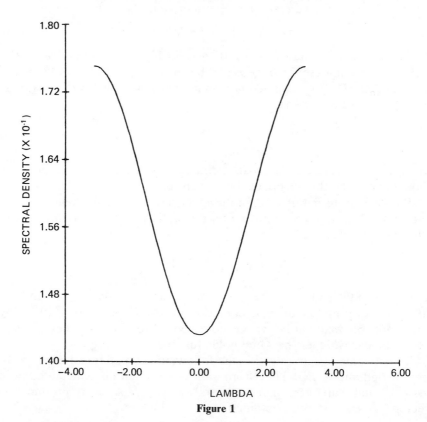

LAMBDA

Figure 1

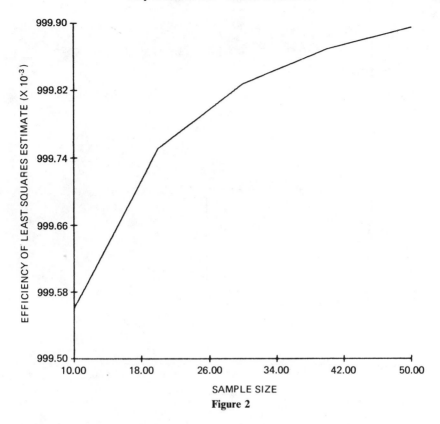

SAMPLE SIZE

Figure 2

   Graphical output of this form was produced for a variety of spectral densities vanishing for $\lambda = 0$, with the same qualitative results.

   Using graphs on log–log paper, it appeared as if $\mathrm{Var}(\hat{m})$ behaved as const $\times$ n^{-3} asymptotically instead of as const $\times$ n^{-1} in the standard case. This evidence was very shaky, to say the least.

   Based on these experiments, it was conjectured that the asymptotic efficiency for $f(0) = 0$ is zero. The form of the optimal c-vectors in these plots led to the conjecture that an optimal estimate [for the case $f(0) = 0$] could be obtained by letting c_t be a second-order parabola as a function of t.

   Both of these conjectures were verified analytically. The second one was expressed as follows: If f is continuous, the limit in (9) exists, and

$$\lim_{\lambda \to 0} \frac{f(\lambda)}{\lambda^2} > 0 \qquad (9)$$

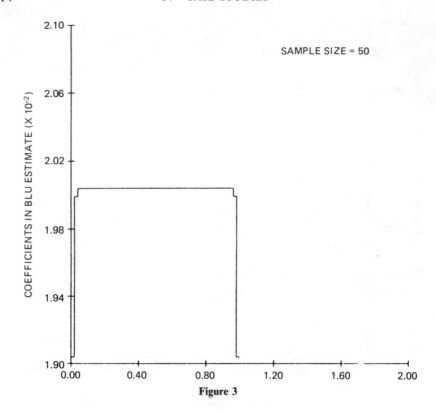

Figure 3

then the estimate with the coefficients

$$c_t = \frac{6n}{n^2 - 1} \frac{t}{n} \left(1 - \frac{t}{n} \right) \tag{10}$$

is asymptotically efficient. Several extensions of this result were obtained later, but they need not concern us in this context. The variance did indeed behave as const $\times$ n^{-3} asymptotically, a fact that played a major role in the analytical treatment of efficiency.

It should be remarked that the role of the computer in this study was not to prove a theorem, only to suggest hypotheses to us. These hypotheses were later studied by standard analytical methods.

We learned several things from this experiment, one of the earliest in the Computational Probability Project. The programming for computing the functions mentioned was easy, but still took a good deal of time to do; debugging and running a program in batch mode is not very efficient in terms of the user's time. It was more time consuming to program the CALCOMP and, more importantly, it was psychologically unattractive to

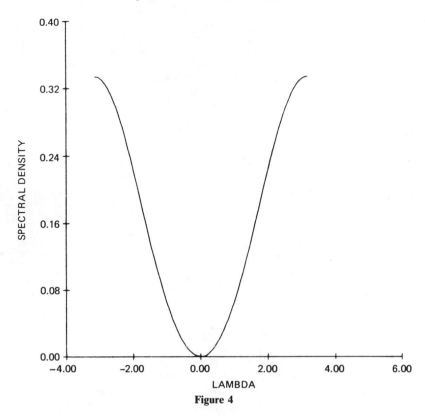

Figure 4

run the plotter off-line. It would have been all right if this had to be done once only, or a few times. As the experiment proceeded, however, and repeated runs were required, it slowed up the interaction between computation and analysis. Today we would do this by other means, to be discussed later.

Another thing we learned was that, since the cost of CPU time for this sort of experiment is low, we could afford to run experiments that were not motivated by the original problem. For example, we computed the relative efficiency of the "parabolic estimate" in (10) for the nonpathological case $f(0) > 0$, and found numerically, to our surprise, that the relative efficiency in this case was not tending to zero as the sample size increased. On the contrary, it stayed fairly close to 1. There must be an analytical explanation of this startling behavior. And indeed there was. A theorem was proved that the relative efficiency had the limit $\frac{5}{6}$, quite close to 1!

A technical problem, typical for experiments of this type, deserves a brief discussion. How should one pick the class F of spectral densities f to

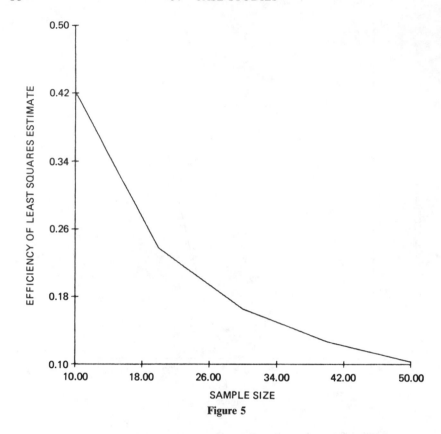

Figure 5

be used computationally? Of course, one could choose some wide class F of continuous nonnegative functions and then compute the covariances r_k by numerical quadrature of the Fourier integral.

But this is bad experimental strategy. We would then have had to keep track of the numerical error in the quadrature and find out how this error propagates through the experiment. This is an elementary exercise in numerical analysis, but, and this is the point, it would take time and effort away from what should be the main task: *to learn about the mathematical problem at hand.*

To avoid such inessential technicalities one could pick, for example, the family F as consisting of functions of the form

$$f(\lambda) = |z - 1|^{2\nu} |Q(z)|^2, \quad z = e^{i\lambda}$$

where ν is a nonnegative integer and Q a polynomial. It is clear that if we allow a fairly high order for Q the resulting family F should be adequate.

We would then get the r_k with little computing and no error from the

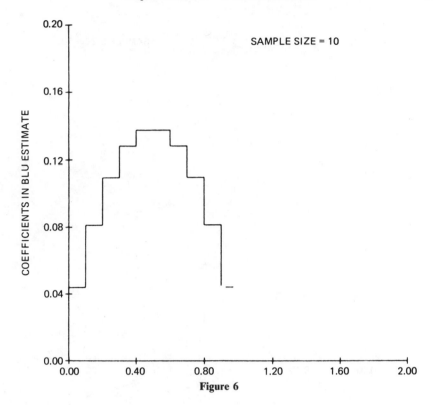

Figure 6

coefficients in

$$f(\lambda) = \sum_{\nu=-r}^{r} A_{\nu} z^{\nu}$$

To get the coefficients A_{ν} amounts to a single convolution and we could do this, for example, by the APL program CONVOLVE discussed in Section 12.1.

This was not the way we actually did it; our way was more arduous, but, as always, it is easier to give advice than to follow it!

Was the computer indispensable for this work? Definitely not. Sooner or later someone would probably have been led to the same conjecture even without the computational results, although *the $\frac{5}{6}$-theorem* would have been difficult to guess. The point is that the experiments not only speeded up the abduction process, but also had an important psychological effect in that they reinforced our belief in the conjectures. We shall come back to this aspect, the psychological one, repeatedly in the following.

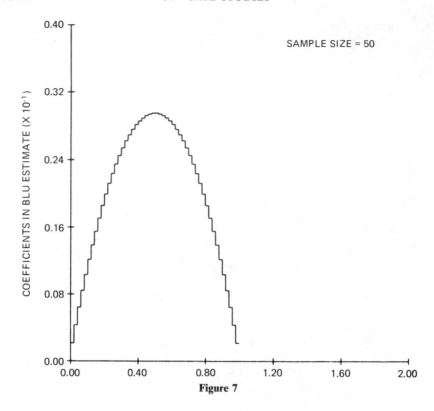

Figure 7

EXPERIMENT 2: From Linear Algebra

Our next case deals with a related problem from linear algebra, but the methodology of experimentation differs in one basic respect: we are not looking for conjectures but for counterexamples to given conjectures.

This problem arose in a study of Phillips's curves in theoretical economics; see Grenander (1976) for more details. It can be stated in a direct form, isolated from the background problem, and in the terminology of the previous example as

Conjecture. *The entries of the vector* $v = R^{-1}e$ *are always nonnegative.*

In the seven economic instances observed, the conjecture was true empirically; n was about 25. A number of special numerical cases were also tried in which the R-matrix was picked more or less arbitrarily. In each case the resulting v-vector had nonnegative entries only.

Empirical evidence of this sort does not, of course, constitute a proof of the conjecture but it strengthens one's belief in it. To study it in general we first prove the following simple result.

Proposition. *If R is as above, and also circulant, the conjecture holds.*

The proof is almost immediate. The eigenvectors of the circulant R are known to be of the form

$$\phi_k = \frac{1}{\sqrt{n}} \left[\exp\left(2\pi i \frac{kt}{n}\right), \quad t = 1, 2, \ldots, n \right]$$
$$k = 0, 1, 2, \ldots, n - 1 \tag{11}$$

so that, if the eigenvalues are denoted $f_0, f_1, f_2, \ldots, f_{n-1}$, we have the spectral representation

$$r_{st} = \frac{1}{n} \sum_{k=0}^{n-1} f_k \exp\left(2\pi i \frac{k(s-t)}{n}\right), \qquad f_k \geq 0 \tag{12}$$

But the inverse of a circulant is also a circulant, always with the same eigenvectors as in (11), so that R^{-1} has as entries the matrix elements

$$r_{st}^{(-1)} = \frac{1}{n} \sum_{k=0}^{n-1} \left(\frac{1}{f_k}\right) \exp\left(2\pi i \frac{k(s-t)}{n}\right), \qquad s, t = 1, 2, \ldots, n \tag{13}$$

Relation (13) makes it possible to calculate the vector v in closed form

$$v_s = \sum_{t=1}^{n} r_{st}^{(-1)} = \frac{1}{n} \sum_{k=0}^{n-1} \left(\frac{1}{f_k}\right) \exp\left(\frac{2\pi i k s}{n}\right) \sum_{t=1}^{n} \exp\left(-\frac{2\pi i k t}{n}\right)$$
$$= 1/f_0 > 0 \qquad \text{for all} \quad s = 1, 2, \ldots, n \tag{14}$$

which not only completes the proof but also indicates that the value of f_0 is the critical one.

The circulant case is, of course, only a special one, but it is well known that it approximates the general case well if n is large. All our efforts to use this to prove the entire conjecture were fruitless, however.

Finally we turned to heuristic search. To get a better understanding of how the vector v varies with R we carried out a computational experiment in which the R-matrix was picked "at random."

A technical difficulty in this connection is that it is not so easy to pick R-matrices at random. We want them to be symmetric, which is easy, and Toeplitz, also easy; see Section 12.6. They also have to be positive definite, which is a property further away from our intuition than, for example, symmetry. One necessary and sufficient way of verifying this property

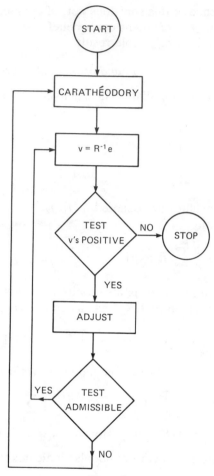

Figure 8

(there are lots of others, equally unattractive) is to use *Sylvester's criterion*

$$\det(R_1) > 0, \quad \det(R_2) > 0, \quad \ldots, \quad \det(R_{n-1}) > 0, \quad \det(R_n) > 0 \quad (15)$$

where R_k stands for the $k \times k$ section of the R-matrix. Determinants are notoriously clumsy to work with both analytically and computationally, so that instead of generating symmetric Toeplitz matrices "at random" and selecting those that satisfied Sylvester's criterion, *we generated them directly by Carathéodory's representation of finite Toeplitz and positive definite matrices:*

$$r_{st} = \sum_{k=0}^{n-1} f_k \cos \lambda_k(s - t) \quad (16)$$

where $f_k \geq 0$, $0 \leq \lambda_k \leq \pi$. We generate the fs as i.i.d. variables $R(0, 1)$ and the λs as i.i.d. variables $R(0, \pi)$. The block in the flowchart (see Fig. 8) doing this is called Carathéodory.

This procedure of generating positive definite matrices directly speeded up the algorithm considerably. We then compute v and test whether all entries are nonnegative. If a negative value is found, the program stops. Otherwise it adjusts the λ-vector as in Fig. 9. The block ADJUST computes the criterion

$$M = \min_k v_k \qquad (17)$$

for two points, say P_1 and P_2 in λ-space. We then move by successive steps in the direction of decreasing M, along a straight line until we get outside the set of admissible λ-values, when we start over again generating two new λ-points at random.

Of course, the λ-values outside the region are not really inadmissible (periodicity!), but we do not want to keep them fixed indefinitely, so this seems a wise choice. A trajectory in λ-space can then look like P_1, P_2, P_3, P_4. It can also change direction as in the trajectory Q_1, Q_2, Q_3, Q_4, Q_5, Q_6.

As a modification of this strategy we also varied the f-vector linearly

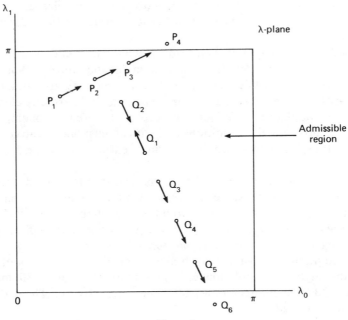

Figure 9

until some component became negative. We then had a $2n$-dimensional phase space.

Using the interrupt feature we can stop the program at any time and print out R, R^{-1}, v, or M, to give us an idea of what is going on. Notice that *we do not use an entirely random search* in our $2n$-dimensional phase space; that would be very inefficient. *Nor do we use an entirely systematic search,* which would be nearly impossible, at least in higher dimensions. Instead a combination of both seemed right.

Executing the program for $n = 5$ for about 30 iterations produced no negative M. We noticed a few cases of v-values close to zero, however. For $n = 4$ about 100 iterations produced no negative vs and we did not even find any close to zero. For $n = 3$, however, after about 30 iterations the program stopped and produced the matrix

$$R = \begin{pmatrix} 1.6036 & 0.6097 & -0.8127 \\ 0.6097 & 1.6036 & 0.6097 \\ -0.8127 & 0.6097 & 1.6040 \end{pmatrix} \qquad (18)$$

with the v-vector $= (1.8940, -0.8167, 1.8940)$ and the *conjecture had been disproved!*

To gain some more experience for other values of n, we then returned to $n = 4$, where about 300 more iterations finally produced a counterexample, and to $n = 5$, where 30 more iterations also gave a counterexample. There was some evidence that a purely random search would have been wasteful.

Armed with hindsight, it is easy to see what we should have done from the beginning: just evaluate the determinants needed for $n = 3$ and let the six variables vary, while keeping the matrix nonnegative definite. We finally carried out the boring and time-consuming but perfectly elementary algebraic manipulations. We found, indeed, that there is a nonempty set of matrices R satisfying the conditions and with at least one negative v-entry. The set is thin; that is why we did not find it earlier during the experimentation.

Before we had the result of the heuristic search it seemed futile to do this exercise in algebra since the result would probably not have been conclusive, and larger values of n would be too cumbersome. *We had been almost sure that the conjecture was correct, so that our main effort went into unproductive attempts to prove it, not to look for exceptions.*

The programming of the search algorithm in APL took about 1 h (to design the algorithm took longer, of course), debugging 30 min, and execution 20 min connect time, all done interactively. The CPU time needed was negligible.

This computational experiment is another example of how the mathematician can exploit the computer as his laboratory to explore hypotheses and use it as a research tool in addition to the traditional deductive method.

We learned from this experiment that the heuristic search could be made fairly quickly, certainly more quickly than purely random or purely systematic search, by employing the analytical structure of the setup, in this case the Carathéodory theorem.

Let us not leave this problem yet. We know that the question has a negative answer, but how negative? In other words, if we normalize $r_0 = 1$, can we give a lower bound for the entries of v? We shall show that no finite bound exists by an example.

The example (18) produced by the experiment gives a hint. The entries in the three diagonals are, after normalization, $\cong 1$, 0.38, and -0.50. The "simplest" entries close to this seem to be $1, \frac{1}{2}, -\frac{1}{2}$ or $\cos \lambda k$, $k = 0, 1, 2$, with $\lambda = \pi/n = \pi/6$. Generally, let us try with

$$M = \{\cos(\pi/n)(s - t); s, t = 0, 1, \ldots n - 1\}$$

which is nonnegative definite with the spectral mass $\frac{1}{2}$ at each of the frequencies $\pm \pi/n$.

This M is singular, but we shall perturb it slightly to the nonsingular

$$R = M + \epsilon I$$

Writing the spectral decomposition of M as

$$M = \sum_{\nu=1}^{n} \lambda_\nu P_\nu = \lambda_1 P_1 + \lambda_2 P_2$$

with the ordering of the eigenvalues such that λ_1 and λ_2 are the two positive ones, the rank of M is two.

Furthermore,

$$R^{-1} = (M + \epsilon I)^{-1} = \sum_{\nu=1}^{n} \frac{1}{\lambda_\nu + \epsilon} P_\nu$$

so that, for small positive ϵ,

$$R^{-1} \sim \frac{1}{\lambda_1} P_1 + \frac{1}{\lambda_2} P_2 + \frac{1}{\epsilon} P, \qquad P = \sum_{3}^{n} P_\nu$$

where we have assumed, of course, that $n \geq 3$.

Our new question therefore amounts to studying the signs of the vector $f = Pe$. But in the present case the eigenvectors corresponding to P_1 and P_2 can be written as [use the fact that $2\lambda n \equiv 0 \pmod{2\pi}$]

$$w_1 = c_1(\cos \lambda t; t = 0, 1, \ldots, n - 1)$$
$$w_2 = c_2(\sin \lambda t; t = 0, 1, \ldots, n - 1)$$

Hence $f \perp w_1$ and w_2, or, written out,

$$\sum_{t=0}^{n-1} f_t \cos \frac{\pi t}{n} = 0$$

$$\sum_{t=0}^{n-1} f_t \sin \frac{\pi t}{n} = 0$$

If all entries of f_t are nonnegative, the second equation implies that $f_1 = f_2 = f_3 = \cdots = f_{n-1} = 0$, and then the first implies that $f_0 = 0$. In other words, there exist constants a_1 and a_2 such that

$$a_1 \cos \lambda t + a_2 \sin \lambda t = 1, \qquad t = 0, 1, \ldots, n-1$$

Writing in polar coordinates $a_1 = \rho \cos \phi$, $a_2 = \rho \sin \phi$, this reduces to

$$\rho \cos(\lambda t - \phi) = 1, \qquad t = 0, 1, \ldots, n-1$$

which is not possible. Hence *no finite lower bound exists:*

$$\inf_{\langle R, r_0 = 1 \rangle} \min_t (R^{-1}e)_t = -\infty$$

EXPERIMENT 3:   An Energy Minimization Problem

In the preceding two cases the mathematical experiments were carried out in order to suggest hypotheses or to provide counterexamples. The next case illustrates a situation where conjectures had already been arrived at in somewhat vague form: one hoped that the experiment would make it possible to formulate the conjecture more precisely and gain better insight into the problem in order to prove (or possibly discard) the conjecture.

Consider a linear chain c of length l made up of generators $a_1, a_2, \ldots, a_l$. For each i, the value of a_i is restricted to d possibilities, which we enumerate just by the natural numbers $1, 2, \ldots, d$. The interaction energy between a couple a_i, a_{i+1} is given by a function $f(a_i, a_{i+1})$. This function need not be symmetric in its two arguments; it can be represented by an energy matrix, denoted by T. It is $d \times d$. Assuming additive energy, the whole chain has the total energy

$$t(c) = \sum_{i=1}^{l-1} f(a_i, a_{i+1}) \tag{19}$$

It will be convenient to use the relative energy

$$r(c) = (l - 1)^{-1}t(c) \tag{20}$$

instead of (19).

The problem is to find a chain that makes the energy as small as possible, then solve the minimum problem

$$t_l = \min t(c), \qquad r_l = t_l/(l - 1) \tag{21}$$

It is clear that the minimum is attained; we have a finite although large number of possibilities. The solution need not be unique. One also feels that the solutions ought to be "regular" in some sense, display some sort of (approximate or exact) invariance with respect to translations.

To study this experimentally we compute the solution of (21) for a set of l-values and display the solution for increasing values of l and a set of possible T-matrices.

It would be awkward, to say the least, to do this by computing $t(c)$ for all c of length l. This complete enumeration would require d^l iterations, which would soon become unfeasible when l became large. To avoid this we used a simple dynamic programming algorithm as follows.

Introduce

$$t_l(i, j) = \min\{t(c)|c \text{ of length } l, a_1 = i, a_l = j\} \tag{22}$$

Then (21) leads to the recursion

$$t_{l+1}(i, j) = \min_k[t_l(i, k) + f(k, j)] \tag{23}$$

We can now compute (23) successively with a modest amount of CPU time. This was done by a very concise APL program.

Let us look at a few solutions. For one choice of T-matrix we got the solution shown in Table 1. We display only enough l-values to indicate the typical behavior. It is clear that here the extremal behavior is just As (except for one at the end).

For another T-matrix the solution is shown in Table 2. It is very similar to the first one.

A third T-matrix led to the solution in Table 3. Here the favored arrangement consists (mainly) of the set $ADBE$ periodically repeated.

TABLE 1

l	r_l	Minimal configuration
2	1	*AB*
3	2	*AAB*
4	2.33	*AAAB*
5	2.5	*AAAAB*
6	2.6	*AAAAAB*
7	2.67	*AAAAAAB*
8	2.71	*AAAAAAAB*
9	2.75	*AAAAAAAAB*
10	2.78	*AAAAAAAAAB*

TABLE 2

l	r_l	Minimal configuration
2	1	*BF*
3	1.5	*DBF*
4	2	*ADBF*
5	2.25	*EADBF*
6	2.60	*EAADBF*
7	2.67	*DBEADBF*
8	2.71	*ADBEADBF*
9	2.75	*EADBEADBF*
10	2.89	*ADBFCCCCC*
11	2.90	*ADBFCCCCCC*
12	2.91	*ADBFCCCCCCC*
13	2.92	*ADBFCCCCCCCC*
14	2.92	*ADBFCCCCCCCCC*
15	2.93	*ADBFCCCCCCCCCC*

The conclusion is that *the conjecture ought to be expressed more precisely* in the form of a limit theorem to allow for the appearance of the boundary effects observed at both ends of the chain. We should claim that r_l tends to a limit as l tends to infinity and that this *limit can be attained asymptotically by a periodic arrangement*. This was then proved using a convexity argument. The two-dimensional analog of this extremum problem is not yet well understood.

A similar experiment, but employing another computer technology, also dealt with *patterns of arrangement, this time in the plane*. Consider a

TABLE 3

l	r_l	Minimal configuration
2	1.00	*BF*
3	1.50	*DBF*
4	2.00	*ADBF*
5	2.25	*EADBF*
6	2.60	*EAADBF*
7	2.67	*DBEADBF*
8	2.71	*ADBEADBF*
9	2.75	*EADBEADBF*
10	2.89	*EADBEAADBF*
11	2.90	*DBEADBEADBF*
12	2.91	*ADBEADBEADBF*
13	2.92	*EADBEADBEADBF*
14	3.00	*EADBEADBEAADBF*
15	3.00	*DBEADBEADBEADBF*

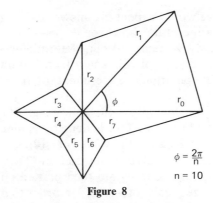

$$\phi = \frac{2\pi}{n}$$
$$n = 10$$

Figure 8

star-shaped pattern of the type shown in Fig. 10. It is generated from line segments of length r_k from the origin and at angle $2\pi k/n$, $k = 0, 1, \ldots,$ $n - 1$. The end points of contiguous line segments are connected linearly and the resulting polygon encloses a set: *the star-shaped pattern.*

Viewing such patterns dynamically, as a function of time t, we want to model their development under the basic hypotheses that contiguous line segments interact with inhibition. The simplest nontrivial model of this is the set of ordinary differential equations

$$\frac{dr_k(t)}{dt} = a - br_k(t) - \sum_{l=1}^{n} c_{k-l}r_l^2(t), \qquad k = 0, 1, \ldots, n - 1 \quad (24)$$

where the subscript of c_{k-l} should be understood as reduced modulo n. The coefficients a and b are arbitrary positive constants, the first one measuring the initial growth rate when the pattern is very small and the second expressing a counterbalancing effect for large patterns. The crucial coefficients $c_0, c_1, \ldots, c_{n-1}$ express the interaction and we shall here only consider the case where only c_1 and c_{-1} are different from zero: nearest-neighbor interaction.

What happens if we start the growth at $t = 0$ with some small patterns of arbitrary shape and then let t increase? We would expect the shape to settle down at some limiting shape and this is indeed so, but with some qualifications.

The singular points of the ODE in (24) are the candidates for such limiting shapes. Note that the r_k should be nonnegative. More importantly, in order that the growth pattern be insensitive to biological noise, always present, we must insist that the singular point be *asymptotically stable*.

Since (24) is of the much studied Lotka–Volterra type, very much in vogue these days, we would expect the answer to be known. We were not able to find it in the literature, however.

An analytical study was carried out, determining such solutions. We need not go into the details of this study [see Frolow (1978)], but shall instead report a computational experiment of interest in the present context.

A program was written to do the following: Start at $t = 0$ with a randomly chosen small pattern. This is done by calling a random number generator repeatedly to generate the values of $r_k(0)$, $k = 0, 1, \ldots, n - 1$, appropriately scaled. Then solve the system in (24) by a standard ODE subroutine. The resulting patterns are then plotted on a TSP plotter on-line for an increasing sequence of t-values in order to observe tendencies. The program was written *intentionally* as an infinite loop with no built-in termination criterion. Instead, the analyst sitting at the plotter decides when the experiment should be terminated and when enough information has been obtained.

One result is shown in Fig. 11. The initial pattern is the inner, some-what deformed, circle. As time increases we observe a rapid growth towards equilibrium. The equilibrium is a circle.

The next result, shown in Fig. 12, is quite different. It started again from a deformed circle seen in the inner part of the figure. The pattern changed rapidly to the large circle in the picture; the intermediate steps are not displayed. This circle was repeated for a long time (on the original it is seen drawn in ink heavier than the rest) and it may have seemed that an equilibrium had been reached. Suddenly the pattern changed into a clearly star-shaped one and remained so during a great number of iterations.

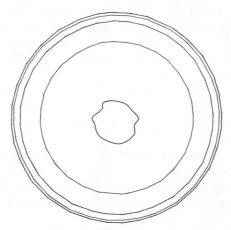

Figure 11

Figure 12

Figure 13

This peculiar behavior was given an analytical solution by analyzing the stability properties of the critical point of the ODE in (24). Both of the limit patterns that seemed to appear during the experiment are indeed real; they correspond to solutions of the equilibrium equations. The first one, the large circle, is not asymptotically stable while the second one is. This explains why the large circle seemed stationary for a while.

A third plot with $n = 50$ is displayed in Fig. 13. The equilibrium is seen clearly. The model used seems to deserve a more complete study.

From this experiment we can learn that it is sometimes valuable to do even the plotting part of an experiment on-line. If it had been done off-line one could easily have missed the first equilibrium. This would have been true, paradoxically, for a high-quality plotter, say a CALCOMP, since the lines drawn repeatedly would not necessarily show up heavier than a line drawn a single time. Also, there is psychological value in being able to watch the formation of patterns dynamically changing in time.

EXPERIMENT 4: Neural Networks (Static Problem)

In the study of formal neural networks, much attention has been given to the idea of a random, or partly random, topology of connections. The neurons are connected at their synapses in a way that does not seem to be well described by deterministic schemes. Controlled randomness appears more likely to offer a good description, although this is still in doubt. To examine the mathematical consequences, we have investigated a model that can be reduced to an eigenvalue problem as follows. The details can be found in Grenander and Silverstein (1977).

Let V be a matrix $\{v_{ik} ; i = 1, 2, \ldots, n, k = 1, 2, \ldots m\}$ where both n and m are large. Here $m = dn$, where d is a natural number called the divergence of the neural network. If $d > 1$, we speak of a diverging network. The entries v_{ik} are independent random variables with mean zero and variances that need not be constant.

For a given input vector, the resulting effect will have the power $\|V^T x\|^2 = x^T W x$, where the matrix $W = VV^T$ is symmetric and nonnegative definite. The asymptotic behavior of W for large networks, say expressed by its eigenvalue distribution, is the main concern at present.

Of course, this eigenvalue distribution is random. As n increases, however, we would hope that it would settle down to a deterministic limit: *that a law of large numbers would be valid for the spectra.*

With a normalizing constant C_n that depends upon n and the variances mentioned above, we can show that

$$E\left[\frac{1}{C_n}\|V^T x\|^2\right] = \|x\|^2 \qquad (25)$$

We can also show that $\text{Var}(C_n^{-1}\|V^T x\|^2) \to 0$ as n tends to infinity.

Hence, with convergence in probability

$$\frac{1}{C_n}\|V^T x\|^2 \to \|x\|^2, \qquad n \to \infty \qquad (26)$$

Arguing heuristically, this would lead to the conjecture that the operator $C_n^{-1}W$ is close to the identity matrix and has an eigenvalue distribution concentrated around the value $\lambda = 1$.

Of course, this is not a valid argument, and before we proceed to a rigorous proof or disproof, we could carry out an experiment. Generate W as described. This takes only a very simple program. Calculate its eigenvalues and display the resulting histogram. We start with no divergence, $d = 1$.

Surprise! The histograms are not at all concentrated around $\lambda = 1$ but are spread out over the interval (0, 4) approximately as shown in Figs. 14a

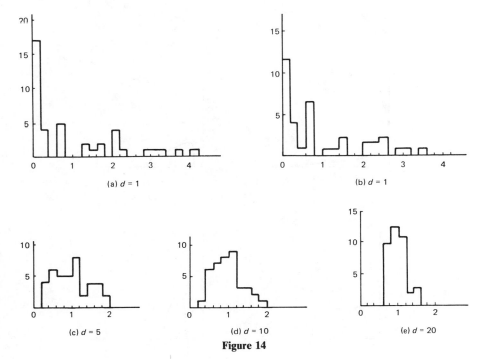

(a) $d = 1$

(b) $d = 1$

(c) $d = 5$

(d) $d = 10$

(e) $d = 20$

Figure 14

and 14b. It is a consolation, however, that the two histograms resemble each other. While the original conjecture is clearly wrong, there is still hope that a deterministic limit exists, although not the one we guessed.

For large values of the divergence [see the graphs in (c), (d), and (e)] the eigenvalue distributions are seen to contract around the point $\lambda = 1$.

Armed with this empirical knowledge, we spent the next several weeks trying to prove the existence of deterministic limits. The proof is long and laborious but the end result is a simple analytical form for the density of the limiting eigenvalue distribution. The particular form of the density need not concern us here; we merely display the density functions in Fig. 15. It also turned out to be true that as $d \to \infty$ the curves become more concentrated around $\lambda = 1$.

During this experiment the plotting played no important role; it could have been done by hand. The main computational work was calculating the eigenvalues. Calculation of the eigenvalues of a symmetric and non-negative definite matrix is one of the standard tasks in numerical linear algebra, and a multitude of algorithms are given in the literature. We had moderate values of n, mostly around the value 40, occasionally up to $n = 100$. For matrices of such sizes the CPU time needed is not a major

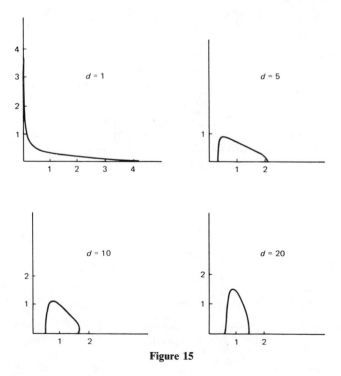

Figure 15

concern and we felt no need for maximum computational efficiency. A program from one of the public program libraries was used with minor modifications. The computation required little effort.

This situation changed when we went ahead to study the eigenvector distribution of W. First, this unavoidably leads to a good deal of number crunching—the CPU-time cost is no longer negligible. Second, we did not have immediate access to a full set of subroutines for this purpose so that we had to spend more time on programming than is usual in our mathematical experimentation. If we had had a well-organized program library we could have saved some of our time and used it more productively for the analytical aspect of the problem. Third, we had no clear conception of what to guess concerning the asymptotic behavior of the eigenvectors. We had established the existence of a deterministic limit for the eigenvalue distribution. Assuming that something similar holds for the eigenvector distribution, how should this be made precise? One should keep in mind that the (normalized) eigenvectors are not uniquely determined, their sign is always arbitrary, and, if the eigenvalues are multiple, the degree of nonuniqueness is even worse.

What we are looking for is, therefore, not the individual eigenvector but the projection operator associated with subsets of eigenvectors. More precisely, let $\lambda_k^{(n)}$, $k = 1, 2, \ldots, n$, be the eigenvalues of the random W-matrix of size $n \times n$, and $P^{(n)}(\lambda)$ the associated spectral family. Then the operator

$$P^{(n)}(\lambda_2) - P^{(n)}(\lambda_1) \tag{27}$$

is the projection down to the subspace spanned by all the eigenvectors $z_k^{(n)}$ for which the corresponding eigenvalues satisfy $\lambda_1 < \lambda_k^{(n)} \leq \lambda_2$.

Correspondingly, let $Q^n(\lambda)$ be a deterministically given approximation, so that $P^n - Q^n$ becomes small in some metric as n tends to infinity. Arguing by analogy with Toeplitz form theory, where the solution of the associated problem is known, we use a distance criterion leading to the convergence statement

$$(1/n) \operatorname{tr}(P^n(\lambda) - Q^n(\lambda))^2 \to 0, \qquad n \to \infty, \quad \forall \lambda \tag{28}$$

Whatever Q^n is, we can test the conjecture (28) by computing two realizations of W and their spectral families $P_1^{(n)}$ and $P_2^{(n)}$. The quantity

$$(1/n) \operatorname{tr}(P_1^{(n)}(\lambda) - P_2^{(n)}(\lambda))^2 \tag{29}$$

should then be small if n is large.

This was programmed and executed for a fine grid of λs and many values of n. The answer left no doubt that (28) is wrong: the criterion in (29) settles down quickly when n is allowed to grow, but not to zero! Discarding our hypothesis we look for alternative ones.

But here we met a fourth difficulty. Once we have computed the $z_k^{(n)}$, say for $n = 100$, what do we do with the numerical result? We have 100 values in $\mathbb{R}^{100}$, that is, a chaos of 10,000 real numbers. It seems impossible to gain any intuition about their behavior; we just lack the means to display high-dimensional data.

No way out of this dilemma is offered by the technology of computer graphics. If the dimension had been, say, 4 or 5 or 6, we might have been able to learn something by displaying two-dimensional projections on the screen of the CRT, rotating the two-dimensional subspace quickly, until some meaningful pattern appeared. For $n > 10$ or so, this is hopeless, since there are so "many" possible rotations.

Realizing this difficulty, which will turn up many times again, we tried to extract low-dimensional information from the experimental results.

We proved the following auxiliary result. Let P_1 and P_2 be generated from a fixed (but arbitrary) rectangular frame by a random rotation. By this we mean that we choose two independent rotations from the population of rotations in $\mathbb{R}^n$ by sampling from the Haar measure over the rotation group. Then with convergence in probability

$$\lim_{n \to \infty} \frac{1}{n} \operatorname{tr}(P^{(1)}(\lambda) - P^{(2)}(\lambda))^2 = 2F(\lambda)[1 - F(\lambda)] \tag{30}$$

where F is the distribution function of the deterministic limit already established. In Fig. 16 we display the right side of (30) for $\lambda \in (0, 4)$ to-

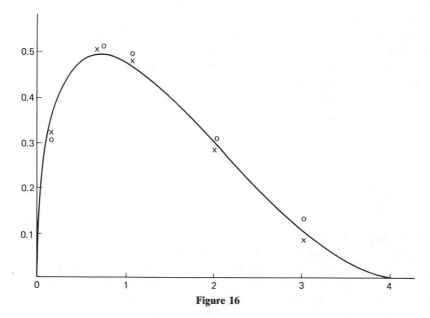

Figure 16

gether with the experimental results from two simulations indicated by crosses and small circles. For details, see Grenander and Silverstein (1977).

The agreement is quite good. This supports the conjecture that the eigenvector distribution is asymptotically completely random in the sense of being generated by the Haar measure of the orthogonal group.

Additional experimental support for this hypothesis, which can be said to be the opposite extreme to our original hypothesis, was obtained later. The complete answer has not yet been obtained, but the most recent analytical result at the time of this writing seems actually to say that the asymptotic eigenvalue distribution is random as mentioned only if a certain moment relation is satisfied; see Silverstein (1980).

We realized through this experiment the importance of selecting low-dimensional information from the mathematical experiment in such a way that it guides our intuition visually. Our present knowledge of how to do this is scanty, to say the least, and must be improved.

Before leaving the problem of the asymptotic eigenanalysis of W, let us mention an unexpected epilogue. In the study of random equations made up of either algebraic equations or differential equations it is believed that their solutions will obey laws of large numbers and central limit theorems. The equations cause couplings between the components of the solution that make it less promising to prove the limit laws by standard techniques from probability theory.

Instead, a technique has been developed [see Geman (1980)] by which the desired results can be derived directly. An early and heuristic version of this technique had a missing link: one had to show that the maximum eigenvalue $\lambda_{\max}^{(n)}$ of a matrix that happens to be generated just like W is uniformly bounded with probability one as $n \to \infty$.

This is not implied by the limit theorem for the distribution of eigenvalues. For $d = 1$ it can be verified that the asymptotic distribution has as its support the interval [0, 4], but this does not mean that *all* the $\lambda_k^{(n)}$ will stay below or equal to 4. Indeed, the statement about the asymptotic distribution of the $\lambda_k^{(n)}$ would be consistent with the possibility that one of them [or a small number $o(n)$ of them] tended to infinity.

At this stage of the work both possibilities received attention: attempts were made to prove the missing link as well as to exhibit the divergence of $\lambda_{\max}^{(n)}$. The psychological circumstances were drastically altered, however, when it was realized that in the earlier experiments, carried out a couple of years earlier and with a different aim, none of the observed eigenvalues had exceeded the value 4 by more than a very small amount.

This strengthened the belief in the conjecture, the analytical effort to prove it was increased, and a positive result was eventually produced.

The computer did not play as crucial a role in this part of the work as in some other cases reported here, but the psychological impact of the empirical findings was not negligible.

EXPERIMENT 5: Limit Theorems on Groups

Computer technology for visual display will certainly be an important tool in mathematical experiments. A good illustration of how this can sometimes be explored is furnished by a study of measures on groups and the corresponding limit theorems.

Let G be a topological group and P a probability measure defined on the Borel sets of G. Take a sample of independent observations of size n from G according to the probability law P and denote the resulting group elements $g_1, g_2, g_3, \ldots, g_n$. Form the product

$$\gamma_n = g_1 g_2 g_3 \cdots g_n \tag{31}$$

A problem that has received a good deal of attention in recent years is the following one. The stochastic group element γ_n has a well-defined probability distribution P_n that can be viewed as the nth convolution power of P:

$$P_n = P^{n*} \tag{32}$$

How does P_n behave asymptotically for large n? Are there analogues to the probabilistic limit theorems on the real line or in other Euclidean vector spaces?

The state of our current knowledge depends very much on the structure of the group G. If G is compact, the problem has been well penetrated and a number of informative theorems have been proved [see Grenander (1963, Chap. 3)]. For commutative, locally compact G we also know a good deal. When we proceed to groups that are locally compact but not commutative, our knowledge is less complete. See, however, the monumental work of Heyer (1977), which contains a wealth of information. *The basic normalization problem has not been settled:* How should we normalize γ_n in (31) in order to arrive at nontrivial probabilistic limit theorems? How should γ_n be transformed by a one-to-one mapping into some other space? In a few special cases we know how to do it, but not in general.

In view of this, it may appear too early to turn to groups that are not even locally compact. There are two reasons for doing this, however.

First, it may be that, again, we can deal with some special case that will tell us something about what possibilities exist and what probabilistic behavior we should be prepared to meet in the absence of local compactness.

Second, a practical motivation arose from some work in pattern analysis. A space X that we shall refer to as the *background space* is mapped onto itself by a random function ϕ called the deformation:

$$\phi: X \rightarrow X \tag{33}$$

Very often in these problems ϕ is bijective and has some continuity in probability. We are interested in what happens when ϕ is iterated:

$$\begin{aligned}
\psi_1(x) &= \phi_1(x) \\
\psi_{i+1}(x) &= \phi_{i+1}[\psi_i(x)], \qquad i = 1, 2, 3, \ldots
\end{aligned} \tag{34}$$

where $\{\phi_i\}$ is a sequence of random functions independent of each other and with the same probability measure P on some function space Φ. This leads obviously to the problem described above.

In some practical cases, X is a Euclidean vector space or a subset of one. We shall look at the case $X = [0,1]$ and

$$\begin{aligned}
\phi \text{ is nondecreasing} \\
0 \le \phi(x) \le 1 \\
\phi(0) = 0, \qquad \phi(1) = 1
\end{aligned} \tag{35}$$

as minimum requirements. We shall make it a bit stronger by assuming ϕ to be continuous in probability. Sometimes we have also assumed that ϕ is strictly increasing so that we get a group as above, not just a semigroup.

We shall use two functions to give a crude characterization of the deformation: the mean-value function $m(x)$ and the variance function $V(x)$:

$$m(x) = E[\phi(x)], \qquad V(x) = \text{Var}[\phi(x)] \tag{36}$$

If the deformation has a tendency to the left in the sense that $m(x) < x$, $0 < x < 1$, we call the deformation *left-systematic;* similar conditions give a *right-systematic* deformation. If $m(x) \equiv x$, which is the most important case, the deformation is called *fair.*

This is not the right place to describe in any depth the study of random deformations of a background space. We shall just indicate how we used the computer to help us look at convolution powers on this group and how this aided us in suggesting some theorems. At this point we should stress the psychological desirability of not just computing the convolution process but of making the *computational results intuitively available* by graphic representation or by other technical means. In this study this was done by crude APL plots and also by using the Calcomp plotter to get better graphic resolution.

A first question is: When do we reach equilibrium distributions and what form do they have? It is not difficult to prove that if the deformation is left(right)-systematic, an equilibrium distribution must be concentrated on the two-point set $\{0, 1\}$. To shed some light on the *fair* deformations we generated a series of Calcomp plots, some of which are given in Fig. 17,

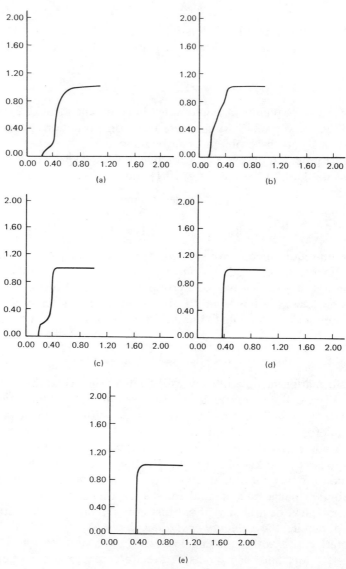

Figure 17

where ψ_n is plotted against x for $n = 10, 20, 30, 40,$ and 50. Other graphs had the same appearance: ψ_n tends to a step function with a single step and the location of the step varies from sequence to sequence but not within a sequence. This led us to the following conjecture:

If ϕ is a fair deformation with a continuous $V(x)$ vanishing only at $x = 0$, 1, there exists a stochastic variable ξ such that

$$\lim_{n \to \infty} \psi_n(x) = \begin{cases} 0 & \text{if} \quad x < \xi \\ 1 & \text{if} \quad x > \xi \end{cases} \tag{37}$$

and where ξ has the rectangular distribution $R(0, 1)$.

It is natural to look at the problem from the point of view of martingales. Actually the proof can be carried out conveniently in a more direct and perhaps more informative way.

Once formulated, the conjecture was not difficult to prove; we just had not asked the right question until the computer experiment showed us the way. Indeed, at first it was something of a shock to see the way the graphs looked with their counterintuitive shapes.

If a (univariate) equilibrium distribution exists for $\psi_n(x)$ for any (fixed) x, the question arises: What can we say about bivariate (and higher-dimensional) equilibrium distributions? Some APL plots were helpful: We show two plots of the bivariate empirical distribution over $(0, 1) \times (0, 1)$ in Fig. 18a and b. It is apparent that the points tend to cluster on the 45° diagonal. This suggested the following conjecture.

If the univariate limit is unique and does not depend on x, then any bivariate limit distribution is singular and has all its mass on the diagonal of the unit square.

The proof is quite simple. Several other related computer experiments were carried out but will not be discussed here.

We have only let the computer help us in making guesses here. These guesses have then been verified (or discarded, occasionally) by using deductive reasoning and the standard analytical tools.

The two-dimensional extension of this problem is not well understood yet; we do not know what the limits signify.

Some caution is needed in regard to the interpretation of topological concepts, such as those occurring above, on the computer. It could be argued that since the machine is a finite service, it is meaningless to implement concepts like continuity or compactness on it. This is true only on a superficial level. The argument misses the point that qualitative properties—without complete correspondence to the computer—can be

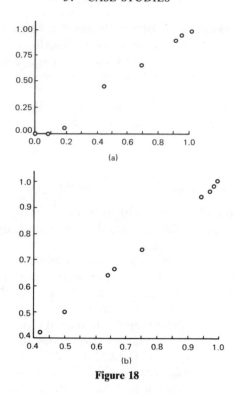

Figure 18

expressed approximately in discrete terms resulting in a meaningful correspondence. But this requires a "translation" of a quite sensitive nature. It takes a good deal of experience before the experimenter acquires such insight and it is easy to be led astray.

EXPERIMENT 6: Pattern Restoration

In the next example the purpose of the computational experiment is not to suggest conjectures or to generate empirical support for hypotheses. The aim was instead to provide numerical information concerning the performance of a given restoration method.

Let I denote a pure point spectrum made up of a finite number of spikes $A_k \delta(x - x_k)$ of frequency x_k and amplitude $A_k > 0$. For the following it will be natural to consider I as a measure on the real line.

This measure, which is not known to the observer, is deformed by two mechanisms. The first widens the spectrum by solving

$$\frac{\partial p(x, t)}{\partial t} = Lp(x, t) \tag{38}$$

where L is a linear operator (on the x-variable) with initial values $p(\cdot, 0)$ corresponding to the measure I. A typical case is $L = \frac{1}{2}D^2$, when (38) reduces to the heat equation, but other choices of L also appear in applications. Set $t = w = $ the widening parameter.

The second deformation consists of adding random noise to $p(x, w)$. The result will be denoted $I^{\mathscr{D}} = I^{\mathscr{D}}(x)$ and is what the observer can see. He wants to restore I.

A naive approach would be to try to solve (38) backwards for w time units, but this will not work. The basic difficulty is brought out clearly if we discretize (38) for both time and space. This will lead to a vector equation

$$p(t + 1) = Mp(t) \qquad (39)$$

with p now taking values in some $\mathbb{R}^n$. When we try to solve this equation backwards, i.e., invert M, we find that typically M has an eigenvalue equal to or close to zero. This makes inversion impossible or at least numerically unstable.

This shows that the problem belongs to the family called "incorrectly posed" as discussed, for example, by Tihonov (1963). A detailed analysis, too long to be described here, has been given by Town (1978). It is based on the following restoration method.

Calculate the Fourier (sine) transform

$$\hat{p}_\mu = \sqrt{\frac{2}{n}} \sum_{x=1}^{n} I^{\mathscr{D}}(x) \sin \mu \phi_x, \qquad \mu = 1, 2, \ldots, n \qquad (40)$$

where $\phi_x = x\pi/(n + 1)$. Form

$$\hat{q}_\mu = \frac{\cos^w \phi_\mu}{\cos^{2w} \phi_\mu + \epsilon} p_\mu, \qquad \mu = 1, 2, \ldots, n, \qquad \epsilon > 0, \qquad (41)$$

and use

$$I^*(x) = \sqrt{\frac{2}{n}} \sum_{\mu=1}^{n} \hat{q}_\mu \sin \mu x \qquad (42)$$

as the restored spectrum.

This procedure can be shown to possess optimality properties, for example, in the Bayesian sense, but it suffers from the defect that w is usually not known to the observer so that he would have to try several values.

To find out how this works, a computational experiment was carried out in which the deformations were simulated, the algorithm in (40)–(42) was implemented with a small modification (all by APL programs), and the resulting restoration I^* was plotted using a simple (x, y) plotter.

It then became clear that the behavior of the method depends crucially on the relation between ϵ and the value of w chosen. If w is too small, not enough "deconvolution" is done and I^* will be a smoothed form of the original point spectrum. As w is increased, the behavior of I^* suddenly becomes erratic and I^* often takes negative values, which is physically meaningless. After a few trials have been observed the experimenter develops a feeling for when w is approximately correct. Some rule of thumb should be of value when applying the restoration method.

An early version of this experiment was carried out at a time when we did not have access to any sophisticated interactive system. The only interactive capability was in terms of BRUIN, a locally developed programming language reminiscent of BASIC. It was barely adequate for this purpose, quite clumsy for the experimenter. A program was written to experiment with this method of image restoration. The several iterations, typical of mathematical experiments, did not develop well, partly because of the by now well-established fact that programming languages of this type are not suitable for our purpose. *They force the attention onto the programming effort and away from what should be the principal objective: the mathematical analysis.*

Even worse was the psychological effect of the nonrobustness of the system. Frequent breakdowns, sometimes with files irretrievably lost, and long interruptions made the experiment painfully slow; there was a feeling of losing momentum. Under such circumstances it is better to stay away from mathematical experiments.

This experiment would have been impossible, or at least very expensive, unless we had had access to the fast Fourier transform (FFT) for computing the numerical Fourier transform. As is well known, this reduces the amount of CPU time needed by a factor $n/\ln n$, a drastic decrease in cost. The need for efficient algorithms will now and then arise in mathematical experiments, and it should be pointed out that *fast algorithms may sometimes be constructed even in situations where the naive algorithm seems to be the only way of organizing the computing task.*

To emphasize this point let us briefly mention a quite different experiment, but one in which the same point comes up. In the study of random sets we were considering closed and bounded convex sets in the plane. For computing purposes these sets were described as polygons with a finite but possibly large number of corners.

Say that we generate a sequence C_1, C_2, C_3, . . . of random convex sets with the same distribution and independent of each other. The Minkowski sum $A + B$ of two sets is defined as the set of all points $z' + z''$ with $z' \in A$ and $z'' \in B$. Form now the Minkowski sum $S_n = C_1 + C_2 + \cdots + C_n$ and normalize it by changing the scale by the factor $1/n$.

The resulting set

$$\mathbf{M}_n = (1/n)(C_1 + C_2 + \cdots + C_n) \tag{43}$$

is also convex and we want to find out if it tends to a limiting set as $n \to \infty$. In other words, is there a law of large numbers for random convex sets? Limit should here be understood in the sense of Hausdorff distance between sets and interpreted either in probability or with probability one.

The experiment led straight to the goal: there was no doubt of convergence. The graphs were very informative and increased our belief in the conjecture.

For an analytical approach it is reasonable to describe all the sets by their support functions and embed these functions in the space $C([0, 2\pi))$. We then reduced the problem to the law of large numbers in a separable Banach space, and we could conclude that convergence indeed takes place.

How does one compute the Minkowski sum of two polygons, each with, say, n number of corners? The naive way would be to form the n^2 sums of all pairs of corners and then calculate the convex hull of these points. This is very wasteful.

A related problem is: When a random set of n points is given, how does their convex hull behave for large values of n? Again we need to compute the convex set of many points.

We could organize an algorithm as indicated in Fig. 19. Approach the

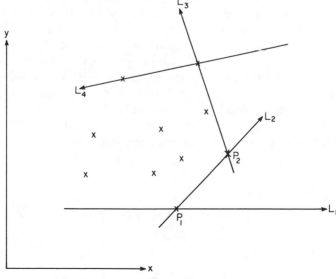

Figure 19

discrete point set from below by half-planes with boundaries parallel with the x axis. They hit the set with boundary L_1. Store its parameters (special rule if more than one point is hit by L_1; similarly in the following). Consider now half-planes with boundaries passing through P_1. Rotate them in the positive direction until the next point P_2 is hit and continue in this way.

This algorithm is correct but also very wasteful of CPU time. It can easily take $O(n^2)$ iterations. When these experiments were done we did not realize that many problems in computational geometry could be solved by efficient algorithms. For the convex hull, for example, the number of steps can be reduced to $O(n \ln n)$, making it feasible for the experiment at a much lower cost or on a larger scale, than if the naive algorithm had been used. Many similar instances could be quoted.

In our experience of mathematical experiments we have only occasionally felt the need for "fast algorithms." The two examples mentioned above are exceptions. We believe, however, that experimenters in the future may feel this need more strongly. For example, in experiments with finite algebraic structures, semigroups, rings, and fields, naive algorithms are prohibitively slow and must be speeded up.

EXPERIMENT 7:   Modeling Language Acquisition

Let us consider one of the problems in the mathematical modeling of language acquisition—learning syntax. To fix ideas let the formal language $L = L(G)$ be generated quite simply by a finite-state grammar G over a terminal vocabulary $V_T = \{a, b, c, \ldots\}$ and a nonterminal vocabulary $V_N = \{1, 2, \ldots, F\}$. In V_N we use the convention that a derivation of a sentence must begin in state 1 and end in state F. We also need a set of rewriting rules that we give in the form $i \xrightarrow{x} j$: go from state i to state j writing the terminal symbol x.

As an illustration consider the diagram in Fig. 20 with $F = 7$. This simple grammar generates, for example, the sentences

$$aac$$

using the rewriting rules

$$1 \xrightarrow{a} 2, \qquad 2 \xrightarrow{a} 4, \qquad 4 \xrightarrow{c} 7$$

$$acbabb$$

using the rewriting rules

$$1 \xrightarrow{a} 2, \qquad 2 \xrightarrow{c} 1, \qquad 1 \xrightarrow{b} 3, \qquad 3 \xrightarrow{a} 3, \qquad 3 \xrightarrow{b} 5, \qquad 5 \xrightarrow{b} 7$$

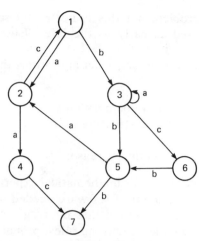

Figure 20

and

$$bcbb$$

using the rewriting rules

$$1 \xrightarrow[b]{} 3, \qquad 3 \xrightarrow[c]{} 6, \qquad 6 \xrightarrow[b]{} 5, \qquad 5 \xrightarrow[b]{} 7$$

The set $L = L(G)$ of all sentences that can be generated in this manner is the language for given G.

For a complete performance description we must also associate probabilities to the rewriting rules. From state 3, for example, we can go to states 3, 5, and 6, and we should specify the corresponding probabilities p_{33}, p_{35}, and p_{36}. Of course, $p_{33} + p_{35} + p_{36}$ must equal one. Similar rules hold for the other states.

With this *syntax-controlled probability model* our present problem takes the following form. Given a set of sentences $s_1, s_2, \ldots, s_n \in L(G)$, how can we learn G? Note that the terminal vocabulary V_T is considered known in the present context but V_N and the set of rewriting rules are not.

We can express the learning task somewhat differently. The learning takes place in two phases: listening and speaking. During the listening phase the learner "hears" one more s_k. During the speaking phase the learner produces a sentence and is told whether it is grammatically correct or not. We include no semantics in this set up; the problem of learning semantics in a formal sense is being studied at this time and will be reported elsewhere.

We want to construct an algorithm such that we can prove that it converges to the true grammar G as $n \to \infty$. At first glance this may appear to be a formidable problem but this is not really so. Indeed, such algorithms were proposed as early as the late 1960s by the author and others.

What is more difficult is to construct algorithms that are "natural" in the sense that

(a)   they require moderate memory only,

(b)   they are tailored to the structure of the sort of grammars allowed, and

(c)   the amount of computing is reasonable.

This problem is well suited for mathematical experiments. To carry out such an experiment the amount of software needed is not negligible.

First, we need a setup program that interrogates the user and asks him what are the vocabularies, rewriting rules, probabilities, and possibly other parameters to be set. The responses are collected and organized into suitable data structures: scalars, vectors, rectangular arrays, and lists of names.

Second, we must have access to a simulation of a given grammar. This simulation takes the result of the setup program and generates one sentence in accordance with the syntax-controlled probability model each time the simulator is executed.

Third, we need a program to collect data from the experiment, analyze them, and display the results graphically. The graphs are displayed using a typewriter terminal to produce crude plots.

Finally we have the main program. The three auxiliary ones mentioned above were written at the beginning of the experiment, but we rewrote and modified the main program during the experiment when our experience enabled us to improve it. Several algorithms were constructed and we sketch only one of them. It is intended for the subtask of discovering word classes; the general task can be dealt with similarly. It was organized into eight steps as follows:

Step 1.   Initialize by creating a single class.

Step 2.   Produce one more sentence s_k.

Step 3.   Select a terminal symbol x from s_k and call its current prototype y. Go to Step 4 if $x = y$, otherwise go to Step 5.

Step 4.   Select another terminal y from the (current) class containing y. Replace x by 2 in the sentence. Treat 2 as x is treated in Steps 6–8. Go to Step 2.

Step 5.   Test the grammaticality of the sentence with x replaced by y. If it is not grammatical go to Step 7.

Step 6.   If it is grammatical call a subroutine that "strengthens belief" of the hypothesis that *x* is in the same word class as *y*. Go to Step 2 or stop.

Step 7.   Move *x* to the next class if there is one, or else go to Step 8.

Step 8.   Create a new (current) word class with *x* as its single element and prototype.

The reader can find a full description of this and related algorithms in Grenander (1978, Chap. 7).

To prove that this algorithm (and some extensions of it) converges in finite time to the true grammar is elementary. The speed of convergence is trickier and systematic experiments were carried out to improve the speed. For this purpose the analysis program mentioned above was useful and we finally arrived at algorithms with performance that was deemed acceptable. No claim was made that they were optimal.

This experiment differs in several respects from the earlier ones. We were not aiming for *the* solution (of an optimality problem) but for reasonably good ones. The choice of good ones was to some extent arbitrary as is often the case in applied mathematics.

Here the graphical displays did not motivate the use of any sophisticated computer graphics technology. Simple typewriter plots were sufficient since the degree of accuracy was low in any case.

The programming was done in APL. It is sometimes stated that whereas APL may be a good programming language for numerical work it is not suitable for nonnumerical computing such as, in this case, string manipulations. *It is our experience that this is not true.* The primitive functions in APL include several that are just what is needed for manipulating strings. The dyadic "iota" is one of them and indexing in APL is very convenient for the present purpose. The CPU time that was required was never excessive in these experiments.

For a more ambitious experiment things may possibly be different. The limitation in APL to rectangular arrays introduces some waste of memory resources. It would have been desirable to be allowed to operate on arrays of arrays, for example. Since APL is implemented by an interpreter (we did not have access to any compiler), the fact that all algorithms used involved a fair amount of looping would have led to a waste of CPU time for larger runs.

For such experiments it would have been better to reprogram the algorithm in, say, LISP and execute the compiled programs. To do this conveniently one should have access to at least *remote job entry* of LISP programs from the terminal with short turn-around time.

EXPERIMENT 8:   Study of Invariant Curves

The next experiment is from ergodic theory. Let us consider an area-preserving mapping $M: \mathbb{R}^2 \to \mathbb{R}^2$ defined as the composition of two mappings M_+ and M_-. To introduce M_+ and M_- we let $N: \mathbb{R}^2 \to \mathbb{R}^2$ be given in polar coordinates (r, ϕ) by

$$N: (r, \phi) \mapsto (r, \phi + \alpha(r)) \tag{44}$$

where the twist $\alpha(\phi)$ is a C_1 function to be specified later. Let M_+ be N-translated to have its center on the x-axis at the point with rectangular coordinates $(\epsilon, 0)$. Similarly, let M_- have its center at the point $(-\epsilon, 0)$.

Since N is obviously area preserving, the same is true of M and this fact can be used for an experiment to study invariant curves, periodic points, ergodic behavior, and so on. In such an experiment [see Braun (1977); for terminology used see, e.g., Nitecki (1971)] the first function was chosen as $r^{-4/3}$ and the constant ϵ as 0.02.

It should be noted that M has many fixed points on the y-axis, corresponding to points that are fixed both for M_+ and M_- and to points $(0, y)$ that are mapped by M_+ into $(0, -y)$. Some of these points are elliptic, with their islands of invariant curves surrounding them, and some are hyperbolic. It has been verified numerically that the stable and unstable manifolds of some of these hyperbolic points intersect homoclinically.

In the experiment a starting point (x_0, y_0) was selected and the iterates $(x_n, y_n) = M^n(x_0, y_0)$ were computed for $n = 1, 2, \ldots, 2500$. This takes only a very simple program. To visualize the solution the points were to be plotted and this leads to an algorithmic problem of a certain significance; its solution is due to McClure (1977).

When the plotter is engaged in plotting the point set, the pen moves to (x_n, y_n), and drops down, lifts up, and moves on to the point (x_{n+1}, y_{n+1}). If the points come in a chaotic order, as, for example, in the ergodic case, the mean length of the pen's movement between points can be expected to be approximately a constant times the diameter of the plotting region; the constant could be considerably larger than zero but, of course, smaller than one. This will lead to a lot of unnecessary movement and the plotting procedure will be slow. It was estimated that about 45 min would be required *for each choice of starting point* (x_0, y_0).

To avoid this waste, the set $\{(x_n, y_n), n = 1, 2, \ldots, 25\}$ was first broken up into bins. On the x-axis, N intervals $(a_i, b_i]$ were chosen and the ith bin was given as

$$B_i = \{(x_n, y_n) | x_n \in (a_i, b_i]\}; \, i = 1, 2, \ldots, N \tag{45}$$

In each bin the points were sorted monotonically according to their y_n values. Then the plotting was executed one bin at a time and in the new order in each bin.

A rough optimality study showed that N should be chosen as approximately $\sqrt{N}$. The resulting saving was considerable.

Some of the plots produced in this experiment are shown in Figs. 21–25. Figure 21 looks like an imperfectly plotted circle. A closer examination shows, however, that this is not so and it suggests an invariant curve. The presence of the many holes speaks against this, since, in an invariant curve, the iterates should be equidistributed. With 2500 iterations the little holes should not be present. The reproduction is unfortunately not quite clear.

It is believed that (x_0, y_0) in the first case was a periodic point of very high order. In Fig. 22 this is even more apparent.

Admittedly, this reasoning is not entirely convincing, since the first figure also had some little holes. This is quite similar to the difficulty mentioned earlier in interpreting finite data in infinite terms such as continuity and smoothness. A plotting device of higher accuracy, for example an HP plotter, would have been helpful here. The next figures are easier to interpret. In Fig. 23 the invariant curves are seen to disintegrate and the

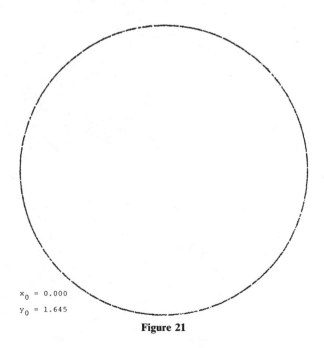

$x_0 = 0.000$
$y_0 = 1.645$

Figure 21

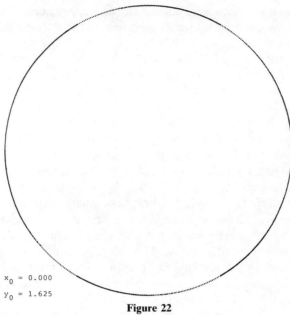

$x_0 = 0.000$
$y_0 = 1.625$

Figure 22

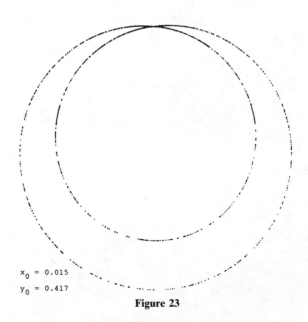

$x_0 = 0.015$
$y_0 = 0.417$

Figure 23

intersection of the "curves" is a hyperbolic fixed point; the stable and unstable manifolds at this point intersect homoclinically.

In Fig. 24 we see a not very pronounced tendency to instability, and in Fig. 25, finally, we see a clearly ergodic behavior.

Case 8 is interesting because it illustrates both the need for careful algorithmic analysis (to a good but not truly optimal algorithm) and because it brings out a difficulty of interpretation of the numerical results. It takes intuitive insight as well as some numerical feeling to be able to interpret the computer output. One hopes, of course, that these would be acquired and learning facilitated during the course of the experiment.

A somewhat related experiment has been reported in the literature [see Brahic (1971)]. It goes back to the following question formulated by S. Ulam.

Consider a ball moving vertically between the floor and a player who hits it repeatedly and tries to make it move faster and faster. More precisely, look at a mass point in vertical motion between a fixed horizontal plane and another parallel plane moving up and down according to a periodic function $f(t)$ of time. At the planes, elastic collisions take place. No other forces act on the particle. Given f, will the velocity function $v(t)$ be bounded or can one find some f for which this is false?

Brahic designs an experiment where $v(t)$ is computed at each collision

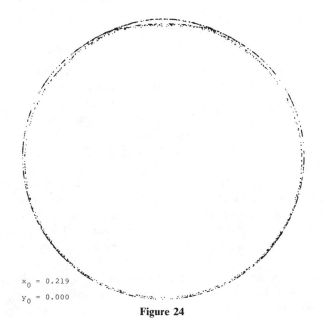

$x_0 = 0.219$

$y_0 = 0.000$

Figure 24

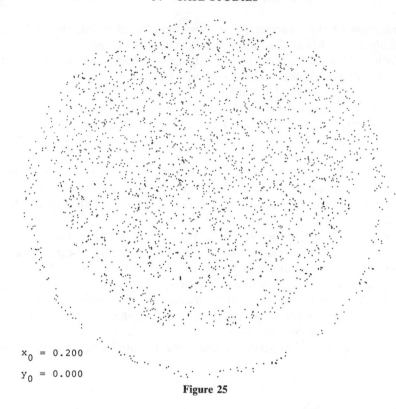

$x_0 = 0.200$

$y_0 = 0.000$

Figure 25

and plots the values for a large number of iterations. The resulting plots are beautiful and sometimes indicate the existence of ergodic zones, sometimes that of invariant curves, and, in a very particular situation, the mass point seems indeed to be "infinitely accelerated." The last statement should be compared, however, with Moser (1973, p. 58). The computing seems to have been done in batch on an IBM 360/65. It involved up to a million values.

The famous FPU paper describes an early and highly suggestive computer experiment [see Fermi *et al.* (1955)]. They consider a linear chain of coupled anharmonic oscillators. The coupling forces were assumed to be, for example, a cubic polynomial. What happens after a long period of time? Will equipartition of energy take place over the various modes of vibration?

The surprising fact is that the experimental result clearly points in a different direction. Instead of equipartition occurring, the energy seems to remain at low Fourier modes, a curious observation that has generated a good deal of thinking among mathematical physicists in the years since the experiment.

This brings us to the crucial question: What is a *successful mathematical* experiment? Clearly, if the experiment truly increases the analyst's insight to the extent that he can formulate and prove a nontrivial statement, then it is a success. But what if it leads to discarding his original hypothesis? Well, this must also be counted as a success; negative knowledge is also of value.

An experiment that does not generate hypotheses that are proved or counterexamples that change the picture is a failure. *A good mathematical experiment must have analytical consequences.*

EXPERIMENT 9: Neural Networks (Dynamic Problem)

In Experiment 4 we studied a problem in modeling a neural network statically at birth, and we shall now consider the corresponding dynamic problem where the synaptic strengths vary with time. These changes, representing long-time learning in the model, follow a nonlinear difference or differential equation.

The model building in Silverstein (1976) led to the basic model, expressed in vector–matrix form,

$$M(t + 1) = \rho M(t) + \sigma[y(t) + \alpha x(t)][y(t) + \alpha x(t)]^{\mathrm{T}}$$
$$x(t + 1) = M(t)[y(t) + \alpha x(t)] \tag{46}$$

In (46) $M(t)$ is an $n \times n$ matrix, depending upon time t. Its entries $m_{ij}(t)$ represent the strength of a synaptic junction between the ith and the jth neuron. The vector-valued $x(t)$, with values in $\mathbb{R}^n$, is the state vector, representing short-term memory. The other vector-valued function $y(t)$ is the input to the network at time t.

There are three global parameters in (46): the forgetting coefficient ρ, the modification coefficient σ, and the amplification factor α.

In (46) we have chosen a discrete-time model—a difference equation. This is convenient for computer simulations since we need not worry about discretization errors. In principle, we could just as well have chosen a continuous-time model—a differential equation, which is of a type more familiar to most analysts. As far as the modeling alone is concerned, the distinction is not important.

For the input $y(t)$ at least four types of behavior should be studied: constant input, periodic input, the case where $y(t)$ is a stochastic process, and systematic behavior (therapy).

For constant input $y(t) \equiv y$ over a long time interval it has been shown that the system has at most three critical points, only one of which is stable [see Silverstein (1976)]. This is done by standard methods of phase portrait analysis.

This sort of qualitative information is certainly useful, but we would also like to know how the phase portrait looks in quantitative terms. To learn about this the following experiment has been carried out. To be able to display the result easily we set $n = 1$; more about this later.

A program was written that solves (46) numerically for a given initial value $x(0)$, $M(0) = 0$, and collects the result in a $2 \times T$ matrix where the tth column means the 2-vector $(x(t), M(t))$; recall that $n = 1$ for the moment. This is then plotted on-line on a TSP plotter. If the user wishes he can ask for a continuation of the same trajectory as long as it is still inside the plotting area on the paper, or he can ask for a trajectory starting at another initial point $(x(0), M(0))$.

A typical result is exhibited in Fig. 26. The trajectories look fairly smooth; the successive changes are small from t to $t + 1$. An exception is for negative values of m, when the discrete behavior is more pronounced. It is doubtful, however, whether negative values have any significance.

The two unstable critical points are seen clearly in the upper half-plane to the right and to the left as the arrows indicate. The stable point is not as easily seen; it is just a little to the northeast of the origin.

It is of great help in computing and displaying the phase portrait to do this on-line. This also gives the user some feeling for the speed with which the system moves along the trajectories, and, in the present case, where the stable critical point was situated.

In Fig. 27 two solutions are plotted for periodic inputs, one coming from above and the other from below. Here the discretization is more

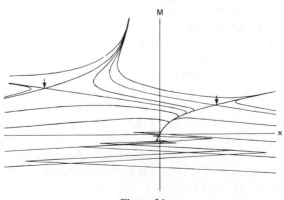

Figure 26

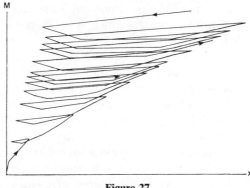

Figure 27

pronounced, and a smaller time step would have been better. Anyway, the tendency to a limit cycle is fairly clear.

For a stochastic process input we can expect less regular behavior for the trajectories, as seen in Fig. 28. Three of the trajectories shown seem to escape from the region in the figure, while the other six tend in general toward the stable critical point. Of course, they do not converge to any point in the plane, but the stochastic 2-vector $[x(t), M(t)]$ can be shown analytically to converge (weakly) in distribution. The diagram gives a rough idea of the spread of the limiting distribution.

The fourth type of input, where the purpose is to choose $[y(t);$ $t = 0, 1, \ldots, T]$ such that the $[x(t), M(t)]$ vector is brought back as quickly as possible to the stable region, has also been studied in a similar manner by Silverstein (1977a), but we shall not discuss the results here.

For $n > 1$, the case of real interest, we can compute the solution almost as easily as for $n = 1$. Of course, it will take more CPU time but no essential difficulty is added. The problem is to *represent high-dimensional*

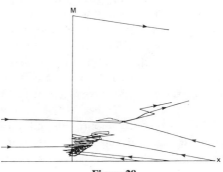

Figure 28

solutions in such a way that it helps our intuition. Note that we would like to "see" not only the state vector $x(t)$, which is in $\mathbb{R}^n$, but also the memory matrix $M(t)$, which is in $\mathbb{R}^n \times \mathbb{R}^n$.

Of course we could plot individual components of $x(t)$ and $M(t)$ or other linear functions of $x(t)$ and $M(t)$, but this would not be of much help. Perhaps it would be better to display $\|x(t)\|$ together with $\|M(t)\|$, since these norms would be more informative, or to display other characteristics of the matrix $M(t)$. We do not have any satisfactory way of choosing really essential functions to be plotted at this time.

In these experiments it would have been useful to have access to some interactive software for drawing phase portraits. Such a program should not be completely automatic, since this would be wasteful in that it would not exploit the experimenter's geometric intuition.

Instead the drawing should be done (interactively, of course!) in steps. The analyst should indicate a point, the trajectory should then be drawn through the point in both directions until it hits a boundary or forms a closed curve or settles down close to a limit cycle. This will involve a lot of testing and requires a good deal of CPU time.

The analyst should be able to point to what he thinks is a singular point and get the linearized analysis in a neighborhood of the point with eigenvalues computed. He should also be able to indicate some systematic procedure for drawing many trajectories together. We do not have such software but hope that it will be made available soon.

EXPERIMENT 10:   High-Dimensional Geometry

In some of the above experiments we have stressed the importance of being able to display the numerical results graphically. The need for some form of computer graphics is of course accentuated in geometric research.

Unless the space in which we are working is of very low dimension, say two- or three-dimensional, this requires more advanced computer technology than in most of the case studies presented so far. This unfortunate circumstance, which is likely to hamper extensive computer experimentation of this type, is due to the fact that it is *difficult to search higher-dimensional spaces systematically* for whatever effect we are looking for.

If the space is $\mathbb{R}^n$, or can be conveniently embedded in $\mathbb{R}^n$, we must, of course, have access to the transformations that are intrinsic to the underlying geometry. This could be the translation group, the affine group, or another elementary transformation group.

But this is not enough. To be able to "see" the phenomenon we must be able to display the results in two-dimensional form, for example, by displaying projections to random planes. Perhaps we may want to display other types of projections. Slicing data can also be informative: we select the data belonging to the region between two parallel planes and project the result. More generally, we select those elements of the computed data satisfying a given criterion and display the result.

It is not difficult to program such computing tasks. The hard part is to write a master program that loops through a set of selected operations in a more or less "intelligent" fashion. It could be completely systematic, which is seldom feasible, or it could exploit heuristic three-searching methods developed in artificial intelligence. Our experience in this is too limited to allow us to be very optimistic about future prospects.

Instead, one may have to *rely on the geometer's own intuition;* in other words, to employ interactive programs. The geometer would be sitting in front of a CRT on which successive pictures are flashed in accordance with his commands. To be able to do this requires much more sophisticated computer technology than we have discussed so far.

The CRT display must be fast and of high quality with little flickering and a high raster density. It must be possible to get the successive pictures rapidly. This requires a lot of number crunching; geometric computing is typically very demanding as far as CPU time is concerned. It also requires a fast channel to the central computer, implying practically that the terminal be close to the computing laboratory.

Let us mention an example from Banchoff and Strauss (1974). In complex 2-space consider the algebraic curve $z^3 = w^2$, $z = x + iy$, $w = u + iv$. This can be viewed as a real surface in $\mathbb{R}^4$, which could be parametrized by $z = \zeta^2$, $w = \zeta^3$, where $\zeta = re^{i\phi}$.

This defines a mapping of the disc into complex 2-space by ($r^2 e^{i2\phi}$, $r^3 e^{i3\phi}$) or in real form ($r^2 \cos 2\phi$, $r^2 \sin 2\phi$, $r^3 \cos 3\phi$, $r^3 \sin 3\phi$). This was displayed on the CRT screen after applying various transformations.

One result was Fig. 29, which shows a fivefold symmetry. It is the result of rotating by 45° in the x–y plane and then by 45° in the x–v plane. This symmetry had not been expected, but once it had been seen, it was verified analytically.

Many other examples were given in Banchoff (1978). A Vector General scope was used, driven by a small computer, the META 4, connected to the large computer. Such advanced and costly equipment cannot be expected to be generally available to the mathematical experimenter at present. As the cost decreases, however, one can hope that computer graphics of this sophistication will become economically feasible for more and more experimenters.

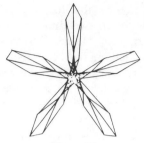

Figure 29

EXPERIMENT 11:   Strategy of Proofs

In another study we were investigating a sequence of functions

$$f_1(m; x_1),   f_2(m; x_1, x_2),   f_3(m; x_1, x_2, x_3),   \ldots ;$$
$$m = 1, 2, \ldots \qquad (47)$$

with the domain

$$-\infty \leq f_n(m; x_1, x_2, \ldots, x_n) < +\infty \qquad (48)$$

The task was to study the first m-value, say m_n, for which the maximum was attained,

$$f_n(m; x_1, x_2, \ldots, x_n) = \max_m f_n(m; x_1, x_2, \ldots, x_n) \qquad (49)$$

and to show that m_n could not grow faster than $o(n)$.

This has to be qualified by a measure-theoretic statement that we omit here; it is not necessary to know how the functions were found in order to understand the following design of the experiment.

It was easy to see that

$$f_n(1; x_1, x_2, \ldots, x_n) \equiv 0 \qquad (50)$$

and that m_n was always attained. The growth of m_n was harder to get at analytically and, to begin with, we were uncertain about what to expect.

To gain some insight into this an experiment was performed in which m_n was calculated for $n = 20, 40, 60, 80, 160$ for various choices of the x-values, as well as for a few very large n-values of the order 500. The results were plotted and also examined by the method that will be presented in Part III.

It became clear that the ratio m_n/n decreased toward zero as n was made large. The convergence was remarkably slow, however, and there was some empirical evidence that it behaved as $O(1/\ln n)$.

This certainly strengthened our belief in the conjecture, and even made it more precise. This is, of course, of psychological value, but what is more important, the computer printout *helped us to plan a strategy for proving the conjecture analytically*.

Note that in (48) we allow the value $-\infty$. This value can in fact be attained and, if it is attained for a certain value M_n, then it will also be attained for larger M_n.

But during the execution of the experiment the computer will be unable to handle this value, or one which is very large and negative. Execution will be halted and an error message (in this case DOMAIN ERROR) will be printed. It was noticed that this happened already for moderate values of m. A closer examination of this showed that M_n/n also seemed to tend to zero.

It was therefore decided that the proof should be organized as follows. Prove that $M_n = o(n)$. If this succeeds, just use the fact that $m_n \leq M_n$ to arrive at the conjecture $m_n = o(n)$.

This strategy, somewhat modified, succeeded! This experiment shows that it may be possible to learn from the experiment what is a reasonable conjecture, and also to be led to a method of proof.

CHAPTER 4

Other Areas

As this project continues we hope to be able to extend it to other areas of mathematics. At present we can make no claim that the same approach will yield anything substantial in these areas. A few comments are in order, however, based only on second-hand knowledge from reading the literature.

The computer is a finite device. This is sometimes a limitation when dealing with "continuous" problems. On the other hand, it makes it reasonable to expect that it should be useful in algebra, for example.

Many reports can be found in the literature on "computational algebra." Most of it is of the "production computing" type and does not concern us here. We refer the reader to Leech and Howlett (1970), Beck and Kolman (1977), and Churchhouse and Herz (1968). The aim in most of this work seems to have been concentrated on (1) the representation of algebraic objects by good data structures, and (2) algebraic algorithms. What data structures are suitable for representing and enumerating various algebraic structures? Can one get fast algorithms for carrying out certain algebraic tasks? Can one find lower bounds for the amount of computing that is needed?

In this context one should also point to the combinatorial algorithms: given a graph, find its connected components quickly; compute the chromatic polynomials efficiently; determine a spanning tree. Many other similar problems could be mentioned. See Auslander, Dunham, and North (1976) for many examples.

Such problems are connected with the theory of algorithms, in particular the study of complexity of computing tasks. Anyone doing experiments involving graphs and other finite structures should learn about this rapidly expanding area of research. Seemingly straightforward problems often turn out to have surprisingly efficient algorithmic solutions; see, for example, Nijenhuis and Wilf (1978).

The same is true for geometric computations. Plotting level curves efficiently is not as easy as it sounds. How quickly can one calculate whether a given set is connected via ϵ-chains? What numerical accuracy should be used in calculating winding numbers? The monograph by Shamos (1976) is a good reference for some questions in geometric computing.

It may well be that some areas in mathematical research do not lend themselves to computational experiments. In order to handle, for example, large matrices without special structure, so much CPU time and/or memory may be required that it should not be done interactively and in the way described in Chapter 2. Instead one should probably look for carefully designed algorithms, perhaps write the code in FORTRAN, and execute the compiled program in batch mode.

The same may be true for large partial differential equations. With one time and three space dimensions, interactive computing in a language implemented through an interpreter is probably impossible or prohibitively expensive. This is the reason why we have not mentioned mathematical experiments in nonlinear partial differential equations, for example, in fluid dynamics. Such experiments require a quite different computational set up from the one we assume here.

Problems that require extensive statistical analysis or large-scale linear programming are probably better dealt with in batch mode.

Many finite problems of a combinatorial nature also belong to the same group, where batch processing is the only feasible solution.

Admitting that this may be true, we would like to emphasize, however, that it may be helpful to start with a small-scale experiment to pinpoint the algorithmic difficulties and to get an idea of the numerical complexity. After such a pilot experiment one would strive for a more efficient code when the full-scale problem is attacked.

CHAPTER 5

Software Requirements

Having presented the case studies in Chapter 3, we can now sum up the requirements for software. They concern *the programming language, its implementation, the supporting system, and program libraries.*

We have already argued that APL is the only rational choice of *programming language* when the purpose is mathematical experimentation. APL generates great enthusiasm among its practitioners, but also an almost theological hatred among its detractors. The main arguments against it are inefficiency, inflexibility, and the fact that it is difficult to read and modify.

It is true that *efficiency* in terms of CPU time expended sometimes leaves something to be desired. The reason is that the standard implementation is not by compilation but by interpretation. This means that each statement in the code has to be interpreted each time that it is executed. A program that loops a lot will, therefore, lead to much interpretation, which can far exceed the amount of "useful" computing.

In passing, we remark that this feature is a characteristic of the implementation rather than of the language. If and when compilers become generally available for APL, this issue will disappear.

Another answer to this criticism is that the structure of APL, especially the array organization, allows the programmer to avoid explicit looping most of the time. If someone insists on programming in FORTRAN/ALGOL style, and some programmers do, then he will no doubt waste a good deal of CPU time. For this reason it is important that the user be instructed from the beginning in exploiting the array capability and primitive functions to their full power.

In addition it must be remembered that our main concern, at least for small- and intermediate-scale experiments, is efficiency in terms of the user's time, not in CPU time; the latter is often quite modest.

The argument that APL is *inflexible* also has an element of truth. Certain data structures are difficult to simulate, for example, trees and arrays of arrays. The output was also severely restricted in earlier forms of APL; the access to "format" has helped considerably. Occasionally a user will also run into trouble due to his limited ability to control the precision. On the whole, these drawbacks do not amount to much.

We would give greater weight to the criticism that APL is difficult to read and modify. Recall the quip: "APL is a write-only language." It is indeed infuriating to run into one of those clever and very long one-liners that solve a substantial computing task. Such code can be almost impenetrable for a reader and hopeless to modify. The same thing can be said about many-line programs where the flow of control is governed by lots of "go to" or "conditional go to" statements hidden in the middle of the statements. This will cause difficulties, especially when the author of the program has provided neither sufficient external documentation nor internal documentation by "describe" vectors or comments in the code.

To avoid opaque code like this one must insist that the programmer fully document the code. Furthermore, he should avoid being "too clever by half." It is better to write straightforward, easily understood code when we are dealing with software for experiments where no program is ever quite definite but will have to be updated and modified repeatedly. Achieve simplicity by avoiding spurious elegance!

Our experience of the various *implementations of APL* has been generally positive, except for an early version implemented on an 1130. The computer was too limited to be useful, both in storage, size of admissible arrays, and extent of the set of primitive functions. At present we favor the VS APL implementation since it allows us to use "execute" and "format," but this may change if more advanced implementations become available.

As far as the supporting *computer system* is concerned, the user will not be really aware of it as long as it is operating well. Assuming that it is robust, with long average intercrash time, and that the load is reasonable, the experimenter need not worry about the system. If the load increases so that response time increases noticeably, the early stage of an experiment will be done under psychologically unfavorable conditions. We are not referring here to the execution phase of the experiment; it is not very serious if it takes some time before the result of a substantial computation is returned. But if the average response time, say for a trivial "edit" command, is more than a second or two, the interactive user will experience frustration. This means that the load on the system is too heavy.

During our work we have had access to many available *programming libraries*. They were certainly useful but we know of no library tailored to the specific demands of mathematical experimentation. This is, of course, the reason why we have developed our own, which will be presented in

Part IV. To some extent it overlaps with existing libraries but its logical organization as well as its content, especially designed for experiments, should make it useful for other mathematicians. We have aimed for "clean" code, comparatively easy to read and modify, based on at least moderately efficient algorithms and with informative documentation including the underlying mathematics.

CHAPTER 6

Hardware Requirements

Our experience from the project has also led to certain demands for hardware: *a central computer or stand-alone computers, terminals, and computer graphics devices.*

The first question is whether to rely on terminals connected to a *central computer or to use a smaller independent computer.* The latter obviously has the attraction that there will be no interference with other users and no peak hours to avoid. This has to be balanced against the severe limitations in memory size, access to public libraries, and the possibility of convenient exchange of files with other users.

A stand-alone computer with, say, 64 kbytes in primary memory plus a large secondary memory, say, in the form of a floppy disk, would suffice for small and medium-size experiments. An attractive compromise is to use a stand-alone computer that can also be connected to a central machine; then the user would have the best of two worlds. What we have in mind is something like the IBM 5200 or more recent versions of it.

In the beginning of our project our work spaces were only 32 kbytes. Later this was increased to 64 kbytes, and, under a virtual machine, to about a megabyte. One seldom needs that much primary or virtual primary memory. It is easy to get spoiled!

The terminal or single computer can be hooked up to the central computer by telephone. We have also been using direct line hookup, which avoids dialing up each time the user is going to log on. The saving in time for the user may seem negligible, but it should be remembered that our sort of computing is typically done in short bursts separated by long "think time." Then the saving is of some psychological importance. This is not a major issue, however. Anyway, direct hookup is economically feasible only when the distance to the central computer is short.

The channel capacity of the line is also a parameter of some interest in cases where nontrivial amounts of information are to be transmitted. This seems to be the case in experimental mathematics, mainly when extensive plotting or other types of visual displays are required; more will be said about this later.

The *choice of terminal* depends on many considerations; an important one is the need for hard copy. A typewriter terminal produces hard copy, of course, but it is rather slow, not cheap, and can be quite noisy. It should not be installed where the noise would disturb other people close by. Thermal printers, for example, the Texas Silent, are quiet and fast. The one mentioned also has the advantage of being truly portable. If the terminal only displays programs and results on a CRT, which is fairly fast and cheap, we get no hard copy unless a secondary device of photoelectric type is also available or we take Polaroid pictures of the screen.

The terminal should have the full set of APL characters. Systems where APL characters are represented via ASCII or EBCDIC code together with some translation convention are clumsy and should be avoided if possible.

Modern terminals often display the text in such a way that the last few characters are hidden by the printing device. For the Texas Instruments Omni terminal one can have the device move away horizontally, while on the Diablo 1600 it moves away vertically. Editing programs can be bothersome in such cases, especially if insertions and deletions are displaced one or more characters. This seems difficult to avoid, however, at least at present.

The maximum length of the line printed is also of interest. If the terminal allows only short lines, say less than a hundred characters, this need not be serious for writing code, but it is a serious limitation when we want to display large arrays: tables or crude plots.

Our experience is that *hard copy is usually absolutely necessary*. Also, the copy produced by Xerox-like devices electronically linked to the terminal is of low quality, especially for geometric output, which is of paramount interest; however, the quality may be different for other devices that we have not tried. Our current view is that a typewriter terminal is preferable.

We now come to the very important question of *graphics facilities*. We have occasionally been using sophisticated computer graphics, say the IBM 2250 (early) and the Vector General (more recently). It would be unrealistic to require that the mathematical experimenter should have access to such advanced and costly technology. It is more reasonable to ask that x–y plotters be available and compatible with the programming language used by the experimenter, preferably APL. The compatibility is needed to avoid messy interface problems. Much of our work has been done using a simple TSP plotter or, more recently, Tectronix and HP plotters. The latter has the attractive feature of programmable choice of color.

A compromise would be to use a typewriter terminal for plotting. The plots will be quite crude, which is sometimes acceptable, unless the terminal has the capability to produce medium-quality plots by presenting a variety of extra symbols (slanted strokes, for example) in addition to the standard ones. The IBM 5200 mentioned above can be supplemented by a small bidirectional printer offering such a capability.

When computational experiments are done by a group of three or more people, and if it is desired to do this on-line with the members of the group present, the difficulty arises that the output at the terminal is not very visible to all in the group. We tried to solve this by installing a TV camera in front of the terminal and displaying the results on a 25-in. TV screen. This did not work out to our satisfaction, however. Either the TV camera was focused on a small subset of the printing area of the paper (then the text was readable, but only a part of geometric displays was seen on the monitor) or the camera focused on the whole width of the paper (the graphs could then be displayed in their entirety but the text symbols were barely visible).

The same problem arises when teaching experimentation on-line. We did not arrive at any satisfactory answer to how to deal with this technical problem, but it should be mentioned that one can send the signal with text (data) directly to the screen without any TV interface. In this way the quality of the picture will be acceptable.

One technological advance that is likely to have a profound impact on mathematical experiments is the development of microcomputers. As they become more powerful and cheaper they will make mathematical experimentation available for a much larger group of mathematicians than at present. Except for some experience (very encouraging!) with the IBM 5000 series we have not yet explored the potential of microcomputers for experiments.

CHAPTER 7

Practical Advice

The only way to learn how to carry out mathematical experiments on the computer is by actually trying some; *it cannot be learned by just reading about it*. Some advice will facilitate the learning, however; so we give some hints here. The book by Freiberger and Grenander (1971) contains further advice, although its main thrust is toward using the computer for teaching rather than research.

The first psychological barrier to be breached by the prospective experimenter is to learn the programming language, of course, in addition to the protocol of logging on/off and such minor details. Although APL is much closer to the spirit of mathematics than other languages, it takes some time to memorize the strange symbols and many conventions. Most mathematicians will find this stage of the learning process boring; it involves a good deal of rote learning.

Some textbook or manual on APL is needed and we mention the widely used book by Gilman and Rose (1976). A shorter but useful text is Hellerman and Smith (1976), which, however, is limited to the early 360 version. The IBM description of APL mentioned in the bibliography is complete but very tersely written; it is best as a reference. To help the experimenter in learning APL from the point of view of mathematical experiments we have included Part II. It also contains programs or parts of programs that may be of use for the experimenter.

After the elements of the language have been learned, it is recommended that the experimenter try out some simple versions of his own problems from his area of research. He should not hope for too much success initially, but it will make him aware of programming problems and ways of solving them that will be useful later on.

After a few weeks he will know enough to be able to begin experiments. It is tempting to try to write programs directly, while sitting at the terminal—composing at the keyboard. Our experience is that this should be avoided; it does not save time; just the contrary. The conventional way is better: draw a flowchart of how control is going to be passed from block

68

to block, try it on a trivial numerical example where the answer can be checked, and do the debugging step by step. The only exception is a very simple program with no "go to" or "conditional go to"; then the more experienced user can compose the code at the keyboard.

Any large program should be broken up into *modules,* each one to be written as a separate "function" and then connected by a master program. This will lead to a slightly longer and possibly less efficient program, but this is more than compensated for by the gain during the debugging–editing phase. It will gain time for the user if not CPU time; this is quite in accordance with the general approach outlined earlier.

Other instances will be mentioned in the following where one intentionally sacrifices CPU time for storage efficiency.

Write using *nmemonics for naming variables and functions.* This will make the names longer than necessary. In APL it takes time to interpret each character so that this recommendation leads to some waste of CPU time. It is usually worth it afterwards, perhaps months or years later, when the function to be executed needs some rewriting.

If the program produces much output data, it is advisable to include statements that print text that laconically describes what the numbers, strings, arrays, etc., represent. Otherwise it may be difficult for the experimenter to remember what the output signifies. Explicit formatting of the output is seldom needed.

When results are plotted one should include code to produce coordinate axes with tie marks to be able to see what scale is used a few weeks later when the scaling has been forgotten. Alternatively, one can make the program supply alphanumeric printout carrying this information.

When a larger experiment is planned in which we are looking for the presence or absence of a particular phenomenon, it is often best to first write a program that executes a particular experiment for fixed parameters. To this we add a *test program* that verifies whether the phenomenon we are looking for is there or not, and a *master program* that iterates the first two through a set of values that need not be predetermined. Instead it may be possible to achieve more informative results by letting the design of the next subexperiment be guided by the previous outcomes. If this is done, it is usually worth the trouble to include a fourth program that collects the empirical evidence and organizes it into readable form close to the user's intuition. The latter often implies that graphical display should be used.

When designing the program one sometimes has the choice of exploiting APL's array capability, which will avoid expensively repeated interpretation of the code. APL is excellent for *looping* (implicit looping), as long as one knows in advance (approximately) how much looping is required. If not, then the amount of (explicit) looping must be restricted, perhaps by batching some of the computing. Instead of repeating the loop again and again, we rewrite the inner loop to do a batch of subtasks. This will lead to some more computing than would be otherwise, but, since the amount of interpretation decreases by an order of magnitude, this is often time efficient.

On the other hand, if we avoid excessive looping by using a large array, we may instead run into *memory-size problems*. It is wonderful to have access to, for example, four-dimensional arrays, but they take up a lot of storage. This is the well-known trade-off between time and space in computing.

A serious difficulty arises in mathematical computer experiments when we are trying to *make infinite inference from finite data*. As mentioned earlier, this will be the case when we try to input in numerical terms statements like "continuity," "belong to L_1 or L_2," and "analyticity." It is difficult to advise the experimenter how to do this; only his own experience will help him to do it well.

It would be wrong to assert, however, that we cannot handle infinite structures exactly on the computer. Say that we look at a structure of numbers involving irrationals, for example $n_1 + \sqrt{2}\, n_2$. Dealing with vectors (n_1, n_2) we can make exact calculations; we can do the same for other extensions of $\mathbb{Z}$ or in dealing with the functions of a continuous variable x, $x \in \mathbb{R}$. Say that the functions are embedded in a vector space of finite dimension $\mathbb{R}^n$ or $\mathbb{C}^n$; then, again dealing with vectors $(x_1, x_2, \ldots, x_n)$, we can make exact linear computations. Many more cases of this sort could be added.

We now know enough about the design, execution, and analysis of mathematical experiments to approach APL carefully and in terms of experimental mathematics.

PART II

A MATHEMATICIAN'S GUIDE TO APL

CHAPTER 8

Language for Mathematical Experiments

1. MATHEMATICAL LANGUAGES

Ordinary mathematical notation is based on many peculiar conventions, sometimes inconsistent with each other, and often lacking in logic. It is firmly anchored in the mathematician's mind, which is why we seldom reflect on its peculiarities.

Since this statement is likely to provoke the mathematical reader to deny the allegation, we had better support it by examples. Functions are often written $f(x)$, or $f(x, y)$, and so on; the argument(s) is (are) enclosed in parentheses after the f or whatever symbol names the function.

But we write $\sin x$ (the parentheses can be left out); or x^2, where the function name appears up to the right of the argument; or $x + y$ rather than $+(x, y)$; or, even worse, $\sqrt{x}$. Sometimes the function name comes after the argument, as in the notation $n!$ for factorial. Worst of all, perhaps, is the notation for ratios:

$$\frac{3 + x}{5 + 3x}$$

The lack of consistency in function notation is, of course, not serious, since we are well versed in the art of interpreting formulas. It becomes serious, though, if the formulas are to be understood by a mechanical device, a machine such as the computer. Then complete precision is needed and we should not rely on the "intelligence" of the machine, at least not very much.

All programming languages have to be based on a stringent syntax. The usual languages, such as FORTRAN, BASIC, and ALGOL, have syntactic structures that have little in common with mathematical notation. This presents a psychological obstacle for the mathematical experimenter, not so much in learning the syntax (that is comparatively easy, actually easier than that of APL) but in using the language fluently. This may be one reason why mathematical experiments on the computer have not yet become widely accepted as a research tool by the mathematical community.

It is the author's conviction, based on a dozen years' experience, that APL is the only programming language today that is suitable for this task. Before we begin a detailed study of its syntax and usage, let us briefly review what the experimenter requires of the language.

2. REQUIREMENTS FOR THE LANGUAGE

We shall insist on three features for a language in order that it be acceptable for experimentation in mathematics.

The language shall be based on primitives close to the fundamental concepts of mathematics.

This implies that the language primitives shall include such notions as propositions, sets, vectors, functions, as well as the usual mathematical operations on them. For psychological reasons the language syntax must remind the user of the mathematical structure in which his experiment is immersed.

Other consequences are that the syntax must allow common operations such as function composition, subscripting of vectors and matrices, vector algebra manipulations, and, of course, the standard Boolean and arithmetic operations.

A second requirement concerns the ease of writing and modifying programs.

The language must have a syntax that facilitates fast and correct translation from the mathematical formulation into program statements.

This should also hold for making changes in a program that has already been written. Even if one has access to an extensive program library, our experience has shown that *it is only rarely that an experiment can be carried out using only library programs with no modification.* Most experiments seem to require tailor-made programs. It is also a matter of speed. The traditional way would be that the mathematician would tell a more or less professional programmer about his problem and the programmer would go away, then return the next week with his program. Let us be optimistic and assume that no misunderstandings would occur. Then the program would be correct, it could be executed, and the mathematician would have his result.

But this is too slow. One has to *keep up the momentum of research,* and this can only be achieved, realistically, if the mathematician writes his own program immediately and directly debugs and executes it. Interactive computing is therefore imperative. After all, we write formulas ourselves; we do not use professional "formula writers" in mathematics!

A third requirement has to do with the implementation of the language.

The implementation should be interactive and include tools for fast and efficient debugging and execution.

Note that we have not included any requirement that the implementation should be highly efficient as regards CPU time. In mathematical experiments "efficiency" should mean efficient use of the mathematician's time; the CPU-time requirements are seldom serious. In the rare instances when they are, we may have to turn to other languages and perhaps to execution in batch mode.

3. THE CHOICE OF APL

The only language that we know of at this time that satisfies our requirements is APL. The following lists its advantages:

(i) It is directly based on "good" mathematical primitives.

(ii) Once its syntax has been mastered it is extremely easy to use for doing mathematics.

(iii) Implementations exist that allow for easy editing and debugging.

Let us also consider some of the criticisms that have been raised against it.

(i) It is difficult to learn, compared to, say, BASIC or FORTRAN. This is true; it takes a while to memorize its primitives with the somewhat strange notation. But when the user has absorbed some of its basic features and begins to understand its highly coherent logical structure, he has access to a formidable arsenal of mathematical operations, much more powerful than other programming languages.

(ii) It is difficult to read. It is true that it can be painfully hard to read and analyze other people's codes. Sometimes it may be hard to understand *one's own code* six months after it was written.

This can be avoided, at least in part, by writing many short, clear statements rather than a few long and clever ones. It also helps enormously to include explanatory *comments* in the code itself as well as *internal documentation* in the form of character vectors stored in the work space used.

(iii)   It is inefficient. It is true that the common interactive implementations can lead to excessive CPU-time costs unless precautions are taken against it. The usual implementations are in the form of *interpreters,* not *compilers.* An interpreter translates one statement at a time and immediately executes it. A compiler translates the whole program into machine code, optimizes it, and then executes the whole object code. This implies that one and the same APL statement may have to be translated 10,000 times if the flow of control loops through the statement that many times. This is, admittedly, wasteful, but it can be argued that APL programs can often be written with little or no (explicit) looping. This is due to its array capability by which we can, for example, manipulate matrices and vectors without (explicit) looping. The user should try to avoid looping by exploiting APL's array capability.

When discussing efficiency one has to make up one's mind about what sort of efficiency one is thinking about: either *CPU-time* (and storage) *efficiency* or *user efficiency*. The latter means that the user's time is treated as a valuable commodity.

For most mathematical experiments our experience has shown that APL is highly user efficient.

4. HOW TO LEARN APL

The best way to learn APL is by using it, not by reading about it. Therefore, the reader is encouraged to sit down at a terminal as soon as he has read a few of the following sections and try some of the examples in the text. Do not be afraid of experimenting with all sorts of statements; sometimes you will be surprised by the result. You will make lots of mistakes in the beginning, but you will learn quickly.

Very soon you will become aware of the great potential of APL, but the real power emerges first in Chapters 10 and 11. It may be a good idea to take a quick look at Chapter 12, for defining your own functions, right after Chapter 9. This will enable you to save your statements in defined functions. Then you should return to Chapter 10.

Try problems from your own area of mathematics. In the beginning you may be frustrated because you do not yet have access to computational tools that you need. This is good, however, because it will show you the need for tools that you will soon encounter.

Do at least some of the exercises following the sections. Usually there are many answers. Some are given in the Appendix "Solutions to Exercises," but they are given only as indications of how one may attack the problems. You can probably think up better solutions.

When you become more skilled you will actually be able to write a whole program directly while sitting at the terminal. This is like "composing at the keyboard" for the musician. It is tempting to do this, but bitter experience has shown that it is usually safer, and not more time consuming, to first sketch the program with its control flow, careful naming of variables, and construction of algorithms, using pencil and paper.

A final word of advice before we start with the syntax: *Do not think of APL programming as an activity essentially different from mathematics. It is just its continuation by other means.*

CHAPTER 9

Mathematical Objects—
Data Structures

1. HIERARCHIES OF OBJECTS

Historically mathematics has evolved starting from simple objects and going to more general structures on increasing levels of abstraction. The simplest objects are perhaps the set $\mathbb{N}$ of natural numbers; then rational numbers, real numbers $\mathbb{R}$ and complex numbers $\mathbb{C}$. Other objects are truth values $\mathbb{B} = \{\text{TRUE}, \text{FALSE}\}$ and abstract symbols $A, B, C, \ldots$, treated just as characters.

These objects are then used to form sets and vectors, so that we arrive at $\mathbb{R}^n, \mathbb{C}^n, \mathbb{B}^n$, and so on. On these structures we introduce basic operations such as addition; multiplication by scalars; or, for sets, unions and complementation; or, for Boolean structures, conjunction, disjunction, and negation; or, for characters, concatenation.

On the next level of abstraction we consider functions between the structures already arrived at. For example, we consider linear maps $\mathbb{R}^n \to \mathbb{R}^m$, which can be represented as $n \times m$ matrices with real entries. Or we form the outer product from the product space $\mathbb{R}^n \times \mathbb{R}^m$ to $\mathbb{R}^{n+m}$ by multiplying all combinations of entries of the n-vectors with the m-vectors.

Still another level of abstraction consists of defining operators on spaces of functions. This process seems to be without end.

When we are going to compute in terms of objects on any level of abstraction the first problem that arises is *how to represent* in the language a certain type of object by some *data structure*—some construction that can be manipulated by the machine. Since mathematical objects are often infinitary in nature (for example, $\mathbb{R}$), but the machine is a finite device, we can expect to run into difficulties. We shall be forced to use approximations, and this will sometimes lead to difficult decisions on which representation to choose. See, however, the comment on p. 70.

The second problem is *how to specify* a data structure that corresponds to the object that we have in mind. Perhaps this can be done by a special assignment; for example, that the object should be given the value 13 exactly or $\sqrt{2}$ approximated to seven decimal digits. More often we need a more *general way of generating data structures,* however, as will be illustrated in the next sections.

Third, we need *names,* to be able to refer to certain objects and distinguish them from one another. These names should be chosen so that it is easy to remember which name corresponds to which object, *mnemonic naming is desirable.*

Fourth, we must *express the basic operations* on the objects in a way that is close to ordinary mathematical notation but subject to the technical constraints we have to obey, some of which have already been mentioned.

The *syntax* of the programming language tells us what combinations of symbols are legal. Learning syntax involves a good deal of memorizing, but the prospective user will soon discover that the many conventions in APL are not as arbitrary as it may seem at first. Instead, they cohere logically to each other. The best way to learn the syntax is by working at the terminal almost from the beginning; reading about it is not enough.

The *semantics* of the language defines what a legal statement means in terms of a computing task. Semantics also involves a fair number of conventions that the user must know, some of which may seem strange at first: for example, the order of execution and embedded assignments (see below).

The *implementation* of the language, which can vary from installation to installation, describes how the computing task is organized in the machine. We shall not devote much space to this since a user needs this knowledge only rarely. A user should have some idea, though, about the accuracy with which the basic computations are made, how much memory is needed to store the objects, and the approximate speed of operation for the machine he is using. The following assumes that VS APL is the implementation but most of it is independent of the implementation.

2. SCALARS

The values in $\mathbb{B}$, the base set for *propositional calculus,* are represented as TRUE = 1 and FALSE = 0; each one takes a single bit of storage.

On $\mathbb{B}$, the *monadic* (one argument) primitive function "~" means negation, so that ~1 means 0, and ~0 means 1.

On $\mathbb{B} \times \mathbb{B}$ we have the *dyadic* (two arguments) primitive functions ∨ for disjunction and ∧ for conjunction. This makes 𝔹 into a Boolean algebra, a statement that will be extended much further below. We have the usual truth table

X	Y	$\sim X$	$X \vee Y$	$X \wedge Y$
0	0	1	0	0
0	1	1	1	0
1	0	0	1	0
1	1	0	1	1

We also have some other primitive Boolean functions that are not needed as often, such as Nand ⩞ and Nor ⩡. They are typed by first typing ∧, backspacing, then ∼, and similarly for ⩡. Other Boolean functions will be mentioned later.

To make a variable, say X, take the value 0 we use a direct *assignment*

$$X \leftarrow 0$$

It requires, of course, one bit of memory to store X. If we also assign $Y \leftarrow 1$ and type in the statement

$$\sim (X \wedge Y)$$

followed by hitting the return key, the terminal will execute it and respond by typing the result

$$1$$

Here we used the names X and Y to denote logical variables. A name can be any string of alphanumeric characters starting with a letter and with no blank in it. For example, INTEGRALS and GREENFUNCTION are legally correct names.

In other words, we use the set CHAR consisting of all the characters on the keyboard. We treat CHAR as the base set for CHAR*, the monoid of finite strings with concatenation as the operation. The identity element is the empty string.

Say that we want to assign the variable NAME the value A, where A stands just for the alphabetic symbol. This is done by

$$NAME \; \leftarrow \; 'A'$$

enclosing the symbol required in single quotes. This requires one byte of memory. When we execute the statement

$$NAME$$

the result will be just

$$A$$

without quotes.

Consider now the set $\mathbb{Z}$ of integers. Let us assign

$$X1 \leftarrow 18$$

and

$$X2 \leftarrow {}^-3$$

Each integer requires four bytes of storage = 32 bits; one byte means 8 bits. Note that the negative sign, which appears at the upper left of 3, is part of the constant; it is not the (monadic) function that forms the negative of a number and will be mentioned later.

On $\mathbb{Z}$ we have the usual arithmetic functions "+," "−," "×" with dyadic meaning. Hence the assignment

$$X1+X2$$

will give

$$15$$

This makes $\mathbb{Z}$ the additive group (w.r.t. +) with 0 as the unit element, and, also the multiplicative semigroup (w.r.t. ×) with 1 as the unit element.

Warning. The internal computing in the machine for VS APL has a range of about 16 decimal digits, which puts a limit on the range of integers that can be precisely dealt with.

We have also monadic + meaning no change, monadic − (so that -3 gives $^-3$) changing the sign, and monadic ×. The latter is a bit unexpected. It means the sign function

$$\text{sign}(x) = \begin{cases} 1 & \text{if } x > 0 \\ 0 & \text{if } x = 0 \\ -1 & \text{if } x < 0 \end{cases}$$

On $\mathbb{Z} \times \mathbb{Z}$ the functions "=," "≠," "<," "≤," ">," and "≥" are meant as *propositions* with truth values, not as relations. The value computed by such a *function* is thus the truth value, so that the statement

$$15 > {}^-2$$

results in

$$1$$

while (when $X1$ is 18 and $X2$ is $^-3$)

$$X1 \; \ne \; X2$$

is computed as

$$1$$

Since the infinite set of reals can only be approximated in the internal computing done by the machine, attention must be given to the tolerance with which two numbers are identified. The relevant concept here is the *comparison tolerance* $\Box CT$; the symbol $\Box = $ "quad" will be discussed later in a more general sense. If two numbers differ by less than $\Box CT$ times the larger absolute value of the two, they are treated as identical. Unless the user issues a specific command (to be discussed later) the quantity $\Box CT$ is set to $1E^-13$.

Hence, if I and J are scalars, the *Kronecker delta* can be written as

$$I \; = \; J$$

Note.   The integers 0 and 1 also mean the truth values, not just the integers, but this ambiguity should not lead to any misunderstanding.

This may be the right place to tell the reader that APL expressions are *evaluated from right to left, but, of course, with the reservation that expressions enclosed in parentheses will be executed first.* Hence we get the slightly surprising result of

$$10-5+2$$

as

$$3$$

but, of course,

$$(10-5)+2$$

gives us the value

$$7$$

Warning.   If possible, avoid noncommuting dyadic functions. For example, to compute $x - 1 - a$ (in usual notation) the APL formula

$$X-1+A$$

is all right, but

$$^-1 + X - A$$

is easier to read, and this sort of writing leads to fewer mistakes.

We now turn to the field ℝ of real numbers. It takes 8 bytes = 64 bits of storage per real number; all computations are done in double precision. They will appear in decimal form or in the *exponential* form, so that $1369 = 1.369 \times 10^3$ means the same as 1.369E3. On ℝ we have all the monadic functions $+, -, \ldots$ defined in the way that naturally extends their definition, already discussed on ℤ, and also their dyadic form to ℝ × ℝ.

In addition to these functions we have monadic "÷," meaning reciprocal, so that

$$\div 4$$

gives

$$0.25$$

and its dyadic version

$$15 \div 3$$

gives

$$5$$

Warning. In APL the statement "0 ÷ 0" is legal and means 1, and there are similar more or less arbitrary conventions for other undetermined expressions (see manual).

Other primitive functions are "floor," typed as ∟. Monadically it means "integral part of" and dyadically "minimum":

$$\lfloor 1.21$$
$$1$$

and

$$21 \ \lfloor 5$$
$$5$$

Similarly, monadic "ceiling," typed ⌈, means the smallest integer at least equal to the argument, while its dyadic meaning is of course "maximum."

The absolute value of the scalar X is written $|X$, which is logical notation when "$|$" is thought of as a monadic function and with right–left order of execution in mind.

Dyadic $|$ is most useful and means residue, so that $X|Y$ is the remainder of Y modulo X; therefore

$$3 \mid 23$$

is

$$2$$

while

$$3 \mid {}^-4$$

gives

$$2$$

Warning.   For the special case where X is zero, the result is defined as Y.

The elementary transcendental functions are introduced for powers and logarithms.

$\star X$	e^x
$A \star X$	a^x
$\circledast X$	$\ln x$
$A \circledast X$	$\log_a x$

Note that $\circledast$ is formed by typing "$\circ$" (large circle on the keyboard over O), backspace, $\star$.

Let us exemplify what we have learned so far by computing Pythagoras' theorem. In a triangle with a right angle denote the hypothenuse by SIDE3 and the other two sides by SIDE1 and SIDE2. Then

$$SIDE3 \leftarrow ((SIDE1 \star 2) + SIDE2 \star 2) \star .5$$

For trigonometric functions, with Y in radians,

$(-X)\circ Y$	X	$X\circ Y$
$(1-Y*2)*.5$	0	$(1-Y*2)*.5$
ARC SIN Y	1	*SIN Y*
ARC COS Y	2	*COS Y*
ARC TAN Y	3	*TAN Y*
$(-1+Y*2)*.5$	4	$(1+Y*2)*.5$
ARC SINH Y	5	*SINH Y*
ARC COSH Y	6	*COSH Y*
ARC TANH Y	7	*TANH Y*

Mnemonic: 1, 3, and 5 for the odd functions, 0, 2, 6 for the even ones, negative values for inverse functions.

Warning. For the many-valued inverse functions, a principal branch is selected [see APL Language (1976, p. 24)] that may cause trouble unless it is kept in mind.

The monadic "o" means multiply by π so that

$$\circ 1$$

results in

$$3.141592654$$

The *factorial* is denoted by "!," where ! is written by typing first ' (the quote over K on the keyboard), backspace, and period. It extends to $\Gamma(X + 1)$ when X is nonintegral. Similarly the dyadic ! computes the *binomial coefficient* y take x, i.e., $\binom{y}{x}$. It also extends to nonintegral values of x and y in the same way as monadic!

Now consider the random-number generator. Monadic "roll," say $?N$, gives a pseudorandom number (actually computed by a multiplicative congruence generator) uniformly distributed over the integers 1, 2, . . . , N. The symbol $?$ can be found above Q on the keyboard: mnemonic, Q for query. At each new execution the results are supposed to be (pseudo)stochastically independent.

Dyadic "roll" or "deal," say $X?Y$, should, strictly speaking, be presented later since it is vector-valued. It produces a sample of X elements sampled with uniform probabilities *without replacement* from the integers

$1, 2, \ldots, Y$. We could get, for example, by executing $3?5$

$$2 \quad 5 \quad 1$$

and executing the *same* statement another time, for example,

$$4 \quad 3 \quad 5$$

Warning.   The random-number generator is initiated by setting a random link to some value. A user who wishes to get reconstructable data should, therefore, set the random link himself rather than leave this to the system [see APL Language (1976, p. 51)].

Remark.   Complex scalars have not yet been included in APL so that the set $\mathbb{C}$ is not directly available. We must deal with them as 2-vectors, which will be discussed below.

Exercises

1.   Verify that the fractional part of X can be computed as $1|X$ and that the truth value of "x is even" for an integer x can be obtained as $0 = 2|X$.

2.   Verify that $X|Y$ can be obtained as the APL expression

$$Y - X \times \lfloor Y \div X + X = 0$$

3.   Write an APL statement that computes the function $f(x, y)$:

$$f(x, y) = \begin{cases} x/y & \text{if } y \neq 0 \\ x & \text{if } y = 0 \end{cases}$$

4.   Write an expression for the Beta function

$$B(x, y) = \int_0^1 t^{x-1}(1 - t)^{y-1} \, dt = \frac{\Gamma(x)\Gamma(y)}{\Gamma(x + y)}$$

5.   Derive an expression for a random variable X, taking the values 0 and 1, and with

$$P(X = 1) = m/(m + n)$$

where m and n are natural numbers.

6.   Verify that

$$A \div B + C \div D + E \div F$$

means a simple continued fraction.

7.   Verify that

$$A - B - C - D - E - F$$

means an alternating series.

3.   VECTORS

Using truth values, integers, or real numbers, vectors of arbitrary (finite) dimension can be formed by direct assignment

$$V \leftarrow 3.6 \quad 4.75 \quad {}^-2$$

We have separated the components by blanks but commas may also be used. If the components are given by symbols (names) with already defined values, commas must be used as in

$$VECTOR \leftarrow V1,V2,V3,V4$$

Vectors can also be formed from characters but no blanks or commas are used to separate the components. The reason is that character vectors are often used as text and a blank is treated as a special character. The vector should be enclosed by single quotes

$$V \leftarrow \text{'THE RESULT IS='}$$

Note that the blank (the empty set) is itself treated as a character here.

We shall often think of vectors as (discrete) functions. The ith component, or in APL notation $V[I]$, is then the value associated with the argument i.

In this way we arrive at the structures $\mathbb{B}^n, \mathbb{Z}^n, \mathbb{R}^n$ as will become clearer in a moment. We shall first mention two ways of generating certain vectors that will be used as building blocks later. We often need the vector $(1, 2, \ldots, n)$ and one of the most useful functions, monadic iota, does this directly

$$X \leftarrow \iota N$$

so that $\iota 1500$ generates the vector $(1, 2, 3, \ldots, 1500)$. This function is also called the *index generator.*

To represent a vector, all of whose components are equal, we use the function *reshape,* dyadic ρ. Mnemonic: ρ is situated over R on the keyboard; R for reshape. The statement

$$N \rho X$$

generates an N-dimensional vector, all of whose components are equal to X. Of course, N must have been defined as a natural number and X as a scalar. Hence $V \leftarrow 100\rho1$ generates the 100-vector $(1, 1, 1, \ldots, 1)$.

Given a vector V, how do we find its dimension? This is done by monadic ρ, the *shape function,* so that

$$\rho V$$

gives

$$100$$

when V is defined as above.

The behavior of ρ in the following special cases may seem peculiar but plays an important role in APL. If X is a scalar, then ρX denotes the empty set. Since the empty set is needed sometimes, for example, in order to generate blanks, it is of interest that we can get it by writing, say, $\rho1$. If X is the empty set, ρX is defined as zero. Thus we get $\rho \rho1$, meaning zero!

We come to an important convention, one of those that make APL so natural for the mathematician. Given a vector V, the result of a primitive monadic function f to V is simply the vector obtained by applying f to each component of V. For example, applying "floor" to a vector

$$\lfloor \ 3.1 \ 2.95 \ 5$$

means the vector

$$3 \ 2 \ 5$$

Similarly, a primitive dyadic function f applied to two vectors $V1$ and $V2$ means the vector whose components are computed componentwise by f applied to $V1$ and $V2$. For example, if $V1 \leftarrow \iota3$ and $V2 \leftarrow 3\rho2$ we have

$$V1 \ + \ V2$$

equal to

$$3 \ 4 \ 5$$

and

$$V1 \ \lfloor \ V2$$

equal to

$$1 \ 2 \ 2$$

In order that this convention make sense we must demand that the two vectors have the same dimension. One exception, however, very convenient it will turn out, is that one of them may be allowed to be a scalar.

Hence

$$9 \times V1$$

means

$$9 \quad 18 \quad 27$$

and

$$2 * V1$$

means

$$2 \quad 4 \quad 8$$

As an example, we mention formal manipulations of polynomials. All polynomials of degree n

$$p(x) = a_0 + a_1 x + \ldots + a_n x^n$$

with coefficients from the real field can be represented by $(n + 1)$-vectors of the form

$$POL \leftarrow A0, A1, \ldots, AN$$

The sum and difference of polynomials as well as multiplication by scalars is, of course, isomorphic with the same operations on the corresponding vectors. Multiplication of polynomials is a different story; it corresponds to convolutions, which we are not yet ready to compute.

How about differentiation of polynomials? This is very easy: the derivative is represented by just the n-vector

$$DER \leftarrow POL[1 + \iota N] \times \iota N$$

Similarly, the indefinite integral, with the arbitrary additive constant set equal to zero, is given by the $(n + 2)$-vector

$$INT \leftarrow 0, POL \div \iota N + 1$$

If we want APL statements valid for polynomials of variable degree, so that N is not given in advance, we should use instead

$$DER \leftarrow POL[1 + \iota^- 1 + \rho POL] \times \iota^- 1 + \rho POL$$

and the analogous expression for the integral.

We now return to look at some mathematical structures that we can handle by the primitive functions introduced so far.

Operating on n-vectors over integers $\mathbb{Z}^n$ and using dyadic $+$ together with dyadic $\times$ with one argument an integer and the other an n-vector, we get $\mathbb{Z}^n$ viewed as a module over the ring of integers with 0 as additive unit and 1 as multiplicative unit.

Operating on n-vectors of real numbers with dyadic $+$ and dyadic $\times$ with scalars we get $\mathbb{R}^n$ viewed as a vector space, but we do not yet have the inner product of a vector space.

With $n = 2$ we get the complex field $\mathbb{C}$ if we interpret the first component as the real part and the second component as the imaginary part. We use here only dyadic $+$ and with multiplication of two vectors

$$Z1 \leftarrow U1,V1$$
$$Z2 \leftarrow U2,V2$$

defined as

$$Z \leftarrow ((U1 \times U2) - V1 \times V2),(U1 \times V2) + U2 \times V1$$

Note the parentheses. Results are similar for division. The absolute value of $Z1$ would be given by

$$ABS \leftarrow ((U1*2) + V1*2)*.5$$

Indexing of vectors in general is done by brackets, so that $V \leftarrow 2 + \iota 5$ is a 5-vector with $V[1]$ equal to 3 and $V[5]$ equal to 7.

Warning.   The index range in APL always starts at 1 (unless the system has been instructed to start at 0; see below) and this can cause confusion when the "natural" starting value is another one.

To change the index origin, use the systems variable (the conventions vary between different implementations)

$$\Box IO \leftarrow 0$$

and to change it back say

$$\Box IO \leftarrow 1$$

Many other systems variables, all preceded by $\Box$, will be encountered later.

This is extended to using several index values. If V is an n-vector and W is a vector consisting of m numbers, each one among 1, 2, 3, . . . , n, then $V[W]$ is an m-vector consisting of components of V, possibly with repetitions. For example,

$$(3 + \iota 5)[2\ 2\ 1\ 4]$$

means

$$5\quad 5\quad 4\quad 7$$

The reader should convince himself that $V[\iota\rho V]$ is just V itself when V is a vector.

To illustrate this sort of indexing by an elegant example, let V be an n-vector. We want to generate a random permutation of V, which can be done simply by the statement

$$V[N?N]$$

Warning. Parentheses indicate order of execution while brackets indicate components.

Another mathematical structure is obtained by considering all (finite-dimensional) character vectors with dyadic "," as the rule of composition. It is the free semigroup over the given character set with the empty vector playing the role of unit. Note that ' ' (with one blank between the quotes) is not a unit since a blank is treated as a character.

Warning. The scalar 1 is not the same as the 1-vector with the component 1. Monadic ρ applied to it results in the empty set in the first case, in the number 1 in the second.

Exercises

1. Let $[A, B]$ stand for the "discrete interval" of the integers $A, A + 1$, $A + 2, \ldots, B$. Find an APL statement using monadic "iota" to generate this vector.

2. Generate the vector $A, A + K, A + 2K, \ldots, A + NK$.

3. To generate a vector $(v_1, v_2, \ldots, v_n)$ with the step function behavior

$$v_i = \begin{cases} 0 & \text{for} \quad 1 \le i \le m \\ 1 & \text{for} \quad m < i \le n \end{cases}$$

use the monadic iota together with dyadic $<$.

4. Generate the character vector $AA \cdots ABB \cdots B$ of n As followed by m Bs.

5. Generate the $3n$-vector of form $ABCABC \cdots ABC$ where A, B, C are the letters. One could use, for example, monadic "iota" together with dyadic $|$, or better, use dyadic ρ, which automatically extends a vector periodically.

4. SETS

To be able to handle sets we shall often employ vectors. As mentioned in the preceding section it is sometimes advantageous to think of vectors (and other arrays) as functions of the index. In other cases it is more natural to think of them as sets, although this should be taken with a grain of salt. The vector $V \leftarrow 3\ 5\ 4\ 5$ is not the same as the set $\{4, 5, 3\}$. Vectors are ordered multiples, sets are not; vectors allow repetitions, whereas this is irrelevant for sets.

With this in mind let us consider sets of elements from some finite universe U. Without loss of generality we can choose $U \leftarrow \iota N$; this only means labeling the elements by the first N natural numbers.

Let V be a subset of U. How do we find out whether some given element I from U belongs to V? This is done by the dyadic function ϵ, *membership*. Mnemonic: it is situated over E (epsilon!) on the keyboard.

The statement

$$I \epsilon V$$

gives the truth value of the proposition. It is immediately extended to the case where the left side is a vector, say W, and

$$W \epsilon V$$

results in a vector of dimension ρW with zeroes and ones indicating the truth values of the respective propositions $W[I]\epsilon V$ for $I = 1, 2, \ldots, \rho W$. For example, with

$$V \leftarrow \quad 3 \quad 5 \quad 8$$

$$W \leftarrow \quad 2 \quad 3 \quad 8 \quad 1 \quad 3$$

we get $W \epsilon V$ evaluated as the 5-vector

$$0 \quad 1 \quad 1 \quad 0 \quad 1$$

How do we handle the usual operations on sets: union, intersection, complementation, and inclusion (viewed as a proposition)? Consider the union of two sets $V1$ and $V2$. We would, of course, use the expression $V1$, $V2$, but this could lead to repetitions of elements, even though we had made sure that $V1$ and $V2$ had no repetitions. We need some way of "cleaning" a vector by which repetitions are removed. To be able to do this we must compare all components with all other components, and then remove possible copies.

This comparison can be done with considerable elegance in APL, using one of its most powerful concepts, the operator "outer product."

This will be discussed in Chapter 11 so that we have to postpone the treatment of set operations till then.

Warning. The empty set comes in many forms in APL (as well as in other discourses). For example, we can form

$$(0 \quad 2) \rho 0$$

which is an Alice in Wonderland array with 0 rows and 2 columns, all of whose entries are zero!

Exercises

1. Think of the arbitrary vector W as a function of index. Write the truth value for the proposition that the equation $W = $ constant C has no root over the given domain of the function.
2. If V is a 4-vector and W is an arbitrary vector, obtain an APL expression (a bit clumsy; you will do better later on) for the number of elements in V that belong to W.
3. V is a vector of integers. Write a statement for the truth value of the proposition that V has at least one component equal to 1 and 2 or 3.

5. MATRICES

To define large matrices, which are two-dimensional arrays, directly on the typewriter is a bit awkward since the typewriter operates essentially in a linear fashion. We shall use dyadic ρ, *reshape,* but with a vector as left argument. It is easiest to show by an example how it works.

Executing

$$2 \quad 3 \quad \rho \quad \iota \quad 6$$

one will get the matrix with two rows and three columns

$$\begin{matrix} 1 & 2 & 3 \\ 4 & 5 & 6 \end{matrix}$$

Executing the statement

$$4 \quad 2 \quad \rho \quad 7 \quad 2 \quad 3$$

one will get the matrix with four rows and two columns (7   2   3 will be periodically extended)

$$
\begin{array}{cc}
7 & 2 \\
3 & 7 \\
2 & 3 \\
7 & 2
\end{array}
$$

Executing

$$M \leftarrow V1 \rho V2$$

where $V1$ is a 2-vector of natural numbers and $V2$ is an arbitrary vector, the result is a matrix with $V1[1]$ rows and $V1[2]$ columns in which $V2$ has been inserted in row after row, *if necessary by extending V2 periodically.*

Notice that if the last assignment is executed, nothing is printed at the terminal; the effect is to store the matrix under the name M.

If we then ask for the *shape* of M by monadic ρ we get by executing

$$\rho M$$

the result (four rows and two columns)

$$4 \quad 2$$

The primitive functions, both monadic and dyadic, apply to matrices in complete analogy with what was said for vectors in Section 3. If M is as above and

$$N \leftarrow 4 \ 2 \ \rho \ \iota \ 8$$

so that N is

$$
\begin{array}{cc}
1 & 2 \\
3 & 4 \\
5 & 6 \\
7 & 8
\end{array}
$$

then

$$M + N$$

gives

$$
\begin{array}{cc}
8 & 4 \\
6 & 11 \\
7 & 9 \\
14 & 10
\end{array}
$$

and

$$M \times N$$

results in

$$
\begin{array}{rr}
7 & 4 \\
9 & 28 \\
10 & 18 \\
49 & 16
\end{array}
$$

Warning. $M \times N$ does not mean the matrix multiplication of linear algebra; it will be treated in Chapter 11.

Character matrices are handled in the same way. Execute

$$M \leftarrow 2 \ 19 \ \rho \ 'A'$$

which gives us a 2×19-character matrix, all of whose entries are A. We now fill it in with text. The first row is indicated by 1 followed by a semicolon, so that the first row is

$$M[1;] \leftarrow 'RUNGE \ KUTTA \ METHOD \ '$$

and for the second-row vector

$$M[2;] \leftarrow 'GIVE \ INITIAL \ VALUES'$$

Note that the first-row vector has an added blank at the end; this is to conform to the length of the second row.

This has changed M from the earlier assignment. If we execute

$$M$$

we get the character array

$$RUNGE \ KUTTA \ METHOD$$

$$GIVE \ INITIAL \ VALUES$$

A matrix, say MATRIX of size $n \times m$, has row vectors MATRIX[I;], column vectors MATRIX[;J], and has the element MATRIX[I;J] in the Ith row and Jth column.

Alternatively one could have used

$$M \leftarrow 2 \ 19 \ \rho \ 'RUNGE \ KUTTA \ METHOD \ GIVE \ INITIAL \ VALUES'$$

It is extremely important to be able to form submatrices (and subvectors and subarrays, in general). If A and B are vectors of arbitrary dimension, A with elements from $\iota(\rho MATRIX)[1]$ and B from $\iota(\rho MATRIX)[2]$, then MATRIX[A;B] denotes a matrix with ρA rows and ρB columns. For example, with N given above

$$N[2 \ 3 \ ; \ 1 \ 2 \ 1]$$

means

$$\begin{matrix} 3 & 4 & 3 \\ 5 & 6 & 5 \end{matrix}$$

Note that A and B can very well contain repetitions, as above, where $A \leftarrow 2 \quad 3$ and $B \leftarrow 1 \quad 2 \quad 1$, and with any order between the elements.

The transpose of a matrix is formed by monadic transpose, "$\lozenge$," formed by overstriking the large circle o by the left slash. Hence

$$\lozenge N$$

is evaluated as the 2×4 matrix

$$\begin{matrix} 1 & 3 & 5 & 7 \\ 2 & 4 & 6 & 8 \end{matrix}$$

Also, monadic *ravel,* indicated by a comma ",", changes a matrix back to vector form by reading the components in the first row, then the second, and so on. Hence

$$,\lozenge N$$

gives the 8-vector

$$1 \quad 3 \quad 5 \quad 7 \quad 2 \quad 4 \quad 6 \quad 8$$

If M is a nonsingular square matrix of real numbers, then the inverse, traditionally denoted by M^{-1}, has the APL expression

$$MINVERSE \leftarrow \boxminus M$$

It is formed by overstriking the quad symbol $\square$ by the division sign $\div$.

Similarly, the traditional expression $M^{-1}N$ is in APL represented by

$$X \leftarrow N \; \boxminus \; M$$

but, interestingly enough, this expression can be meaningful for nonsquare and for singular matrices and has a most attractive interpretation.

This is how it works. Say that N is $n \times k$ and M is $n \times l$ (they must have the same number of rows) and that the columns of M are linearly independent. Then X means the matrix with l rows and k columns, which is defined as the *least-squares solution of*

$$\sum_{i,j} \left(N_{ij} - \sum_r M_{ir} X_{rj} \right)^2 = \min$$

This is unconventional and deserves some further remarks. If N is a column vector, $k = 1$, then the least-squares solution X will be an l-row vector. This will be useful when one wants to compute, for example, polynomial approximations to given functions or observed data, or more generally for approximation theory in L_2-norm. We shall return to such questions in terms of linear algebra in Chapter 11.

Exercises

1.   Derive APL expressions and execute them to solve the system of equations

$$
\begin{aligned}
3x_1 - 2x_2 - 7x_3 \quad\quad &= 12 \\
x_2 + x_3 - x_4 &= 3 \\
x_1 - x_2 \quad\quad + x_4 &= 8 \\
x_1 + x_2 + x_3 + x_4 &= 9
\end{aligned}
$$

Use ⊟ to do this.

2.   For simple graphic display at the terminal (without using a plotter) it is often a good idea to use character matrices. Write an expression that displays a chessboard, B for black and W for white. Use the vector 'BW' and the residue function.

3.   Think of a given real matrix M as a function of the two subscripts. To plot the region in which this function is greater than some threshold T one could form the character matrix (with the blank in the statement!)

$$(' \ * ')[1+M>T]$$

Execute this statement for some matrix M, for example,

$$M \leftarrow 10 \ 10 \ \rho \ \iota \ 7$$

and $T\leftarrow3$.

4.   For a given real matrix M, write a statement that calculates the symmetrized matrix $\frac{1}{2}(M + M^T)$.

6.   OTHER ARRAYS

With a somewhat unorthodox terminology, an APL scalar is said to have rank 0, a vector is said to have rank 1, and a matrix has rank 2. We also speak of arrays of higher rank = $\rho\rho A$ = the length of the shape vector. We shall discuss rank 3 here, and it will be obvious how this extends to higher values. In data analysis, high-rank arrays are powerful for storing and analyzing data with multiple categorization; analysis of variance and covariance should also be mentioned in this connection.

Tensor analysis, say, in differential geometry, can be handled very elegantly by APL arrays of higher rank.

Consider the assignment of the character array

$$ALPHA \leftarrow 2 \ 3 \ 4 \ \rho \text{'}ABCDEFGHIJKLMNOPQRSTUVWX\text{'}$$

and execute ALPHA. It results in two plane arrays separated by a blank line

$$ABCD$$
$$EFGH$$
$$IJKL$$

$$MNOP$$
$$QRST$$
$$UVWX$$

Note that each plane has shape 3   4.

In analogy to what is true for matrices, the element in the Ith plane, Jth row, Kth column is written ALPHA[$I;J;K$]. Notice the two semicolons separating the three *axes*, which is APL-ese for subscripts (coordinate axes).

Leaving out one or more of the subscripts means that its (their) total range should be used. For example,

$$ALPHA[1;;]$$

is the first of the two planes shown above and

$$ALPHA[;2;4]$$

is simply the 2-vector *HT*.

APL is *array oriented*, which is just what one needs in many mathematical experiments. Rank-1 arrays are related to vector spaces, rank-2 arrays to linear operators, and higher-rank arrays can sometimes also be given natural interpretations as linear operators: for example, arrays of rank 4 can be used to parametrize linear operators from $\mathbb{R}^2$ to $\mathbb{R}^2$, or from $\mathbb{C}$ to $\mathbb{C}$.

The data structures that we have described, together with others that will be derived from them, are usually enough for our needs. There are, however, a few others that deserve some comment.

Any sufficiently experienced user of APL will ask whether it is possible to form, say, a matrix where each element is itself a matrix, or more generally, *arrays of arrays*. Unfortunately, the answer is no. In future extensions of APL this feature may be incorporated, but at present it has to be done by direct programming by the user. While this is no doubt inconvenient, one can live with it.

One way of handling this is by *list structures* using pointers. In the first matrix we denote the element matrices by natural numbers or by other labels. These labels are associated with particular matrices by direct statements somewhere in the software developed by the user. We may write

$$ARRAY \leftarrow 2 \; 3 \; \rho \; 1 \; 2 \; 2 \; 1 \; 3 \; 4$$

and use as pointers assignments of the type

$$MATRIX1 \leftarrow 2 \; 2 \; \rho \; 2 \; 4$$
$$MATRIX2 \leftarrow 2 \; 2 \; \rho \; 15$$

. . .

The execute and format functions (to be described later) will be useful in this context.

Such list structures employing pointers can often be used; it is slightly cumbersome but after some experience the user will not find it too bad.

A data structure, actually related to the above, that is sometimes needed is the *tree*. In Iverson's original description of APL [see Iverson (1962)] trees were envisioned as APL structures; they may also be implemented in the future. Other graphs, such as partially ordered sets, are also missing and have to be treated by programming.

Exercises

1.   The monadic transpose ⍉ that was used in the preceding section for matrices also applies to arrays of higher rank. Its effect is to reverse the order of the axes. What is ⍉ALPHA, where ALPHA was given above? Try to write out what it should be and compare the result by executing ⍉ALPHA.

2.   A real array A of rank 3 has its first dimension equal to 4. Write a statement that computes a rank-2 array (matrix), each element of which is the maximum of the respective elements in the four planes.

3.   Compute a rank-3 array with 2 planes, 2 rows, and 25 columns, with random entries uniformly distributed. In the first plane all entries should be distributed over the range 1–10; in the second plane over 11–20.

CHAPTER 10

Functions—APL Statements

1. PRAXIS OF WRITING APL CODE

So far we have used only the simplest primitive functions, leaving some of the more powerful ones to this chapter. In spite of this we are already able to form expressions for a multitude of mathematical tasks, and we have reason to pause and reflect upon what we already know and to learn more about the *programming utilities.* They do not belong to the syntax of the language, strictly speaking, but are part of the system in which the language is implemented.

Only one sort of utilities will be treated here: *error messages;* others will be presented in Chapter 12.

When we type in a statement and execute it, the result will sometimes not be what we had intended but an error message. Something is wrong and the system points to an error (there may be others not yet detected).

Error messages come in different forms. Say that we try to execute

$$(X + (Y - Z) * 2 \div U$$

This will result in a *syntax error;* the system will respond by

$$SYNTAX\ \ ERROR$$

$$(\underset{\wedge}{X} + (Y - Z) * 2 \div U$$

Obviously the parentheses do not match. We have forgotten one after 2. On the other hand, the expression that we intend to mean $x + yz$ in mathematical notation,

$$X + YZ$$

will result in VALUE ERROR; we have forgotten the multiplication symbol $\times$ between Y and Z and the system thinks that we refer to a variable named YZ, but the value of YZ has not been defined.

To avoid difficulties with parentheses the following hint may be of

100

help. We are going to type a fairly long and nested expression. Type it first with no or few parentheses, but leave lots of blanks between all the symbols. When this is done back space and insert matched pairs of parentheses and let the typing element move back and forth until all parentheses needed are typed. This procedure is allowed since it does not matter in what order the symbols are typed, only the order in which they finally appear in the statements matters.

Next consider *domain errors*. This is what we will get if we try to execute

$$NUM \div DEN$$

if the variable DEN has already been assigned the value zero (except if NUM also happens to be zero; see p. 83); a similar error occurs for other functions when the argument does not belong to the function.

With arrays we sometimes get error messages that indicate that arrays do not fit each other. Say that

$$V1 \leftarrow 1\ 2\ 3$$
$$V2 \leftarrow 4\ 5$$
$$V3 \leftarrow (2\ 2)\rho\ 1\ 2\ 3\ 4$$

Then

$$V1 + V2$$

will give a *length error* as in

$$LENGTH\ ERROR$$
$$V1 + V2$$
$$\wedge$$

indicating that the two vectors cannot be added since they are of unequal length.

The statement

$$V2 + V3$$

will yield a *rank error* as shown by

$$RANK\ ERROR$$
$$V2 + V3$$
$$\wedge$$

We have tried to add arrays of ranks 1 and 2, respectively; a vector and a matrix. This is not allowed.

If we try

$$V1[4]$$

or

$$V1[0]$$

we will get an *index error*. The indices used are outside the range of the indices in the array. This will happen now and then when we believe that the range is one larger or one smaller than is really the case. The error is then at the end points of the domain as in the two examples just given.

A very common message is for *value error*. The expression

$$((X*2)+Y*2)*.5$$

looks fine but it can happen that it is not executable. We may get

$$VALUE\ ERROR$$
$$((X*2)+Y*2)*5$$
$$\quad\quad\quad\quad\wedge$$

indicating that the value of Y has not been defined. Then we have to give Y a value and then execute our function again. Value error can also occur when we try to assign a value whose magnitude is too large (see Section 9.2).

A different message is RESEND, which does not mean that the user has made an error, but that the transmission has been faulty. Usually the remedy is to just try again!

2.   NUMERICAL MIXED FUNCTIONS

The primitive functions used so far have the property that the shape of the result is the same as the shape(s) of the argument(s). The remaining primitive functions do not have this property and are called *mixed*. They are a bit harder to understand and to learn since they do not resemble the ones used in mathematics as closely as do the previous ones. They come in five classes: *numerical, structural, selection, selection information,* and *transformations*. We shall start with the numerical ones.

The *decode function* "⊥," which is situated over B on the keyboard, computes values from a radix vector. By this we mean the following: The expression

$$R \perp X$$

where both R and X are vectors, means the number for which X is the representation in a number system with radices given as the elements in R.

For example, with all radices equal to 10, the statement

$$10 \ \ 10 \ \ 10 \ \bot \ 3 \ \ 1 \ \ 4$$

means the number 314. As usual, a scalar (or 1-vector) argument is automatically extended to a vector of the appropriate length, so that the above statement is equivalent with

$$10 \ \bot \ 3 \ \ 1 \ \ 4$$

Or, in base 2, we have

$$2 \ \bot \ 1 \ \ 1 \ \ 0 \ \ 1$$

equal to $1 \cdot 2^3 + 1 \cdot 2^2 + 0 \cdot 2^1 + 1 \cdot 2^0 = 13$.

Our way of measuring time has the radix vector $R \leftarrow 24 \ \ 60 \ \ 60$ so that

$$24 \ \ 60 \ \ 60 \ \bot \ 1 \ \ 0 \ \ 11$$

yields the value 3611, meaning that 1 hour and 11 seconds equals 3611 seconds.

As an application of considerable usefulness let us evaluate a real polynomial with coefficient vector COEFF. The coefficients should be given in descending order of the powers of the variable, contrary to the arrangement we chose in Section 9.3. If X is a real scalar, then

$$X \ \bot \ COEFF$$

yields the value of the polynomial for the argument equal to X. For experiments involving polynomials this is very useful.

The decode function is also defined for other arrays. We shall not go into this now, since it would obscure the presentation. The same procedure will be chosen for the most part in the rest of this chapter. Scalars and vectors will be discussed but not higher-rank arrays, for which the reader should consult the APL manual.

The encode function "T," situated over N, is in a vague sense the inverse to $\bot$. We have

$$2 \ \ 2 \ \ 2 \ \ 2 \ \top \ 13$$

equal to the 4-vector 1 1 0 1.

On the other hand,

$$2 \ \ 2 \ \ 2 \ \top \ 13$$

means the 3-vector 1 0 1 and

$$2 \quad 2 \quad \top \quad 13$$

means the 2-vector 0 1.

To generate the set of all binary vectors of length R we can execute

$$2 \quad \top \quad {}^{-}1 + \iota 2 * R$$

It yields a matrix with 2^R rows, each row representing one binary vector.

The matrix function ⊞ is also, strictly speaking, a mixed numerical function. For didactic reasons we introduced it already in Chapter 9 so that it need not be discussed here.

Exercises

1.   Consider the function e^x. Expand it in a Taylor series up to powers of order N. Find an APL statement that evaluates this finite series. Compare the result with that of $*X$.

2.   In the series

$$y = \sum_{k=1}^{\infty} \frac{x_k}{2^k}$$

let the x_k be independent random variables taking the values 0 and 1 with probabilities $\frac{1}{2}$ each. Then it is well known that the random variable y is uniform on $(0, 1)$. Use this to generate (approximately) a random variable with a uniform distribution on $(0, 1)$ by truncating the series and using the primitive functions $?$ and $\bot$.

3.   Modify the definition of y in Exercise 2 to generate a random variable whose probability distribution is given by the singular Cantor measure on $(0, 1)$. Hint: let the x_k still be i.i.d. but now taking values 0 or 2, and set

$$y = \sum_{k=1}^{\infty} \frac{x_k}{3^k}$$

3.   STRUCTURAL MIXED FUNCTIONS

We have already encountered some of these in Chapter 9. We know what monadic ρ, the shape function does: it gives us the shape of a given array. Dyadic ρ, the reshape function, takes an array (we only considered scalars and vectors) and changes it into the shape indicated by the left argument.

Monadic , means ravel: it changes an array into a vector. Dyadic , stands for concatenation of scalars and vectors as we have seen but its extension to matrices (and more rarely to higher-rank arrays) is of interest. If A and B denote matrices of size $l \times n$ and $l \times m$, then the statement A,B produces the $l \times (n + m)$ matrix

$$\boxed{A \mid B}$$

If we want to concatenate matrices along rows to give the form (assuming, of course, the same number of rows now)

$$\boxed{\begin{array}{c} A \\ \hline B \end{array}}$$

the axis operator [ ] must be used (see Chapter 11).

A simple and attractive function is *rotate,* that is, monadic "φ." Type the large circle and overstrike it with the |, the absolute value sign. Applied to a vector it gives a vector consisting of the same elements but read in the opposite order. Hence the statement

$$\phi\iota 6$$

gives the result when executed

$$6 \quad 5 \quad 4 \quad 3 \quad 2 \quad 1$$

Applied to a matrix it reverses the order of the columns (remember: always operate on the *last subscript* unless the axis operator appears). If

$$M \leftarrow (2 \quad 3)\rho\iota 6$$

so that M means

$$\begin{array}{ccc} 1 & 2 & 3 \\ 4 & 5 & 6 \end{array}$$

then

$$\phi \ M$$

is

$$\begin{array}{ccc} 3 & 2 & 1 \\ 6 & 5 & 4 \end{array}$$

If we want to reverse a matrix along the rows instead of the columns, the axis operator [ ] should be used, as will be described later.

Dyadic φ, called *rotate,* performs cyclic permutations. For example, with $V \leftarrow \iota\ 5$ the statement

$$2\ \phi\ V$$

means the cyclic permutation of V by two steps

$$3\quad 4\quad 5\quad 1\quad 2$$

Also, when M is a matrix

$$K\ \phi\ M$$

where K is an integer means the matrix obtained by a cyclic permutation with K steps applied to the columns of M.

Warning.   It is easy to forget in what direction the cyclic permutations should be performed. To be safe, try them on short vectors and check that you get what you intended.

Dyadic rotate is quite useful when dealing with periodic functions approximated by functions on the finite cyclic groups. The same is true for functions on tori of higher dimension than one.

Finally we have the transpose function "$\lozenge$," obtained by overstriking the large circle by the left slash, upper case of the lowest, farthest right key. Applied monadically to a matrix M, it does exactly what it says, it forms the transpose M^T, changing columns into rows and vice versa. This has already been mentioned.

Dyadic transpose is a little peculiar if A is an array and V a vector of natural numbers. Then

$$V\ \lozenge\ A$$

means the array formed from the elements of A such that the Ith subscript of A becomes the $V[I]$th subscript of A. For example, with

$$A \leftarrow 2\quad 3\ \rho\iota 6$$

then

$$1\quad 1\ \lozenge\ A$$

means the 2-vector of diagonal elements

$$1\quad 5$$

Note that the dyadic

$$2\quad 1\ \lozenge\ M$$

applied to a matrix M is equivalent to the monadic

$$\lozenge\ M$$

Exercises

1. Given an $n \times n$ matrix M, find the vector consisting of the diagonal elements of M in their natural order. Try several ways of doing it.

2. Write a statement that produces a $2 \times N$ matrix of which the first row is the first N natural numbers and the second row is the same backwards.

3. Generate a table of the function $\ln x$ for $x = 1, 2, \ldots, N$. Give it the form of a $N \times 2$ matrix, with the first column the x values and the second column the results.

4. Generate a table of the functions $\cos x$ and $\sin x$ for $x = (2\pi/N)$, $(4\pi/N), (6\pi/N), \ldots, 2\pi$ arranged as an $N \times 3$ matrix. The first column should contain the x argument.

5. Given a real vector with entries v_k, $k = 1, 2, \ldots, n$, generate a vector of the same length with the entries as the second differences

$$v_{k+1} - 2v_k + v_{k-1}$$

where the subscript is treated as periodic with period n. Be careful with the noncommuting dyadic $-$. Apply it to

$$V \leftarrow A + (B \times \iota N) + C \times (\iota N) * 2$$

where A, B, C are real constants. Boundary effects!

6. If D stands for the difference operator

$$D = \frac{f(x + 1) - f(x - 1)}{2}$$

derive an APL expression for the higher-order (periodic) difference operator

$$a + bD + cD^2$$

where a, b, c are either real constants or represent (periodic) functions, i.e., vectors.

7. Rotate extends to vector-valued arguments so that, with M a matrix and V a 2-vector,

$$V[1]\phi[1]V[2]\phi M$$

means the matrix formed from M-elements, such that the row subscript is shifted $V[1]$ steps and the column subscript is shifted $V[2]$ steps. This statement employs the axis operator [1] to be discussed later. Use this to construct (with periodicity) the discrete Laplacian

$$m_{k+1,l} + m_{k-1,l} + m_{k,l+1} + m_{k,l-1} - 4m_{k,l}$$

4. MIXED FUNCTIONS FOR SELECTION

We have already seen how indexing can be used to compute subarrays (possible with repetitions) of given arrays, using semicolons to separate axes (subscripts).

Two quite simple functions, "take" and "drop," are used to keep or to delete some elements of given arrays. Drop, denoted by "↓," located above U on the keyboard, leaves out the first elements for positive values of the left argument or the last elements for negative values of the left argument.

Hence

$$↓ \ 2 \ ι5$$

means

$$3 \ \ 4 \ \ 5$$

while

$$^{-}2 \ ↓ \ ι5$$

means

$$1 \ \ 2 \ \ 3$$

To illustrate what happens for matrices, let

$$A ← (2 \ 5)ρ(ι \ 10) * 2$$

so that A is

$$\begin{array}{ccccc} 1 & 4 & 9 & 16 & 25 \\ 36 & 49 & 64 & 81 & 100 \end{array}$$

then

$$0 \ 3 \ ↓ \ A$$

is (drop no row but the first three columns)

$$\begin{array}{cc} 16 & 25 \\ 81 & 100 \end{array}$$

and

$$0 \ {}^-2 \ \downarrow \ A$$

means (drop no row but the last two columns)

$$\begin{array}{ccc} 1 & 4 & 9 \\ 36 & 49 & 64 \end{array}$$

Generally, if the left argument is a vector V it means that the first (last) $|V[I]$ subscript values along the Ith axis should be dropped. Hence the statement

$$1 \ \ 2 \ \downarrow \ A$$

results in the 3-vector

$$64 \ \ 81 \ \ 100$$

The function "take," denoted by "$\uparrow$" located above Y, operates in the same way except that it keeps certain indicated elements, rather deletes them. For the character matrix M of rank 2 and shape 2   30

> *ASSIGN VALUES TO X AND Y AND Z*
> *AS COMPLEX NUMBERS*

becomes after executing

$$1 \ \ 18 \ \uparrow \ M$$

the character vector of shape 18

> *ASSIGN VALUES TO X*

which could also be obtained by

$${}^-1 \ \ {}^-12 \ \downarrow \ M$$

The functions "compress" and "expand" are more essential for most mathematical experiments. A "compress" statement

$$V/A$$

uses right slash "/" and assumes that V is a vector with Boolean entries, i.e., consists of 0s and 1s. It forms a new array consisting of all the entries in A for which the last subscript corresponds to 1-values in V.

Warning. The length of the V-vector must be equal to the last dimension of the shape of A. There is one exception, however: if V is 1, we get just A, and when V is 0, we get just the empty set.

For example,

$$1 / X$$

means X, while

$$0 / X$$

is the empty set. This will give us a useful device for conditional branching to be studied in Chapter 12.

Another example, a bit more complicated, starts from a real vector V. We want to create a subvector of all the V elements that are positive. To this end form the Boolean vector

$$B \leftarrow V > 0$$

If

$$V \leftarrow {}^{-}5 + \iota 8$$

so that V means

$$^{-}4 \quad ^{-}3 \quad ^{-}2 \quad ^{-}1 \quad 0 \quad 1 \quad 2 \quad 3$$

then B is

$$0 \quad 0 \quad 0 \quad 0 \quad 0 \quad 1 \quad 1 \quad 1$$

Hence

$$(V > 0) / V$$

gives us as desired the vector consisting of the positive elements

$$1 \quad 2 \quad 3$$

In a study of formal languages **VOCABULARY** is a character matrix with 500 rows and 18 columns starting with

```
TABLE
THE
HOUSE
PROPERTY
AFTER
PEDESTRIAN
WALK
SEE
     . . .
```

We want to compute the subarray that consist of all rows not starting with

'P. Consider the logical vector

$$VOCABULARY[;1] \neq 'P'$$

This is the Boolean vector for the row subscripts that do not begin with a P. The result is therefore obtained by

$$\lozenge(VOCABULARY[;1] \neq 'P')/\lozenge VOCABULARY$$

and will begin with

$$TABLE$$
$$THE$$
$$HOUSE$$
$$AFTER$$
$$WALK$$
$$SEE$$
$$\cdot \quad \cdot \quad \cdot$$

The double use of $\lozenge$ can be avoided by use of the axis operator.

Say that we want to find all factors of a natural number N. Introduce the vector

$$V \leftarrow \iota N$$

of the N first natural numbers; they are the possible candidates. But

$$N \div V$$

will be an integer only when N is divisible by whatever entry is in V. Therefore

$$1 | N \div V$$

is a logical vector with zeros when division is possible. Make the nonzero elements into 1s by applying the sign function $\times$ monadically. The desired result is then achieved by

$$(\sim \times 1 | N \div V)/V$$

or written out explicitly

$$(\sim \times 1 | N \div \iota N)/\iota N$$

Another illustration of the use of "compress" may be helpful. Consider the Galois field GF(p), for p an arbitrary prime number, and let it be represented by $x = 0, 1, \ldots, p - 1$. It is easy to write statements for adding and multiplying two elements $x, y \in$ GF(p). For addition we say simply

$$P | X + Y$$

and similarly for multiplication

$$P \,|\, X \times Y$$

But how about division?

One way would be to see for what $z \in \mathrm{GF}(p)$ we have $yz = x$ by the APL statement

$$X = P \,|\, Y \times 0 , \iota P - 1$$

This is the truth value vector for the component in $y \times 0, y \times 1, \ldots ,$ $y \times p - 1$ to be equal to X. Therefore we get the division result by the compression

$$(X = P \,|\, Y \times 0 , \iota P - 1) / 0 , \iota P - 1$$

A more elegant way, with some drawbacks, however, is to use Fermat's theorem $y^{p-1} \equiv 1 \pmod p$ so that $z = y^{p-2}x$ is a solution since

$$yz \equiv y^{p-1}x \equiv x$$

Therefore the division result could be computed as

$$P \,|\, (Y \ast P - 2) \times X$$

If p is large, then the numbers produced by the last statement will be too large. They then cannot be handled by the computer so that we have to return to the first solution. Sens morale: the elegant solution is sometimes less useful than the straightforward one.

Remark. "Drop" and "take" are really special cases of "compress" but with less control over the deletions. The former are very handy to use when their power is sufficient.

"Expand" is denoted by "$V \backslash A$" where V is a Boolean vector. It expands A by filling in places with 0s if A is numeric, or with blanks if A is a character array. The filled-in places correspond to the values of the last subscript of A that correspond to 0s in V. The number of 1s in V must equal the length of the last subscript of A.

To make this clear,

$$1 \quad 0 \quad 0 \quad 1 \quad 1 \backslash 3 \times \iota 3$$

means

$$3 \quad 0 \quad 0 \quad 6 \quad 9$$

and with

$$X \leftarrow 3 \quad 3 \,\rho\, 'ABCDEFGHI'$$

the expression

$$1 \quad 1 \quad 0 \quad 0 \quad 1 \backslash X$$

gives the character array (with five columns!)

$$
\begin{array}{ccc}
AB & & C \\
DE & & F \\
GH & & I
\end{array}
$$

Exercises

1. Write a statement for the vector of natural numbers from 1 to N that are not divisible by K. Use the compress function and iota.

2. For a real vector X write an APL expression for the subvector whose entries are integers. Use compress and the residue function.

3. A real-valued function of a real argument has been computed and represented as a vector V. We want to restrict the domain of the function to the set when its values are between A and B. Write an expression for the resulting values on the contracted domain.

4. A real function is given on the "discrete interval" $[1, N + M]$, N and M integers. We want to compute a function on the discrete interval $[1, N + M + L]$ by the definition

$$
f(x) = \begin{cases}
f(x) & \text{on} \quad [1, N] \\
0 & \text{on} \quad [N + 1, N + L] \\
f(x - L) & \text{on} \quad [N + L + 1, N + L + M]
\end{cases}
$$

Get an expression for doing this when the function is represented by a real vector F of length $N + M$.

5. MIXED FUNCTIONS FOR SELECTION INFORMATION

We have already met, and used repeatedly, the index generator monadic ι that gives us the first natural numbers. We now also encounter dyadic ι.

Say that f is a function; for simplicity, the argument x taking on the values $x = 1, 2, 3$. We represent, as usual, the function f by an APL vector F with ρF equal to 3.

We often need the location of the maximum (maxima) or minimum (minima). Well, it is easy to find the value of the maximum:

$$MAX \leftarrow F[1] \lceil F[2] \rceil F[3]$$

but for what index (x-value) is it attained?

This is what dyadic ι helps us to do. It finds information about the indices of array elements with a given property. More precisely, for a vector V and arbitrary array A, the expression

$$V \; \iota \; A$$

gives us, for each A-element, the smallest V-subscript I for which the value $V[I]$ is equal to the A-element. If no $V[I]$ is equal to the A-element, the result is $1 + \rho V$. The shape of the resulting array is equal to the shape of A.

For example,

$$5 \; 1 \; 2 \; 8 \; \iota \; 2$$

is

$$3$$

but

$$5 \; 1 \; 2 \; 8 \; \iota \; 7$$

is

$$5$$

Dyadic iota is needed all the time, and also in unexpected situations, but we shall start by simple applications. Say that we want to find the location of the scalar element X in the vector V. Of course, this is obtained as

$$V \; \iota \; X$$

Warning.   If the above statement results in the value $1 + \rho V$, this tells us that X does not appear in V. Also, one should be aware of the possibility that X may appear more than once; we shall only be told about the *first* occurrence. The others must be found by direct programming.

Character arrays can also be searched in this way. For example, if

$$CHAR \leftarrow (3 \; 2) \rho \, 'AB1234'$$

so that CHAR means

$$AB$$
$$12$$
$$34$$

then

$$'2A8' \iota CHAR$$

gives the numeric matrix

$$
\begin{array}{cc}
2 & 4 \\
4 & 1 \\
4 & 4
\end{array}
$$

Now let V be a real 3-vector; we want to compute the location of its largest element. We first find the largest value as above:

$$MAX \leftarrow V[1] \lceil V[2] \rceil V[3]$$

and then get the location as the subscript value given by

$$V \iota MAX$$

Combined, this can therefore be written

$$V \iota V[1] \lceil V[2] \rceil V[3]$$

For higher dimensions this solution is clumsy, and, if the dimension is not specified a priori, it is no solution. We shall return to this later by the "reduction" operator.

Equations can sometimes be solved by using dyadic iota. For example, we want to solve $x^5 + 3x^4 - 2x^3 - 6x^2 + x + 3 = 0$ for integers as possible roots. For example, put

$$V \leftarrow {}^{-}7 + \iota 13$$

Then combine the values taken by the polynomial for the V-values into a vector A. Execution

$$A \iota 0$$

yields

$$6$$

so that we have one root $V[6]$, which is the same as $^{-}1$, and could proceed from there in various ways to find other roots.

The components of the A vector could be obtained, for example, as

$$A[I] \leftarrow V[I] \perp COEFF$$

with

$$COEFF \leftarrow 1 \ 3 \ {}^{-}2 \ {}^{-}6 \ 1 \ 3$$

(see Section 2). This would then have to be executed for $I = 1, 2, \ldots$ up to ρV. It would be convenient to do this in a defined function but we shall not go into this question at this point.

Rather than get involved in this morass of numerical analysis, let us finish discussing dyadic "iota" (but only for the time being) by answering a question we mentioned earlier. We want to remove all duplicates from a vector V already ordered in nondecreasing order and we start by considering

$$W \leftarrow V \iota V$$

If no duplicate existed, W would be just the vector $1, 2, \ldots, \rho V$, which is

$$\iota \rho V$$

A moment's reflection shows that

$$(\iota \rho V) = V \iota V$$

is the truth-value vector for the 1st, 2nd, etc., element of V to be the first occurrence. Hence the cleaned vector is obtained by using compression

$$((\iota \rho V) = V \iota V)/V$$

Two other mixed functions order numeric vectors. "Grade up," denoted "$\blacktriangle$," is obtained by overstriking the $\triangle$, over H on the keyboard, by $|$. It computes the permutation vector of the first N natural numbers that orders a given vector V in increasing order; N is the length of the vector. Similarly, "grade down," denoted "$\blacktriangledown$," gives the permutation that reorders the vectors in descending order.

For example,

$$\blacktriangle \; 55 \;\; 47 \;\; 41 \;\; 72 \;\; 61$$

gives

$$3 \quad 2 \quad 1 \quad 5 \quad 4$$

and

$$\blacktriangledown \; 55 \;\; 47 \;\; 41 \;\; 72 \;\; 61$$

means the permutation vector

$$4 \quad 5 \quad 1 \quad 2 \quad 3$$

Note that $\blacktriangledown \; V$ is the same as $\phi \;\; \blacktriangle \; V$, reversing the order of the components in $\blacktriangle \; V$.

Warning.   These functions do not order the given vector, they only give us a permutation that would order the vector when applied to the subscripts.

It is now clear how a vector V can be ordered ascendingly. We just reorder its components by applying the permutation vector $\triangle V$ to its subscripts so that the numeric V-vector with components 55, 47, 41, 72, 61

$$V[\triangle V]$$

results in

41 47 55 61 72

and

$$V[\triangledown V]$$

results in the vector

72 61 55 47 41

As an application, say that we have observed a sample vector SAMPLE from a real-valued stochastic variable. We want to compute the Kth-order statistic from below. The grade up function immediately gives us the result

$$(V[\triangle V])[K]$$

We would get, for example, the minimum by setting $K = 1$. This would be an inefficient way of computing the minimum since it would involve an unnecessary number of comparisons; we shall return to this in Chapter 11.

Similarly, let us compute the trimmed mean of SAMPLE, leaving out the smallest and largest values. Use the drop function after ordering in increasing order. We get the APL expression

$$^-1\downarrow 1\downarrow SAMPLE[\triangle SAMPLE]$$

for the vector consisting of all observations except the smallest and largest ones. Then these components should be summed; we return to how this can be done later.

Or, say that we want to compute the median of SAMPLE. If the length of SAMPLE is odd it is simply

$$(SAMPLE[\triangle SAMPLE])[.5\times 1+\rho SAMPLE]$$

If ρ SAMPLE is even, we may use the average of the two middle components of the ordered vector, but we leave this to the reader.

The functions grade up and grade down are related to the symmetric group S_N over N objects. Let P be a permutation written as a vector and considered as an element of the group. To get the product of two group elements we compute

$$P_1[P_2]$$

To get the unit element E of the group we use

$$E \leftarrow \iota N$$

and to compute the inverse of an element P we need only execute

$$⍙\ P$$

This gives us an elegant tool to handle finite groups computationally.

Two other functions, $\in$ for membership and $?$ to generate randomness, also belong to the mixed functions that produce selection information. They have already been introduced in Chapter 9.

Exercises

1.   Consider a numeric vector V. Write some expressions to compute the truth value that the maximum element is unique.

2.   Let X be a positive integer. Get an expression that calculates the position of the first nonzero binary digit in the binary expansion of X. Let the number of binary digits used in the expansion be L.

3.   CHAR is a character vector. Use dyadic iota to determine whether CHAR contains any occurrence of the letter T.

6.   TRANSFORMATIONS OF DATA

We now encounter two mixed functions with quite different purposes. They have been included fairly recently in the public versions of APL. This section can be skipped at first reading.

The function "format," denoted " $⍕$," is obtained by overstriking ⊤ by the little ring ∘ located over J. *Monadic format* will transform an array into a character array. Hence its application to a character array has no effect. Applied to a numeric array, it yields a character array of the same shape except that expansion along the last axis usually takes place, since a number usually takes several numerals to be represented.

For example, if the variable X has been evaluated to be 415 then

$$⍕X$$

gives the 3-array, with characters as components,

$$415$$

This is useful, among other things, for *mixed output* in mathematical experiments. Say that we want a program to type out the line THE RE-

SULT IS followed by the evaluated value of *X*. This could be done by the statement producing a character vector

$$\text{'THE RESULT IS ', } \bar{\phi} X$$

This device will be used often in experiments when it is desired to have hard copy preserving the results with text reminding us of what the different numbers mean. Other uses will be indicated later.

Dyadic format, which applies only to numeric arrays, is used for controlling the printing format of the array. If the left argument is a 2-vector [for other cases, see APL Language (1976)], its first component specifies the width of the number field. The second component determines the precision: if it is positive, it is the number of digits to the right of the decimal point; if it is negative, it is the number of digits in the multiplier in the exponential form.

If MATRIX is the numeric 3×2 array

$$
\begin{array}{cc}
12.34 & {}^{-}34.567 \\
0 & 12 \\
{}^{-}0.26 & {}^{-}123.45
\end{array}
$$

then

$$12 \quad 3 \quad \bar{\phi} \quad MATRIX$$

is the character array

$$
\begin{array}{cc}
12.340 & -34.567 \\
.000 & 12.000 \\
-.260 & -123.450
\end{array}
$$

where each column is 12 characters wide and 3 decimal digits follow the decimal point.

On the other hand, using exponential form, the statement

$$9 \quad {}^{-}2 \quad \bar{\phi} \quad MATRIX$$

yields another character array, namely

$$
\begin{array}{cc}
1.2E01 & -3.5E01 \\
0.0E00 & 1.2E01 \\
-2.6E^{-}01 & -1.2E02
\end{array}
$$

The other new function is "execute," denoted by "$\pounds$," whose introduction into the language caused great joy in the APL community. It is applied to a character vector, interprets this as an APL statement, and if it is syntactically correct, executes the statement.

For example, if V is the character vector

$$V \leftarrow \; '5 * 2'$$

then

$$\underline{\pm} \; V$$

yields the numeric answer

$$25$$

Or, if X is the character vector

$$X \leftarrow \; '2 \; 3 \rho 15'$$

the execute statement

$$\underline{\pm} \; X$$

means the numeric matrix

$$15 \quad 15 \quad 15$$
$$15 \quad 15 \quad 15$$

It is often possible to exhibit great ingenuity in program construction by appealing to the power of execute.

Exercises

1.   We want to solve the transcendental equation $x e^x = 1$, $-\infty < x < \infty$, numerically. It is clear that the root must be between 0 and 1. Introduce, with a small step size (large N), the grid vector

$$V \leftarrow \; (\div N) \times \iota N$$

and consider the Boolean vector

$$W \leftarrow \; (1 \geq V \times \ast \; V) \iota 1$$

Use this to approximate the root. Since the solution will overshoot the true value, linear interpolation can be used to improve its accuracy. This solution is computationally wasteful (better ones will be given later), but it is nevertheless of some interest since it illustrates a useful procedure for data that are completely irregular (in contrast to the present case with a smooth function).

2.   Given the universe $U = 1, 2, \ldots, N$, which is used in the following exercises, write a statement that takes two arbitrary subsets A and B (as

vectors) and produces the union of *A* and *B* as a set, with no duplicated values. Use the cleaning procedure in the text.

3. Do the same for the complement of a set *A*; use compression and assume that *A* has no repetitions (if it does, the previously mentioned procedure could be used).

4. Use the membership function ϵ to compute unions and intersections of two vectors (sets) *A* and *B*. The result may have repetitions.

5. Let ALPHA be a 26-vector of characters, the letters arranged in their usual order. Use dyadic *?* to generate a random permutation from S_{26} acting on ALPHA. Given a message in clear, with no blanks, represented as a character vector MESSAGE of arbitrary length, write a statement that codes this by the substitution cypher associated with the permutation and another statement that decodes it back into clear.

6. Compute the permutation vector (over S_N) for the inversion making the elements numbered *I* and *J* change place. Use several assignment statements if you wish.

7. Let *V1* and *V2* be numeric vectors of equal length. We want to reorder them by the same permutation and such that *V2* becomes nondecreasing. Write a statement for the vector that *V*1 becomes after reordering.

8. Let *M* be the numeric matrix

$$3 \quad 3 \quad \rho \quad (\iota 9) * .5$$

Use format to display *M* with a field of width 10 and with the exponential form with four digits in the multiplier.

9. If *R1* and *R2* are numeric scalars, write an APL statement that will print out text for saying that the first root has the value *R1* and the second root the value *R2*.

CHAPTER 11

APL Operators

1. THE NOTION OF AN OPERATOR

When we speak of an operator in ordinary mathematical discourse we usually think of a specific class, for example, self-adjoint operators in Hilbert space. Then the notion of operator is not clearly delimited from that of a function.

In APL, however, the term operator has a clearly delimited sense; there are only five operators at present: *reduction, scan, axis, inner product,* and *outer product*. They are used to modify certain functions, and one can think of an operator as being applied to one or two functions and producing a new one, the modified function.

It is primarily with the operators that APL attains its full power and the reader is encouraged to use them extensively. They are not as easy to learn as the primitive functions already discussed, but with increasing experience it will be found that they can be used to handle sometimes quite sophisticated mathematical constructs.

2. REDUCTION

How do we deal with expressions like

$$\sum_{k=1}^{n} a_k$$

or double sums? What do we do with a formula like

$$\prod_{k=1}^{n} b_k$$

In general, we need some syntactic device for *iterated applications of functions,* whether these are addition, multiplication, or other dyadic

122

functions. If $n = 5$, we can write the sum as

$$A[1] + A[2] + A[3] + A[4] + A[5]$$

but this does not help when n is not a predetermined number.

The solution is offered by the operator "reduction" denoted by "/." Recall that the *function* compression is also denoted by this symbol but this will introduce no serious ambiguity.

If V is a vector and f stands for a dyadic function, the statement

$$f \, / \, V$$

means f iteratively applied to the entries of V, starting with the last two, the result of this and the third last V component as the argument of f, and so on, until we have reached $V[1]$.

This will be easier to grasp by looking at several examples. Take f as dyadic $+$ and A as the vector above. Then

$$+ \, / \, A$$

is clearly the same as the sum above. Hence $+/$ *serves the role of the summation symbol* Σ.

To calculate the zeta function approximately with argument S we can write

$$ZETA \; \leftarrow \; + / \div (\iota N) \star S$$

and to calculate a power series

$$\sum_0^n a_n x^n$$

we can use, with the obvious meanings of the symbols,

$$+ / A \times X \star 0 , \iota N$$

For example, Mittag-Leffler's E_α function

$$E_\alpha(z) = \sum_{\nu=0}^{\infty} \frac{z^\nu}{\Gamma(1 + \alpha\nu)}$$

can be approximated for real arguments by

$$EALPHA \; \leftarrow \; + / (Z \star 0 , \iota N) \div \, ! ALPHA \times 0 , \iota N$$

Before looking at other choices of f, let us remark that, if $f/$ is applied to an array, the reductions are along the last subscript. Hence, if MATRIX is a numeric matrix the statement

$$+ / MATRIX$$

produces a vector of length

$$(\rho MATRIX)[1]$$

whose entries are the row sums, as in

$$\sum_{j=1}^{n} m_{ij}$$

Hence a double sum over a numeric matrix MATRIX can be obtained by two single summations, or, in APL, as

$$+/+/MATRIX$$

$+/+/$ plays the role of the double summation symbol Σ.

Now let f be dyadic $-$; what is the result of

$$-/\iota 4$$

Well, it should mean the same thing as the APL statement (remember that "$-$" is not commutative!)

$$1 - 2 - 3 - 4$$

which is the same as

$$1 - (2 - (3 - 4))$$

which is -2. Note that in ordinary mathematical notation the expression could be written as $1 - 2 + 3 - 4$. Hence $-/$ *means an alternating series.*

The dyadic $\times$ of course gives us the *symbol for product.* We illustrate this by the product

$$\prod_{\nu=1}^{n} \frac{a + b\nu}{c + d\nu}$$

as

$$\times/(A+B\times\iota N)\div(C+D\times\iota N)$$

But this is only a small beginning as far as the usefulness of the reduction operator is concerned. With dyadic "ceiling" we get

$$\lceil /V$$

meaning the largest of the V-components; *it represents the mathematical expression max.*

Similarly, "floor,"

$$\lfloor /V$$

represents min.

We can now find easily where the maximum occurs for a real function (seen as a vector V). We search for the index I where $V[I]$ = maximum (see Section 10.5) and get

$$V \iota \lceil / V$$

The reader should think about what happens when the maximum is not unique.

Now let us look at the Boolean primitive functions. Say that B is a Boolean vector and consider

$$\wedge / B$$

It means clearly *the logical conjunction of all the truth values in B* and corresponds, in the traditional notation of Boolean algebra, to

$$\bigwedge_{\nu=1}^{n} b_{\nu}$$

A similar definition applies to the logical "or," written traditionally as $\vee$. The $\wedge /$ therefore corresponds to the universal quantifier in predicate calculus, while $\vee /$ means the existential quantifier.

Let us now combine reduction with what we have learned earlier.

If $V1$ and $V2$ are numeric vectors of equal length, their *inner product* is simply

$$+ / V1 \times V2$$

The *correlation coefficient,* which is a modified inner product, is then, in terms of the empirical means,

$$M1 \leftarrow (\div \rho V1) \times + / V1$$

$$M2 \leftarrow (\div \rho V2) \times + / V2$$

and in terms of the empirical variances,

$$VAR1 \leftarrow (\div \rho V1) \times + / (V1 - M1) * 2$$

$$VAR2 \leftarrow (\div \rho V2) \times + / (V2 - M2) * 2$$

equal to

$$(\div (\rho V1) \times (VAR1 \times VAR2) * .5) \times + / (V1 - M1) \times (V2 - M2)$$

How do we find out whether $V1[I] = V2[I]$ for some index I? Just by

$$\vee / V1 = V2$$

and, similarly,

$$\wedge / V1 = V2$$

which gives the truth value for the two vectors to be identical. Say that we are interested in the norm of a square matrix MAT. To calculate

$$\max_i \sum_j |m_{ij}|$$

we write, using two reductions,

$$\lceil / + / | MAT$$

Easy, is it not?

Exercises

1. Approximately compute Euler's constant by an APL statement for

$$\sum_{\nu=1}^{n} \frac{1}{\nu} - \ln n$$

   Try various n-values.

2. Find the approximate location of the maximum component of the vector $(v_\nu; \nu = 1, 2, \ldots, n)$ where

$$v_x = \frac{xe^x}{x + e^{2x}}, \qquad 0 \le x \le 1, \quad x = \frac{\nu}{n}$$

3. What does the expression $\div / V$ mean for a numeric V-vector?

4. M is a numeric matrix. Find an APL statement for the truth value that M is a Markov transition matrix.

5. Calculate the truth value for a square numeric matrix M to be (a) symmetric, (b) skew symmetric, and (c) to have at least one integer in each row.

6. Derive expressions for the geometric and harmonic means of a numeric vector V with positive components.

7. Let A and B be vectors with no elements repeated, consider them as sets, and find the truth value for $A \subseteq B$.

8. V is a numeric vector. What is the truth value of the statement: the entries in V are increasing.

9. Let A and B be two sets in $V = \iota N$ without repetitions and with the same number of elements. What is the truth value that $A = B$ as *sets?*

3. SCAN

The next operator "scan" is also applied to dyadic primitive functions f and is denoted "$f \backslash$". It is in a way an extension of reduction. We have seen that, for example, $+/$ stands for the summation symbol; $+\backslash$ computes *all partial sums* so that

$$+\backslash \quad 2 \quad 7 \quad {}^{-}3 \quad 4 \quad 1$$

is the 5-vector

$$2 \quad 9 \quad 6 \quad 10 \quad 11$$

In general $f \backslash V$ means the vector of length ρV obtained by evaluating "$f/V1$", where $V1$ takes on the vectors

$$V[1]$$

$$V[1],V[2]$$

$$V[1],V[2],V[3]$$

$$\cdot \quad \cdot \quad \cdot$$

successively. We shall illustrate this operation, which is often useful when studying how the degree of approximation is increased when more terms are included. Consider the infinite alternating sum

$$\pi = 4 \sum_{\nu=1}^{\infty} \frac{(-1)^{\nu+1}}{2\nu - 1}$$

We get the corresponding partial sums by the expression (notice the $-$ sign rather than $+$)

$$PISUMS \leftarrow 4 \times -\backslash \div {}^{-}1 + 2 \times \iota N$$

Executing this with N set equal to 15 we get the 15-vector (slow convergence!)

4	2.666666667	3.466666667	2.895238095	3.33968254	2.976046176
	3.283738484	3.017071817	3.252365935	3.041839619	3.232315809
	3.058402766	3.218402766	3.070254618	3.208185652	

Another illustration starts from a discrete probability distribution PROB given as a vector of probabilities. Hence the entries in PROB are nonnegative and their sum equals one. For example, the binomial distribution $B(n, p)$, with probabilities

$$\binom{n}{k}p^{k}(1 - p)^{n-k}$$

could be computed by first setting

$$V \leftarrow 0 , \iota N$$

and then

$$PROB \leftarrow (V \, ! \, N) \times (P \star V) \times (1 - P) \star N - V$$

The distribution function associated with any PROB is simply obtained using "scan" as

$$DIST \;\leftarrow\; + \backslash PROB$$

Returning to an arbitrary PROB, how can one generate one pseudorandom number from this probability distribution? Let us first argue in ordinary mathematical notation. If we had access to an $R(0, 1)$ stochastic variable Y, meaning one with Lebesgue measure over $(0, 1)$, then we could set

$$X = \begin{cases} 1 & \text{if} \quad Y \le p_1 \\ 2 & \text{if} \quad p_1 < Y \le p_1 + p_2 \\ 3 & \text{if} \quad p_1 + p_2 < Y \le p_1 + p_2 + p_3 \\ \quad \cdots \end{cases}$$

where p_k stands for the general entry in the PROB-vector. This is all right; it gives, for example,

$$\begin{aligned} P(X = 2) &= P(p_1 < Y \le p_1 + p_2) \\ &= (p_1 + p_2) - p_1 = p_2 \end{aligned}$$

as required.

But the cumulative sums appearing in this construction are just those that are computed by scan. We therefore introduce a Boolean vector

$$Y > + \backslash PROB$$

and

$$X \;\leftarrow\; 1 + + / Y$$

gives the result.

Exercises

1. Consider the transition matrix TRANS of a Markov chain x_1, x_2, x_3, ... Let I be the name of a state I. Write a statement that generates the pseudorandom value of x_{t+1} when we know that x_t has the value I.

2. Write an expression to get a rough but quick idea of the speed of convergence of the infinite product

$$\prod_{k=1}^{\infty} \left(1 - \frac{x^2}{n^2}\right)$$

Try it for some different values of x.

3. For a numeric vector V, what does the following statement mean:

$$(1\downarrow+\backslash V) \ - \ {}^{-}1\downarrow+\backslash V$$

Write the answer in APL form. This is related to the fundamental theorem in calculus.

4. AXIS OPERATOR

We have several times encountered the convention that when some computational operation is applied to an array it is understood to be done over the last axis; i.e., the last subscript. If the array happens to be a matrix, this means the second axis corresponding to columns. This is an unattractive restriction and is removed by the *axis operator,* denoted "$[I]$", where I is the number of the axis over which we want to execute the command.

For example,

$$+/[1]MATRIX$$

means the vector of the column sums of MATRIX, and

$$\wedge/[1]BMATRIX$$

where BMATRIX is the name of a Boolean matrix, gives us a Boolean vector whose entries are the truth values that columns of BMATRIX consist entirely of 1s.

The axis operator applies to reduce and scan, as indicated, and also to reverse, rotate, compress, and expand. It is easy to learn and to use and we need not spend any more time on it here; it will be encountered often. We only note that

$$+/[2]MATRIX$$

means the same thing as

$$+/MATRIX$$

and so on for other primitive functions.

If we want to place the matrix B "below" the matrix A, assuming that they have the same number of columns, it can be done by

$$A,[1]B$$

See also Section 3 in Chapter 10.

Exercises

1.   A numeric matrix M can be thought of as a representation of a function $f(x, y)$ of two real arguments x and y. If we execute M, we get the rectangular array M printed out with the first row first, of course. This happens to be contrary to the standard convention: we usually have the axis pointing upwards (but Descartes did not!). Therefore it is useful to reverse the order of the rows. Write a simple APL statement that reverses the order of the row of subscript M.

2.   A three-dimensional array A has the entries a_{ijk}. Compute the matrix whose entries are

$$\sum_j a_{ijk}$$

3.   When an experiment involves a collection of vectors it is often convenient to organize them into a matrix (sometimes higher-rank arrays are better). Say that our vectors are collected into the matrix MV as row vectors. If we are interested only in the nontrivial vectors, the question arises: How can we remove the others that are just the zero vector? Do this using compression and the axis operator.

5.   INNER PRODUCT

We already saw in Section 2 how to compute the inner product of two vectors A and B. The same result will be obtained in another way by executing

$$A+.\times B$$

but this also means *matrix product* when A and B are two numeric matrices. Of course, their shapes have to conform so that A has the same number of columns as B has rows.

This is only one of the instances of the *general inner product*

$$Af.gB$$

where f and g represent two dyadic functions; in the special case above we had $f = +$ and $g = \times$. If A and B are vectors it could also be written, using reduction, as

$$f \, / A \, gB$$

with a natural symbolism.

Note that in the inner product B could be a vector and A a matrix. The usual matrix–vector product Mx from linear algebra hence appears as

$$M + . \times X$$

The general inner product is a powerful device but its somewhat unconventional form makes it harder to learn and to use than some of the other operators. An example will illustrate its use.

On the universe $U = \{1, 2, 3, \ldots, N\}$ we have defined a relation $\mathcal{R}$. It can be represented by a Boolean matrix R so that I and J are related by $\mathcal{R}$ iff $R[I;J]$ has the truth value 1. Let us calculate $\mathcal{R}^2$, meaning the iterated relation: I and J are related iff there exists an element K such that $I\mathcal{R}K$ and $K\mathcal{R}J$ hold. In other words, we should be able to find at least one path of length 2 from I to J. The truth values of $\mathcal{R}^2$ can then be calculated as

$$R2 \leftarrow R \vee . \wedge R$$

Verify this. What does the following expression mean:

$$R \vee . \wedge R2$$

Generalize it further.

Here is another example. We want to compute in a single computation all the polynomial values for $i = 1, 2$:

$$\sum_{k=1}^{N} a_{ik}^{k}$$

Combine the a-values into matrix A; we then get the results as the vector

$$A + . \star \iota N$$

We now look at a slightly more complicated example. In a finite-dimensional vector space consider a subspace S. Let us span S by vectors that we combine as column vectors into a matrix SUB. Now if X is an arbitrary vector, what is the projection of X onto the subspace S?

Recalling the interpretation of $\boxdiv$ as a least squares solution (minimize distance!), we then get the elegant solution

$$PROJ \leftarrow SUB + . \times X \boxdiv SUB$$

As a last example we consider the use of the general inner product to find character vectors in character arrays. In mathematical experiments this is sometimes required in order to be able to identify names coded as numbers. Say that we have a character array; for example, VOCABU-LARY in Section 10.4 with, for example, 18 columns. A name appears as a character vector NAME with at most 18 entries. What is its coded number, that is, the row subscript in VOCABULARY?

To solve this problem we first make NAME into an 18-vector

$$NAME18 \leftarrow 18 \uparrow NAME$$

which adds the required number of blanks; we need 18-ρNAME blanks. See the complete definition of $\uparrow$ in APL Language, p. 36. What does

$$VOCABULARY \wedge . = NAME18$$

mean?

It operates along the last (second) axis and searches for a fit, all the elements in the Ith row have to equal the respective elements of the entries of NAME18. Therefore the expression is a Boolean vector with 1s only in the entries corresponding to a row of VOCABULARY exactly equal to NAME18.

The result is easily obtained by reduction

$$(VOCABULARY \wedge . = NAME18)/\iota(\rho VOCABULARY)[1]$$

since the shape of VOCABULARY is the 2-vector

$$\rho VOCABULARY$$

so that the number of rows in VOCABULARY is

$$(\rho VOCABULARY)[1]$$

Exercises

1.   Consider the vectors $v^0, v^1, v^2, v^3, \ldots, v^r$ given in component form in $\mathbb{R}^n$

$$v_s^r = s^r, \quad s = 1, 2, \ldots, n$$

Let X represent an arbitrary vector in $\mathbb{R}^n$ and find a statement that computes the projection of X upon the subspace spanned by $v^0, v^1, v^2, \ldots, v^r$. In other words, compute the least squares polynomial approximation of a function given as a vector F on $1, 2, \ldots, n$.

2.   Let DISTANCE be a symmetric numeric matrix with nonnegative entries. Use the general inner product with $\llcorner$ to calculate the shortest distance from state I to state J passing through at most one intermediate state, or with at most two intermediate states.

3.   Consider the quadratic form $x^T Q x$, $x \in \mathbb{R}^n$. Calculate its value for given x.

6. OUTER PRODUCT

The most difficult of the APL operators, but also the most rewarding, is the *general outer product*. You can feel that you begin to master APL when you feel at ease with outer product; it will not happen immediately.

Let us start with the outer product as it is sometimes defined in vector algebra. Let $V1$ and $V2$ be two numeric vectors, then

$$V1 \circ . \times V2$$

(with the little ring ∘) means the matrix whose entry in row I, column J is

$$V1[I] \times V2[J]$$

Generally if f is a primitive dyadic function, then for numeric arrays

$$A1 \circ . fA2$$

means the array obtained by evaluating the f-result of *any* entry of $A1$ combined with *any* entry in $A2$.

For example, if $A = \{a_{ij}\}$ and $B = \{b_{kl}\}$ are numeric vectors than $A\circ . \times B$ means the *Kronecker product* $\{a_{ij}b_{kl}\}$, a four-dimensional array.

This is useful for *generating mathematical constructs*. Say that we are working in a matrix algebra in $\mathbb{R}^n$ and we need the unit element. The question of how to compute it may at first glance appear nonsensical to the mathematician since it is already known. What we really mean is to compute the *numerical representation* of the unit element which is not quite trivial.

It can be done in several ways, one of which is as the outer product with =:

$$UNIT \leftarrow (\imath N) \circ . = \imath N$$

Remember that = is a proposition, not a relation. In the above assignment we consider any entry of $\imath N$, say I, with any entry of $\imath N$, say J. We get a 1 iff I equals J, so that UNIT is indeed the identity matrix.

A similar task is to compute a numeric matrix, say with shape $N \times M$, all whose columns are equal to a vector V. The vectors V and $M\rho1$ are of length N and M, respectively. Therefore the (M, N) matrix

$$V \circ . \times M\rho 1$$

has the entries in the kth row and lth column equal to $V[K] \times 1$ as required.

Or, let us calculate the Hilbert matrix

$$H = \left\{ \frac{1}{i + j}, \; i, j = 1, 2, \ldots, n \right\}$$

Without the outer product this would not be easy; now just put

$$HILBERT \leftarrow \div (\iota N) \circ . + \iota N$$

and we are done.

Let T be a given numeric vector with entries $t_0, t_1, \ldots, t_{n-1}$ and calculate the symmetric Toeplitz matrix

$$\{t_{i-j}, i, j = 1, 2, \ldots, n\}$$

A slight difficulty lies in the indexing; we index vectors starting at 1 rather than at zero; see, however, 0-origin indexing in APL Language, p. 15, describing a device that we shall not use here. Say, instead that T is an N-vector such that $T[1]$ corresponds to t_0, $T[2]$ to t_1, . . . , $T[N]$ to t_{n-1}.

Form

$$W \leftarrow (\iota N) \circ . - \iota N$$

meaning the matrix

$$\begin{bmatrix} 0 & -1 & -2 & \cdots & -N+1 \\ 1 & 0 & -1 & \cdots & -N+2 \\ 2 & 1 & 0 & \cdots & -N+3 \\ \vdots & \vdots & \vdots & & \vdots \\ N-1 & N-2 & N-3 & \cdots & 0 \end{bmatrix}$$

With the other indexing we need instead the matrix U

$$\begin{bmatrix} 1 & 2 & 3 & \cdots & N \\ 2 & 1 & 2 & \cdots & N-1 \\ 3 & 2 & 1 & \cdots & N-2 \\ \vdots & \vdots & \vdots & & \vdots \\ N & N-1 & N-2 & \cdots & 1 \end{bmatrix}$$

But we can compute U from W as

$$U \leftarrow 1 + |W$$

Therefore the result is simply

$$TOEPLITZ \leftarrow T[1 + |(\iota N) \circ . - \iota N]$$

A circulant matrix

$$\begin{bmatrix} c_0 & c_1 & \cdots & c_{n-1} \\ c_1 & c_2 & \cdots & c_0 \\ \vdots & \vdots & & \vdots \\ c_{n-1} & c_0 & \cdots & c_{n-2} \end{bmatrix}$$

can be found by a similar idea since

$$(^-1+\iota N)\circ.+^-1+\iota N$$

means

$$\begin{bmatrix} 0 & 1 & \cdots & N-1 \\ 1 & 2 & \cdots & N \\ \vdots & & & \\ N-1 & N & \cdots & 2N-2 \end{bmatrix}$$

This brings to mind, since we are now operating on the finite cyclic group, the dyadic |, that is, the residue function modulo N. We thus get

$$CIRCULANT \leftarrow C[1+N|(^-1+\iota N)\circ.+^-1+\iota N]$$

where C is the APL vector with entries $c_0, c_1, \ldots, c_{n-1}$.

Let us return for a moment to the relation $\mathcal{R}$ in the preceding section. It was represented by a square Boolean matrix R. Let us compute the truth value that $\mathcal{R}$ be reflexive. It means that all its diagonal entries be equal to 1. Multiplying R by the identity matrix (elementwise, not by matrix multiplication!) we get

$$A \leftarrow R \times (\iota N)\circ. = \iota N$$

Now A has zeros everywhere but in the diagonal. Hence the row sums should all be 1 and their sum must equal N in order that $\mathcal{R}$ be reflexive. Hence we get the truth value sought for as the proposition

$$N = +/+/R \times (\iota N)\circ. = \iota N$$

The reader may find it amusing to get the solution using instead dyadic transpose to get the diagonal elements.

To test whether $\mathcal{R}$ is symmetric is easy, just execute

$$\wedge/\wedge/R = \lozenge R$$

How about a test for transitivity? This is better left to "defined functions" to be treated in the next chapter.

In a mathematical experiment say that we have iterated some procedure and obtained the numeric scalar results $x_1, x_2, \ldots, x_n$ combined into a vector RES. Say that the values are integers from 1 to R. Some of the values appear several times. We want to summarize the experiment by calculating a frequency table: how many xs are equal to one, how many to two, etc.

A simple way of doing this is to first consider the Boolean matrix

$$B \leftarrow (\iota R)\circ. = RES$$

The Ith row and Jth column element tells us the truth value of $X[J]$ equal to I. We therefore get the frequencies as

$$+/B$$

which could be displayed as a table by the statement

$$(2,R)\rho(\iota R),+/B$$

Convince yourself that this works; try it at the terminal. This, and more general versions of the same idea, are useful in many experiments when the results are complicated and motivate a clear display. The outer product can also be used to obtain "quick-and-dirty" plots. When a pilot experiment is under way one is seldom interested in plots of high quality. The important thing is to quickly get enough idea of what the result is. One probably does not want to use the plotter for this—it takes too much time for the impatient experimenter. Instead one can do with typewriter plots. They are crude but easy to get.

Say again that we have obtained results $x_1, x_2, \ldots, x_n$ and that the results are real numbers. Compute minimum and maximum

$$MIN \leftarrow \lfloor/RES$$

$$MAX \leftarrow \lceil/RES$$

and scale the results

$$SCALED\leftarrow(RES-MIN)\div STEP$$

with the step size

$$STEP\leftarrow(MAX-MIN)\div NSTEP$$

into a number NSTEP of classes.

Then SCALED is an N-vector of real numbers between 0 and NSTEP. Round them off to the next highest integer

$$SCALED\leftarrow\lceil SCALED$$

(we can use the same name for the new vector!).

To display this result against the subscript let us form the Boolean matrix

$$B\leftarrow(\phi 0,\iota NSTEP)\circ.=SCALED$$

Note that we applied "rotate" in order to get the ordinates in the graph pointing upwards as is the custom.

To get this into geometric form we can use a 2-vector of characters, for example, '0+'. We want a + where B has a 1, a 0 otherwise. This is obtained by the plotting statement

$$('0+')[B+1]$$

or written out in full as

$$('0+')[1+(\phi\imath NSTEP+1)\circ.=SCALED]$$

Try this on the terminal. Use some simple function such as the quadratic polynomial

$$RES\leftarrow A+(B\times\imath N)+C\times(\imath N)*2$$

Remark. If the experimenter wants a more informative plot, but still only using the typewriter terminal, he could use one of the functions available in many public program libraries. He can then get coordinate axes with tic marks displayed with automatic scaling; he can display several functions at once, and so on.

Warning. The outer product often leads to elegant solutions; its use should be encouraged, but it must be admitted that sometimes the elegance is spurious. An example of that is the following pretty APL statement to find the prime numbers among the first n natural numbers.

Introduce the set M as

$$M\leftarrow\imath N$$

and consider the outer product with the residue dyadic

$$M\circ.|M$$

What does this mean for row I and column J? It means 0 iff J is divisible by I. Hence the Boolean matrix

$$B \leftarrow 0 = M\circ.|M$$

has a 1 in (I, J) iff I divides J. If we sum along columns, i.e., over the first axis, we get the number of Js among $\imath N$ that divide I. If this set of J consists only of 1 and I, we have a prime. Therefore

$$2 = +/[1]B$$

is the Boolean vector with 1s for the primes. We get the primes themselves by

$$(2=+/[1]0= M\circ.|M)/M$$

This looks fine but the matrix generated by the outer product is very large, with n^2 elements, and it will not be possible to handle it unless n is fairly small. Also the amount of computing is wasteful. One should therefore organize the computing differently.

The same is true now and then: the outer product tends to generate large arrays, difficult to keep in storage and time-consuming to handle. If this is so, sacrifice elegance for practicality—it is sad but necessary!

Exercises

1.   Given a square matrix M, compute its trace.

2.   Compute the diagonal matrix whose entries in the main diagonal are given by the vector DIAG.

3.   Compute the subdiagonal matrix $\{a_{ij}, i, j = 1, 2, \ldots, n\}$ for which $a_{ij} = 1$ for $i > j$ and $= 0$ otherwise. Use outer product together with the proposition $\geq$.

4.   For a given square matrix M and real x write an expression for the resolvent, in mathematical notation $(I - xM)^{-1}$.

5.   Consider a numeric N-vector F as a periodic function with period N. Write a statement that produces the vector of all the components of the difference operator

$$c_1 f_i + c_2 f_{i+1} + \cdots + c_r f_{i+r-1}$$

for general r. Use a circulant matrix associated with the c-vector and matrix multiplication.

6.   Given a numeric vector V in $\mathbb{R}^n$, use the outer product to get a simple expression for the projection operator to the one-dimensional subspace generated by V.

7.   On a finite algebraic structure with N elements labeled $1, 2, \ldots, N$ a multiplication table M (as a matrix) is given as a square matrix M with entries between 1 and N. It is easy to write a program to verify if this multiplication is commutative (do it!) but how do we verify associativity?

8.   Consider the finite Fourier series

$$\sum_{k=0}^{n} c_k \cos kx$$

and form the vector of such values for $x = 0, (2\pi/L), (4\pi/L), \ldots, [2(L - 1)\pi]/L$. Write the resulting vector in APL notation.

9.   Let X be a numeric vector with components $x_1, x_2, \ldots, x_n$ and compute Vandermonde's matrix with entries

$$x_k^l, k = 1, 2, \ldots n, l = 0, 1, \ldots, n - 1.$$

10.   Compute the $n \times n$ covariance matrix of the Wiener matrix with a discrete time parameter. Its entries are $\min(i, j)$. Use an outer product.

11.   (a) Compute the $n \times n$ matrix M_{00} with zeroes everywhere except in (N, N) where the value should be one. Also the matrix M_{10} with a single one at $(1, N)$. Continue like this with M_{-10} with the single one at $(N - 1, N)$, M_0 at $(N, 1)$, and M_{0-1} at $(N, N - 1)$. (b) Use these five matrices to compute the Laplacian on the discrete torus in two dimensions.

12.   If V is a permutation over ιN compute the binary orthogonal matrix that corresponds to this permutation.

13.   Let M be a symmetric matrix with eigenvalues $\lambda_1, \lambda_2, \ldots, \lambda_n$ and associated normalized eigenvectors $v_1, v_2, \ldots, v_n$. Let L be the vectors of λs and O a matrix whose rows are the v-vectors. Write an expression that computes the spectral representation of M.

CHAPTER 12

APL Programs—Defined Functions

1. DEFINITION MODE

Assume that we have written a statement, for example, the one in the preceding section intended for a crude typewriter plot

$$('0+')[1+(\phi\iota NSTEP+1)\circ.=SCALED]$$

We execute it and get the plot we asked for.

That is all very well, but if we want to plot the results of another experiment a bit later, we shall have to type in the statement again. Long statements take time to type; this has a negative psychological effect on the experimenter, and one is apt to make typing errors. It would be better if one could save the statement somehow for later use.

One way of doing this would be by assigning it a name as a character vector

$$PLOT \leftarrow '("0+")[1+(\phi0,\iota NSTEP)\circ.=SCALED]'$$

A brief remark about the double quotes: If we want a quote (considered as a character) *inside* a character array, we must represent it by a double quote as was done above. The next time we need the plotting statement, we use "execute" (see Section 10.6). We just type briefly

$$\pm PLOT$$

and get the plot. In other words, in APL we *can save formulas* and execute them just as we are used to doing in mathematics.

This is fine, but suppose that we want to save many statements and join them together to implement an algorithm. We then write a program consisting of all these statements, which in APL is called just a *function,* or, to distinguish it from the primitive functions, a *defined function.* To indicate the beginning and end of the definition we use the symbol "del," denoted ∇, located over G on the keyboard.

140

Warning. There is also a delta, denoted Δ, on the keyboard. It means something different; see Section 4.

A function definition could look like this

$$\nabla \ HILBERTSCH$$
$$[1]NORM \ \leftarrow \ (+/+/MATRIX*2)*.5$$
$$\nabla$$

to calculate the Hilbert–Schmidt norm of a matrix

$$\sqrt{\sum_{i=1}^{n} \sum_{j=1}^{n} m_{ij}^2}$$

Note that we have given the function a name. The function, i.e., program, is automatically saved in temporary memory (how to save it permanently comes later) so that the next time we want to execute it we just type

$$HILBERTSCH$$

The effect is not visible but the result is computed and stored under the name NORM. If we want to have the result typed out, we say

$$NORM$$

and the norm is printed on the next line.

We can have several lines in a program, although APL functions tend to be much more concise than those in FORTRAN, ALGOL, and the other conventional languages. To illustrate how statements are combined into a function, let us calculate N independent and identically distributed stochastic variables, each taking the values 0 and 1 with probability $\frac{1}{2}$ for each. Then compute the partial sums, so that we are studying a stochastic process with independent increments, and print out the result with some text that reminds us of what the numbers mean. We could define

```
     ∇ BERNOULLI
[1]  V ← ‾1+?Nρ2
[2]  SUMS ← +\V
[3]  'THE PARTIAL SUMS ARE ='
[4]  SUMS
     ∇
```

A bit more concisely, but with the same result, using format to produce the numerical result converted into characters, we have

```
     ∇ BERNOULLI1
[1]  'THE RESULT IS = ' , ⍕+\‾1+?Nρ2
     ∇
```

When several statements appear in a function the convention is that *the statements are executed in the order in which they appear* (unless otherwise specified). Each statement is automatically given a statement number: [1], [2], and so on.

To be able to execute BERNOULLI, the variable N must have been assigned a value; otherwise we shall get the error message

```
      VALUE ERROR
      BERNOULLI[1]
[1]   V ← ¯1+?Nρ2
               ∧
```

One way of specifying values is via *arguments*. In mathematics we are used to arguments such as

$$\sin \sqrt{x}$$

or

$$\sqrt{x^2 + y^2}$$

In BERNOULLI we used no *explicit argument*. We are allowed to use one explicit argument, placed to the right of the function name, or two, one placed to the left and the other to the right.

For example, we could define

```
    ∇ BERNOULLI2 N
[1] 'THE RESULT IS =', ⍕+\2×¯1.5+?Nρ2
    ∇
```

To execute it we say simply

```
      BERNOULLI2 15
```

for the partial sums of 15 Bernoulli variables taking values ± 1.

As an example of a function with two explicit arguments, consider the definition

```
    ∇ M BERNOULLI3 N
[1] 'THE RESULT IS = ',⍕+\2×¯1.5+(M,N)ρ2
    ∇
```

which produces M samples, each of length N, of our stochastic process.

In mathematical notation we are accustomed to *composition* of two or several functions, say

$$\sin x^2$$

or

$$(1 + e^x)/(1 - e^{-x})$$

We can do the same in APL without difficulty by using functions with *explicit result*. This is indicated in the function heading by an assignment arrow; for example,

```
     ∇  RESULT ← SAMPLESIZE  K
[1]     RESULT ← ⌊K*.5
     ∇
```

If we say

$$BERNOULLI2 \ SAMPLESIZE \ 16$$

we shall get the result of BERNOULLI2 executed with $N = \sqrt{16} = 4$.

The convention that an APL function cannot have more than two arguments at first appears restrictive. Remembering, however, that the argument names can stand for vectors or larger arrays, this criticism is seen to be essentially unfounded.

Let us write a little program that convolves two vectors. Say that the two vectors are P and Q of length NP and NQ

$$NP \leftarrow \rho P$$
$$NQ \leftarrow \rho Q$$

We want to form a new vector, say R, with entries

$$r_1 = p_1 q_1$$
$$r_2 = p_1 q_2 + p_2 q_1$$
$$r_3 = p_1 q_3 + p_2 q_2 + p_3 q_1$$

We are going to organize the program without explicit looping, using matrix multiplication instead. Consider the vector V:

$$
\overbrace{}^{NP} \qquad \overbrace{}^{NQ}
$$
$$
0 \ \ 0 \ \cdots \ 0 \qquad \cdots \ \ q_3 \ \ q_2 \ \ q_1
$$

that is, using monadic rotate,

$$V \leftarrow (NP\rho 0), \phi Q$$

Now define a matrix in terms of the entries v_k of V

$$
NP + NQ \left\{
\begin{array}{cccc}
\overbrace{\phantom{v_1 \quad v_2 \quad \cdots \quad v_{NP+NQ}}}^{NP + NQ} \\
v_1 & v_2 & \cdots & v_{NP+NQ} \\
v_1 & v_2 & \cdots & v_{NP+NQ} \\
 & & \vdots & \\
v_1 & v_2 & \cdots & v_{NP+NQ}
\end{array}
\right.
$$

that is

$$M \leftarrow ((NP+NQ)\rho 1)\circ.\times V$$

Now shift the rows one step, two steps, . . . , $(NP \times NQ)$ steps with dyadic rotate

$$M \leftarrow (-\iota NP+NQ)\phi M$$

Then the convolution is realized by multiplying the new M-matrix with the vector

$$\overbrace{p_1 \quad p_2 \quad \cdots \quad p_{NP}}^{NP} \quad \overbrace{0 \quad 0 \quad \cdots \quad 0}^{NQ}$$

that is,

$$R \leftarrow M+.\times P,NQ\rho 0$$

except that we should drop the last entry:

$$R \leftarrow {}^{-}1\downarrow R$$

Combining these statements, and making the variables NP, NQ, V, and M local, we get

```
∇    R←P CONVOLVE Q ;NP;NQ;V;M
[1] NP←ρP
[2] NQ←ρQ
[3] V←(NPρ0),φQ
[4] M←(-ιNP+NQ)φ((NP+NQ)ρ1)∘.×V
[5] R←¯1↓M+.×P,NQρ0
∇
```

Here R is the explicit result, P and Q are the two arguments.

Note that statement [4] creates a matrix that will be prohibitively large if NP and NQ are large. We shall then get problems with work space size and may have to use a program that loops explicitly with a →LABEL statement. It will require less space but will be much slower—the usual trade-off between space and time in all computing.

It would be easy to compactify the program into a one-liner. Don't!

Let us now instead simulate sample functions from a Wiener process. After discretization this amounts to calculating partial sums of the type

$$x_1 + x_2 + \ldots + x_t, \quad 1 \le t \le n$$

where (x_k) are independent stochastic variables, all with a normal distribution with mean zero and variance 1. We may scale the result later on if desired.

A common way of simulating a sample of size N of normally distributed stochastic variables, mean zero and variance one, is to form, for each sample value,

$$y_1 + y_2 + \ldots + y_{12} - 6$$

where all the ys are independent and distributed according to a uniform distribution over $(0, 1)$. The rationale behind this algorithm is based on the central limit theorem; the normalized random variable

$$\frac{x_1 + x_2 + \ldots + x_n - nm}{\sigma \sqrt{n}}$$

is approximately normally distributed if n is large. In our case n is 12, $m = \frac{1}{2}$, and $\sigma = 1/\sqrt{12}$. This is easy:

```
    ∇ RESULT ← NORMAL N
[1] RESULT ← ¯6++/ 1E¯6×?(N,12)1E6
    ∇
```

We then get the sample function, as a vector, from applying "scan" with addition:

```
    ∇ SAMPFUNC ← WIENER N
    SAMPFUNC   ← +\NORMAL N
    ∇
```

We have plotted one such sample function in Fig. 1.

Figure 1

Now we do the same thing for a Cauchy process. It is easy to simulate a Cauchy distribution with frequency function

$$\frac{1}{\pi} \frac{1}{1 + x^2}, \qquad -\infty < x < \infty$$

We compute the distribution function (it is an arctan), form its inverse (it is a tan function), and apply it to a uniform distribution over (0, 1). Hence

```
     ∇ RESULT ← CAUCHY N
[1] RESULT← 3○ ○.5E‾6×‾1+?N⍴1E6
[2] RESULT← (‾1*?N⍴2)×RESULT
     ∇
```

In [1] we simulate positive random variables. The ‾1 is added to make sure that we never try to compute tan $\pi/2$. In [2] we symmetrize the distribution by multiplying the values in RESULT by ±1 with random signs.

We now get the sample functions of the Cauchy process by applying $+\backslash$. One result is shown in Fig. 2.

Figure 2

Compare the sample functions in Figs. 1 and 2. The first one is (practically) continuous while the second has large jumps, illustrating the theory very well.

This example could easily be extended for the Wiener process by looking at the Lipschitz property p, $p < \frac{1}{2}$.

Finally, let us write a short program to compute the baker's transformation, familiar from ergodic theory. It maps the unit square $[0, 1]^2$ onto itself by the map $(x, y) \rightarrow (x', y')$:

$$x' = \{2x\}$$
$$y' = (y + [2x])/2$$

In the picture, we press the square "dough" down and then cut it in half, placing the right half on top of the left:

$$\boxed{A\ B} \Rightarrow \boxed{A\ B} \Rightarrow \frac{\boxed{B}}{\boxed{A}}$$

Hence we can write

```
        ∇ RESULT BAKER Z
  [1] X ← Z[1]
  [2] Y ← Z[2]
  [3] XPRIM← 1|2×X
  [4] YPRIM← .5×Y+⌊2×X
  [5] RESULT← XPRIM,YPRIM
        ∇
```

It is easy to experiment with plotting successive points similarly to Experiment 8 in Chapter 3.

Exercises

1. Define a function UNION with an explicit result U and the two explicit arguments A and B that compute the union U of the two sets A and B (viewed as vectors) without repetitions. Include some text saying that the union of the two sets is = ... See Exercise 2 in Section 10.5. Execute it with some choices of A and B and check that it works.

2. Define a function with two arguments N and P that will compute the sum

$$\sum_{k=1}^{n} k^p$$

Include some text that reminds the user of what it represents and what N and P are numerically.

2. FLOW OF CONTROL

In any moderately involved mathematical experiment we need the ability to control the order of execution of statements. The order is not always known in advance and must then be computed as we go along, executing one statement after another.

This is done by the right arrow →, located above the left arrow on the keyboard. A statement like

$$[3] \rightarrow 12$$

means that, once we get to statement 3, we should jump to statement 12. This is an unconditional "go to" statement. Or

$$[5] \rightarrow STATEMENT + 2$$

means that we should go two statements beyond that whose number is already computed and stored as the variable STATEMENT.

Using compression, we can include "conditional go to" as in

$$[8] \rightarrow B/2$$

where B is a Boolean variable. If B is equal to one, $B/2$ is, as we know, equal to 2 so that we should go back to statement number 2.

If B is 0, then

$$B/2$$

is the empty set. What does this mean? The convention is that → *empty set is a command that should be disregarded:* just continue with the next statement, in this case number 9.

Another convention is that the control statement

$$\rightarrow 0$$

means stop execution. This device is often used, especially when we do not know in advance how many iterations are needed.

When planning a large experiment we often do something like this. We are going to execute a defined function, say FUNC, a number of times, say ITER. We store the results each time by calling a function named STORE. After we are done we use another defined function, say ANA-LYZE, to analyze all the accumulated data. Finally we display the result of the analysis by another defined function, say DISPLAY.

We have to keep track of how many iterations have been completed. We do this by a *counter*, say I, which is initially set to 0. The *master program* calling all the defined functions could be given the form

```
      ∇ EXPERIMENT ITER
[1]  I ← 0
[2]  FUNC
[3]  STORE
[4]  I ← I+1
[5]  →(I≤ITER)/2
[6]  ANALYZE
[7]  DISPLAY
      ∇
```

The reader should convince himself that this program does exactly what it is supposed to do.

Remark.   To refer to statement numbers one can use computed *labels*, which are indicated by colons. In the above function, if statement [2] had the form

```
[2]  LABEL: FUNC
```

then the value of the variable named LABEL is computed as 2. We could then replace [5] by

```
[5]  →(I≤ITER)/LABEL
```

leading to the same result. We shall see in Section 12.4 why this may be more advantageous than referring to the absolute statement number 2. In this connection it should also be mentioned that defined functions are allowed to call themselves—to be *recursive*. We illustrate this by an example. Given X and Y as natural numbers, we want to compute their greatest common divisor. Consider the function

```
      ∇ Z ← X GDR Y
[1]     Z ← Y
[2]     → (X=0)/0
[3]     Z ← (X|Y) GDR X
      ∇
```

Verify that it works!

Warning.   Recursive definitions are often elegant but can be inefficient. If the nesting of the recursion is large, the program may stop since the system is not able to handle the depth of recursion. The error message that may result will be discussed later.

Summing up, we can have statements of the following types.

1. Compute and assign to name X; for example,

$$[3] \quad X \leftarrow Y + 2$$

2. Compute and display result; for example,

$$[4] \quad Y + 2$$

3. Go to; for example,

$$[7] \quad \rightarrow \quad Z + 12$$

which also can be given conditional form.

Exercises

1.   Define a function with two explicit arguments A and MAX that calculates the vector with entries v_k

$$v_k = Ae^k + k^2 \qquad \text{for } k = 1, 2, \ldots$$

until v_k reaches the value MAX.

2.   Define a recursive program that generates the Fibonacci series $x_1, x_2, \ldots ,$

$$x_k = x_{k-1} + x_{k-2}$$

starting with x_1, x_2 combined into the 2-vector INIT used as left argument. Compute the vector of length N used as right argument.

3.   Define a function that calculates Bernoulli variables—stochastic variables that take the values 0 and 1 with probabilities $1 - P$ and P. Go on calculating them until their sum is equal to SUM. Use P and SUM as explicit arguments and the resulting vector of Bernoulli variables as the explicit result.

4.   Let f be an increasing continuing function $[0, 1] \rightarrow [0, 1]$ with $f(0) = 0$, $f(1) = 1$, and $f(x) > x$ for $x < \zeta$, $f(x) < x$ for $x > \zeta$. To find the stable fix point ζ, $\zeta = f(\zeta)$, write a program FIX that starts with the value INIT in $(0, 1)$ and calculates

$$x_{k+1} = f(x_k)$$
$$x_1 = \text{INIT}$$

successively until the termination condition holds:

$$|x_{k+1} - x_k| \le \text{EPS}$$

Use INIT and EPS as explicit arguments. Treat f as an already defined function. Try it for some special function, for example,

$$f(x) = 1/2 + 4(x - 1/2)^3$$

Define another function F with one explicit argument X and an explicit result Z for this choice of f.

5. M stands for a matrix of norm less than one. Define a function $F1$ that calculates successive powers M^n until

$$\sum_{i,j} |m_{ij}^{(n)}| \le \epsilon$$

The explicit result should be this M^n. Have the function call another function TEST, defined by you, that computes the sum above and the truth value of the proposition. TEST should have one explicit argument M and an explicit result as the truth value.

3. MANAGING MEMORY

The computing takes place in a *work space,* but you may have several work spaces. They are managed by system commands that are not part of the language but characteristic of the system and the implementation of the language. They can therefore be expected to vary from system to system.

After logging on, you will have access to a work space. You can find its name—its work space identification—by the *system command*

```
)WSID
```

and the system responds by the name of the current work space, say,

```
ODE
```

If the response is

```
THIS IS CLEAR WORK SPACE
```

it means that it has not been given any name. You can name it by

```
)WSID NAME.
```

All system commands are preceded by), right parenthesis.

What is contained in the work space? The names of the defined functions can be found by the system command

```
)FNS
```

with the answer, for example,

$$RUNGEKUTTA1 \ RUNGEKUTTA2$$

To find the names of the variables, type in

$$)VARS$$

which may yield

$$INITIAL \ STEPS \ STEPSIZE \ X \ Y \ Z$$

To delete variables or defined functions, ask for

$$)ERASE \ X \ Y \ Z$$

resulting in the action that the values of X, Y, Z are deleted from memory and their names from the table of symbols.

Suppose that instead you want another work space, say an empty one. This is obtained by

$$)CLEAR$$

and you can give it a name (otherwise it is impossible to save it) by the command mentioned above,

$$)WSID \ NAME$$

If you wish to use another work space, called WS3 for example, that you have already been working with, say

$$)LOAD \ WS3$$

However, if you only need one object, a variable, a function, or a group, say

$$)COPY \ WS3 \ INTEGRATION$$

which will copy the object INTEGRATION from work space WS3 and insert it into the work space in which you are currently active.

You are always storing objects temporarily; they will be lost eventually unless you want to save them permanently and indicate this by the system command (if the WS name is ODE)

$$)SAVE \ ODE$$

This results in the entire active work space ODE being saved permanently.

There are also *system variables*; only two will be described and the reader is referred to APL Language (1976, pp. 47–51) for the others.

One of the system variables that the user needs early is *account information,* denoted by $\Box AI$ (the quad symbol $\Box$ is located above L). It usually means "display" or something related to this. Executing

$$\Box AI$$

will give the user a 4-vector of integers representing an identification, computer time used during this session, connect time since log on, and the time you have had available for typing. All times are given in milliseconds. If you get the answer

$$481 \quad 2379 \quad 18342 \quad 16521$$

it means that you have used 2.379 sec of CPU time. Repeated use of this, together with subtraction of the second components of the vectors, tells you how much CPU time you have consumed between two calls of $\Box AI$. This is useful to get an idea of how efficient your code is.

The system variable *working area,* denoted by $\Box WA$, tells you how much storage is *left* in your active work space. The unit is one byte. If you get

$$8150$$

you know that you cannot define a 20×80 matrix with real numbers as entries since this takes about $20 \times 80 \times 8 = 12800$ bytes.

4. WRITING CODE PRAXIS

Although we have now seen most of the facilities that we shall need (there are a few that will be mentioned soon), it takes time and patience to learn how to exploit them fully. To show how program writing should look, let us consider a simple mathematical experiment. It is intended to simulate a growth model.

On the discrete plane $\mathbb{Z}^2$ with coordinates (i,j) we want to study growth with nearest neighbor interaction. The organism occupies a certain set ORG in $\mathbb{Z}^2$ at time T.

For each coordinate pair (i, j) for which $(i, j) \notin$ ORG, the probability that a cell will be *born* is

$$b_{-1,0}e_{i-1,j} + b_{1,0}e_{i+1,j} + b_{0,-1}e_{i,j-1} + b_{0,1}e_{i,j+1}$$

where the bs are specified constants and $e_{i,j}$ is the indicator function of ORG.

If $(i, j) \in$ ORG, the cell at (i, j) will *die* with probability

$$d_{-1,0}e_{i-1,j} + d_{1,0}e_{i+1,j} + d_{0,-1}e_{i,j-1} + d_{0,1}e_{i,j+1}$$

with constants $d_{-1,0}$, $d_{1,0}$, etc.

Let us consider the finite part of $\mathbb{Z}^2$ for which $1 \le i \le N$, $1 \le j \le N$. For simplicity we shall assume the model to be cyclic with period N in both the i and the j directions. In other words, we operated on the discrete 2-torus. This is not important except for the boundary effects, which will not appear unless the organism grows close to the boundary of the square in which we are operating; this will be avoided by making the square large enough.

At $T = 1$ we start with an initial configuration ORG0. We shall let the simulation run with the number of iterations equal to ITER.

We shall need a main-line program MAIN, similar to what we described in Section 1, and several other defined functions for specific tasks. One of them, PBF, computes the birth probabilities; another, PDF, the death probabilities.

To exploit APL's array organization, ideal for much of mathematics, we shall handle all (i, j) values at once. Of course, current computers work almost entirely sequentially but the program will look parallel, much to our advantage. Therefore the result PB of the PBF function should be an $N \times N$ matrix with real entries and PD will be similar for the PDF function. The organism ORG should be represented as a Boolean $N \times N$ matrix.

We also need a defined function BIRTH that computes a random variable equal to 1 with probability PB and a similar one for a function DEATH. They should also operate directly on $N \times N$ matrices.

We now write the main-line program. Open the definition and type in the code

```
        ∇ MAIN ITER
[1]  ORG ← ORG0
[2]  T ← 1
[3]  PB ← PBF ORG
[4]  PD ← PDF ORG
[5]  BIRTHS ← BIRTH PB
[6]  DEATHS ← DEATH PD
[7]  ORG ← BIRTHS ∨ ORG ∧ ~ DEATHS
[8]  T ← T + 1
[9]  → (T ≤ ITER)/3
        ∇
```

Only line [7] needs explanation. In it we delete cells that are dead by

$$ORG \ \wedge \ \sim \ DEATHS$$

and then add newborn ones by the $\vee$ part of the statement.

We had better save this by the system command)SAVE.

Consider now the definition of BIRTH. It has as explicit result the $N \times N$ Boolean matrix BIRTHS, and the $N \times N$ real matrix PB as its right and only explicit argument. Remember that, if the entry (i, j) has the value one, then BIRTH should not change it. Write the definition

```
      ∇  BIRTHS  ←  BIRTH  PB
 [1]      BIRTHS  ←  PB  ≥  1E¯6×?(N,N)ρ1E6
      ∇
```

Before going any further we had better try this function. To do this, choose small and/or simple values, for example, $N = 5$:

$$N \ \leftarrow \ 5$$

and

$$PB \ \leftarrow \ (N,N)\rho.1$$

so that all birth probabilities are equal to 10%. Executing

$$BIRTH \ PB$$

we get the Boolean matrix

```
          0  0  0  0  0
          1  0  0  0  0
          0  0  1  1  0
          0  1  0  1  0
          0  0  0  0  0
```

It looks all right. Now we do the same for DEATH and put

$$PD \ \leftarrow \ (N,N)\rho.1$$

writing the code

```
      ∇  DEATHS  ←  DEATH  PD
 [1]      DEATHS  ←  PD  ≤  1E¯6×?(N,Nρ1E6
      ∇
```

But executing this does not succeed; we get the error message

```
SYNTAX  ERROR
DEATHS←PD≤1E6×?(N,Nρ1E6
                       ∧
```

We have clearly forgotten the right parenthesis. To remedy this we open
the definition

$$\nabla \; DEATH$$

and are prompted to type the next statement

$$[2]$$

but this is not what we want. Instead we type

$$[1] \; DEATHS \; \leftarrow \; PD \; \leq \; 1E^{-}6 \times ?(N,N)\rho 1E6 \; \nabla$$

Note the close-definition ∇ at the end. The editing can actually be done
without typing in the whole line again. We shall return to this below.
Executing DEATH now gives

$$
\begin{array}{ccccc}
0 & 1 & 1 & 1 & 1 \\
1 & 1 & 0 & 1 & 1 \\
1 & 1 & 1 & 1 & 1 \\
1 & 1 & 1 & 1 & 0 \\
1 & 1 & 1 & 1 & 1
\end{array}
$$

But this is not a likely result; there are too many ones. We would expect to
get on the average $5 \times 5 \times 0.1 = 2.5$ but we find 22 ones. The error is in
the inequality sign; we gave it the wrong direction.

This could be changed as above. We shall choose an alternative
method, illustrating a convenient debugging tool. Type in

$$\nabla \; DEATH[1\square 5]$$

which means: open the definition, display line 1, and be prepared for a
correction at approximately the 5th character of this line. The last number
is not important.

The terminal will respond by

$$[1] \; DEATHS \; \leftarrow \; PD \; \leq \; 1E^{-}6 \times ?(N,N)\rho 1E6$$

Now move the carriage to $\leq$, type in /, meaning delete this character,
followed by 1, meaning leave one space blank. The terminal's response
will be

$$[1] \; DEATHS \; \leftarrow \; PD \quad 1E^{-}6 \times ?(N,N)\rho 1E6$$

Now position the carriage under the blank and type $\geq$. The terminal will
take this change into memory, and goes on by prompting us for the next
statement

$$[2]$$

We do not need any more, so we should answer by closing the definition by typing just

$$\nabla$$

We had better execute the function a few times, perhaps with different arguments, and see that everything looks fine.

Assuming this, we go on to define the function PBF. Here we use dyadic "rotate," recalling that we made the model cyclic. Look up the definition of "rotate" for arrays in APL Language (1976, p. 33). To get the term $e_{i-1,j}$ we should shift the whole ORG matrix 1 step along the i direction. This is done by (with the axis operator [1])

$$^-1\phi[1]ORG$$

To get $e_{i,j+1}$ we do

$$1\phi ORG$$

and so on. This leads to the tentative definition

```
      ∇ Z ← PBF ORG
[1]     Z ← (B[1]×⁻1φ[1]ORG)+(B[2]×1φ[1]ORG)+
            +(B[3]×⁻1φORG)+B[4]×1φORG
      ∇
```

Just to get some values for the constant let us put

$$B\leftarrow4\rho.1$$

and also

$$D\leftarrow4\rho.001$$

Try this on a few examples. Then write a similar function for PDF but, of course, with the B-vector replaced by the D-vector. Then try it, debug it if necessary, and)SAVE. It is a good idea to do)SAVE more often than actually needed to prevent a possible computer breakdown from destroying programs that have been completed but not saved.

We now try to execute, and type

$$MAIN\ 3$$

but this is not possible, as indicated by the error message

```
      VALUE ERROR
      MAIN[1] ORG ← ORG0
                        ∧
```

showing that we have not yet defined the initial shape $ORG0$. Hence let us define ORG0 as, for example,

```
ORG0 ← (N,N)ρ0
ORG0[3;2 3 4] ← 1 3 ρ 1
```

so that

ORG0

```
0  0  0  0  0
0  0  0  0  0
0  1  1  1  0
0  0  0  0  0
0  0  0  0  0
```

meaning that the organism forms a bar of length 3 and width 1.

If you are working on a fairly large program with a complicated flow of control, and if it does not work the way you wish, it may be useful to *follow the execution by a trace control command.*

This is done by setting the trace control vector equal to one or several statement numbers that appear suspicious. For our program MAIN we may say (note the delta, not del!)

$$T\Delta MAIN ← 3 \ 5 \ 8$$

Then execution will proceed as before except that, each time we pass through statement 3, the terminal will type out

$$MAIN[3]$$

together with the result of that statement, and so on for statements 5 and 8.

In this way we can trace the crucial steps, which often facilitates debugging the program.

To negate the trace control command, change the trace control vector to the empty one:

$$T\Delta MAIN ← \iota 0$$

There is also a *stop control* command that can be used to force execution to halt at certain specified statements. See APL Language (1976, p. 68) for a description.

Now let us execute MAIN again. We get no visible result although ORG has been computed three times. Its value has been stored, however. How does it look?

To get a simple graphical display we define

```
∇  SHAPE ← DISPLAY ORG
[1]    SHAPE ← (' o')[1+ORG] ∇
```

We want to insert this after line 7 in the main-line program. This can be done by opening the definition of MAIN by the statement

$$\nabla \ \ MAIN$$

The response is

$$[10]$$

to which we respond by

$$[7.1] \ \ DISPLAY \ ORG \ \ \nabla$$

Now when we execute MAIN we get the shapes dynamically displayed. To separate the picture better and to indicate the number of iterations, we open the definition again

$$\nabla \ \ MAIN$$

and get

$$[11]$$

Now the statements have been renumbered as integers, [7.1] is now [8] and we have 10 statements. Type in

```
[7.1] 'THE ORGANISM IS AT TIME', ⍭ T
[7.2] ⍳0
[7.3] ⍳0 ∇
```

The last two statements separate the pictures by printing (well, not really!) two blank lines.

We now have 13 statements. Let us also do some analysis by

```
∇ GROWTH
[1] SIZEVECTOR ← SIZEVECTOR,+/+/ORG ∇
```

and insert this statement after [7]. We also have to initialize the vector SIZEVECTOR by a statement in MAIN

$$[1.1] \ \ SIZEVECTOR \ \leftarrow \ +/+/ORGO$$

Run this, but now for larger values of N, and experiment with different birth and death probabilities specified by the numeric 4-vectors B and D. Do a)SAVE!

We now want to see the entire main-line program, by

$$\nabla \ \ MAIN[\Box] \ \ \nabla$$

and see the APL code with automatically recomputed statement numbers

```
        ∇ MAIN ITER
[1]     ORG←ORGO
[2]     SIZEVECTOR←+/+/ORGO
[3]     T←1
[4]     LOOP:PB←PBF ORG
[5]     PD←PDF ORG
[6]     BIRTHS←BIRTH PB
[7]     DEATHS←DEATH PD
[8]     ORG←BIRTHSVORGΛ~DEATHS
[9]     'THE ORGANISM IS AT TIME ',⍕T
[10]    ι0
[11]    ι0
[12]    GROWTH
[13]    DISPLAY ORG
[14]    T←T+1
[15]    →(T≤ITER)/LOOP
        ∇
```

The reader will notice that we have used the label LOOP in statement [4]; this made debugging faster.

Executing the program MAIN with $N = 20$ and

$$ORGO←(N,N)ρ0$$

and

$$ORGO[9\ 10\ 11;\ 9\ 10\ 11]←\ 3\ 3ρ1$$

gave the result

```
        MAIN 10
   THE ORGANISM IS AT TIME 1
```

```
        ooo
        oooo
        oooo
```

```
   THE ORGANISM IS AT TIME 2
```

```
        oo
        oooo
        oooo
```

THE ORGANISM IS AT TIME 3

```
   ooo
   oooo
oo  oo
   o
```

THE ORGANISM IS AT TIME 4

```
   oooo
    oooo
oooooo
   o  o
```

THE ORGANISM IS AT TIME 5

```
   oooo
   ooooo
oooo  o
 o  oo
```

THE ORGANISM IS AT TIME 6

```
   oooo
   ooooo
oooooo
 o  oo
```

THE ORGANISM IS AT TIME 7

```
   oooo
   ooooo
ooooooo
   ooo
```

THE ORGANISM IS AT TIME 8

```
     OOOO
    OOOOOO
  OOOOOOO
    OOO
```

THE ORGANISM IS AT TIME 9

```
     OOOO
   OOOOOOO
 OOOO  OO
   O  OOO
```

THE ORGANISM IS AT TIME 10

```
     OOOO
   OOOOOOO
 OOOO  OO
  OO  OOO
    O    O
```

The size of the organism has grown according to the vector

SIZEVECTOR
9 11 10 12 16 17 18 19 20 21 24

In a larger experiment one would probably prefer to print out the display only when T is some multiple of some number, say NUM. This could be achieved by a conditional branch from statement [8] to [14] except when the proposition

$$0 = NUM \mid T$$

is true. One should then also move [12] to [8.1], for example, so that the updating of the variable SIZEVECTOR will also take place when no display is made.

An experiment of this type requires a good deal of planning and the programming takes some time. Let us examine a quite different case by an example that is characteristic of an often useful experimental situation.

Consider the stochastic variables

$$y_k = f\left(\frac{k}{n}\right) + e_k \qquad k = 1, 2, \ldots, n$$

where the e_ks are i.i.d. from $N(0, 1)$ and f is a continuous increasing function. We observe the set $\{y_k\}$ and want to estimate f, especially when n is large.

Note that we do not observe the vector (y_k), but only the *set of values;* no labels are attached to them. Such problems are next to unknown in statistical theory at present, but can be expected to become of great practical interest.

Let us try to estimate, for example, $\alpha = f(\frac{1}{2})$. To do this, order the y_ks in increasing order so that we obtain a vector $z_1, z_2, \ldots, z_n$. Use $\alpha^* = z_m$, $m = [n/2]$ as an estimate of $f(\frac{1}{2})$, or form the statistic

$$\beta^* = \frac{1}{n} \sum_{k=1}^{m} z_k$$

as an estimate of the quantity

$$\beta = \int_0^{1/2} f(x)\, dx$$

The motivation behind the choice of α^* and β^* is that ordering the y_ks will dampen out the influence of the e_k disturbances so that α^* and β^* will converge in probability as $n \to \infty$ to the limits α and β, respectively.

Are these conjectures true? Can they be supported by a mathematical experiment? Let us write a little function to generate the y_ks:

```
      ∇  Z ← DATA SS
 [1]  V ← (÷SS)×ιSS
```

This statement produces a vector of x values spaced by $dx = \div SS$. Now form the e_ks and add $f(k/n)$

```
 [2]  SAMPLE ← (GAUSS SS) + F V
 [3]  PERM ← ⍋ SAMPLE
 [4]  Z ← Z[PERM]
 [5]  □ ← ALPHASTAR ← Z[⌊ .5×SS]
      ∇
```

Let the function $f(x)$ be $x - \frac{1}{2}x^2$

```
      ∇  Z ← F    X
 [1]  Z ← X  -  .5×X*2
      ∇
```

Note that in DATA we do not make PERM a local variable; we want to be able to look at it later on. To be able to handle several *f* functions conveniently, we could have made DATA have as a left argument the name of the function as a character string and then used an "execute" statement in line [2].

We are now going to replicate the experiment several times. The typical way of doing this is by a master program looking something like

```
      ∇      REPLICATE TIMES
[1]   'WHAT IS NAME OF FUNCTION TO BE REPLICATED?'
[2]   NAME ← ▯
[3]   T ← 1
[4]   LOOP: ⍕ NAME
[5]   T ← T+1
[6]   → (T ≤ TIMES)/LOOP
      ∇
```

This will replicate the function called NAMES again and again until TIMES replications have been obtained. The reader may want to save a copy of REPLICATE in a work space for other uses.

Start with *n*, that is *SS*, = 100, and do 10 replications. We get

```
      SS←100
      REPLICATE 10
WHAT IS NAME OF FUNCTION?
DATA
0.411725
0.018296
0.386263
0.367047
0.246805
0.452604
0.238718
0.317233
0.343978
0.311863
```

Convergence does not seem to take place although the sample size *n* is considerable, namely 100.

Increase *n* to 300. We get, with again 10 replications,

```
      SS←300
      REPLICATE 10
```

```
WHAT IS NAME OF FUNCTION?
DATA
0.3712174444
0.3237627778
0.3410217778
0.3751704444
0.343109
0.3573007778
0.4133024444
0.2952267778
0.2929887778
0.365739
```

No convergence is visible and, even worse, the estimate seems to have a considerable negative bias.

This experiment was repeated in different versions and the results confirmed (empirically) the above impression. Also, the second estimate β^* was computed; note that we can get it easily by changing [5] in DATA to

$$[5] \quad \Box \leftarrow BETASTAR \leftarrow (:SS) \times \iota / Z[\iota \lfloor .5 \times SS]$$

The results are negative here too.

Since all the experimental evidence contradicted the conjecture, we decided to try to disprove it analytically. This turned out to be fairly easy, but that is another story.

In this case the programming effort was negligible, about an hour including debugging. Nevertheless, the experiment was of some psychological help for the analyst.

Now we give some general advice about the praxis of writing APL code for mathematical experiments.

(a)   Resist the temptation to be clever and elegant. Write reasonably *short statements* rather than one-liners that do an awful lot of computing. In this way the code becomes more transparent; it can be read and understood by others, and by yourself a month later. Remember Boltzmann's advice: elegance should be the concern only of tailors.

(b)   Include internal documentation in the program. Any number of *comments* can be inserted, indicated by the "lamp post," denoted ⍝, obtained from ∩ overstruck with the small circle ∘. It could look like

$$[5] \quad ⍝ \; COMPUTES \; LARGEST \; EIGENVALUE$$

or

$$[11] \quad \text{\tiny A} \; X \; MEANS \; REAL \; PART \; OF \; Z$$

(c) Let the control flow have *clean topology*. Because of its array handling, APL can often do without lots of "GO TO," conditional, and unconditional statements. When → is used, see to it that you do not get any overlapping, as in

```
 ┌ [3]  →  B/5
 │┌[4]  →  LABEL ←┐
 └→[5]  →  4       │
   →[6]  LABEL:X  ←  (X*2)+*X
```

(d) Write the code in *modular* form, building large programs from simpler modules—defined functions. This will make the debugging simpler and many possible mistakes will be avoided. One has to be careful about naming variables in functions that call each other: if the same name appears in both functions, confusion can occur. Make the names local to avoid trouble, if possible.

(e) Use mnemonics for names. It is very annoying to have to decipher code with variables $Z12$, $Y5$, etc., rather than the mnemonic EIGENVEC, UNIT, etc.

(f) When → is needed, use labels rather than absolute statement numbers. Remember that adding new statements in the middle of a program will automatically change the following absolute statement numbers.

(g) Contrary to the conventional wisdom among orthodox programmers, it may be perfectly all right to recompute what has already been computed. Sometimes memory is more crucial than CPU time!

(h) Avoid cluttering up the work space by scores of global variables. If the variables $X, Y, Z, \ldots$ are needed only for the internal computing of a function FUNCT, this can be indicated in the definition heading

$$[0] \quad FUNCT;X;Y;Z$$

using semicolons. These variables are then *local*, in contrast to *global* ones that continue to be defined. The price we pay for this is that the values of $X, Y, Z, \ldots$ are no longer available after execution.

However, do not be hasty in making variables local. During the construction and debugging of a program it is advisable to keep them all global, so that one can check their values. Only at the end, when the function is to be saved, should one change the heading of the function, making some variables local.

Note. Even local variables are available if the program is interrupted; this helps in debugging.

(i) Save functions in groups, not too large, so that a group contains functions that do related things and/or support each other. If a function always calls another function, then a copy of the second one should be included in the group containing the first. This leads to some waste of storage but this is seldom serious.

(j) Write auxiliary code for doing the data analysis of the results of the experiment. Such code should arrange results into legible form, well-organized tables or graphs. Cosmetic formatting of the results into beautiful tables can usually be dispensed with, though.

(k) If the results of the experiment look peculiar, do not be discouraged. Perhaps you have stumbled upon something essential! To be able to pursue this further, it is often a good idea to let APL functions with an explicit result store the result under the name of a global variable. In this way you can continue the analysis of the result without having to replicate the experiment. This is important especially when the result is a large array.

(l) The printout from the terminal should of course be saved, but it is often necessary to write down comments on it. Such penciled comments could specify parameters and conditions under which the experiment was carried out.

(m) If you run into problems with a large APL function with elaborate branching structure, execute it for simple choices of the parameters. Preferably choose values that make the computing task trivial so that you can check the result.

(n) Do as much of the calculations as possible analytically. Remember that *the computer is useful for our purpose as a supplement to, not a substitute for, analysis*.

Exercises

1. Define a function POISSON that computes the harmonic function in polar coordinates (r, ϕ):

$$u(r, \phi) = \frac{1}{2\pi} \int_{\psi=0}^{2\pi} \frac{1 - r^2}{1 - 2r \cos(\phi - \psi) + r^2} \, b(\psi) \, d\psi$$

for given boundary values. Here $b(\psi)$ should be given as an arbitrary numeric N-vector, corresponding to N uniformly spaced points on the unit circle. Let B, for $b(\psi)$, and Z, for $u(r, \phi)$, be the argument and result.

2.   Consider mappings between the sets $\{1, 2, 3, \ldots N\}$ and $\{1, 2, \ldots, M\}$. Let the mapping be given as an N-vector MAP.

   (a)   Compute the truth value of MAP being surjective. Let the function have an explicit result, a left argument MAP, and a right argument M as above.

   (b)   Same as above, but test for injectivity.

   (c)   Define a function that computes the inverse of MAP for a given argument. The explicit result can be a set, possibly empty.

3.   Given vectors arranged as row vectors in a matrix called ROWS, define a function that computes the Gram–Schmidt orthogonalization. The result should be a matrix whose rows are the generated orthonormal vectors. Allow for linear dependencies in the original vectors.

4.   Given a set of orthonormal vectors in $\mathbb{R}^n$ arranged as rows in the matrix ORTHO. Define a function that computes the projection of a vector X in $\mathbb{R}^n$ to the subspace spanned by the orthogonal vectors.

5.   (a)   Given a discrete function F with two arguments x and $y = 1, 2, \ldots, N$. Compute the truth value that F is superharmonic, i.e.,

$$f(x + 1, y) + f(x - 1, y) + f(x, y + 1) + f(x, y - 1) - 4f(x, y) \geq 0$$

for $2 \leq x, y \leq N - 1$. Think of F as an $N \times N$ matrix.

   (b)   Same as (a), but F should be discretely harmonic, i.e., equality in the above relation.

6.   For a given natural number N compute all permutations of order N.

7.   Compute all permutations of ιN that leave the subvector V of ιN invariant.

8.   Given two finite real power series

$$a(z) = \sum_{k=0}^{n} a_k z^k, \qquad b(z) = \sum_{k=0}^{m} b_k z^k$$

to compute the Taylor coefficients c_k in

$$a(z)b(z) = \sum_{k=0}^{n+m} c_k z^k$$

In other words, form the convolution of two vectors A and B of length $n + 1$ and $m + 1$, respectively. Do this by defining a function with explicit result $(c_k, k = 0, 1, \ldots, m + n)$ and two explicit arguments, the vectors A and B. Let the function loop explicitly to save space; see the discussion in Section 12.1.

9. In function theoretic experiments we need power series in complex form. To do this we have to be able to calculate z^k as a complex number. Define a function POWER that calculates $z_1^k, z_2^k, \ldots, z_r^k$ as a $2 \times r$ real matrix (explicit) result and has one explicit argument equal to k and the other as the $2 \times r$ real matrix

$$\begin{bmatrix} x_1 & x_2 & \cdots & x_r \\ y_1 & y_2 & \cdots & y_r \end{bmatrix}$$

Do it by calling a function MULT written by you that calculates the product of two complex r-vectors, each given as an $2 \times r$ matrix.

10. Consider the finite power series in complex form

$$f(z) = \sum_{k=0}^{n} c_k z^k$$

where $c_k = a_k + ib_k$, $z = x + iy$. Complex numbers are, of course, represented as real 2-vectors with the real part first.

Define a function TAYLOR with two explicit arguments C and Z. Here C should be a $2 \times (n + 1)$ real matrix (use concatenation and the axis operator)

$$C = \boxed{\begin{array}{c} A \\ \hline B \end{array}}$$

and Z a $2 \times m$ real matrix

$$Z = \begin{bmatrix} x_1 & x_2 & \cdots & x_m \\ y_1 & y_2 & \cdots & y_m \end{bmatrix}$$

The explicit result should have the same shape as Z. Use POWER from the preceding exercise.

11. Given a real-valued function of a real argument, $f(x)$. Assume we have defined a function F with vector-valued explicit results and arguments, define a function that approximates the integral

$$\int_A^B f(x)\, dx$$

by the trapezoidal rule using N interval

$$\frac{B - A}{N} \left[\frac{1}{2} f(A) + \sum_{k=1}^{N-1} f\left[A + \frac{k}{N}(B - A) \right] + \frac{1}{2} f(B) \right]$$

The left argument should be the two-vector (A, B) and the right argument the number of intervals N.

12.   In matrix manipulations it is often convenient to be able to refer
to the entry at (i, j) by a one-dimensional subscript k that runs through
one row at a time. For example, a 2×3 matrix can be indexed as in

$$\begin{pmatrix} 1 & 2 & 3 \\ 4 & 5 & 6 \end{pmatrix}$$

Define a function CODE with explicit result K, left argument IJ equal
to the (i, j) vector, and right argument SHAPE equal to the (m, n) vector
describing the shape of the matrix. Define an inverse function DECODE
with explicit result (i, j), left argument k, and right argument (m, n).

13.   Say that we have a monoid with elements labelled $1, 2, \ldots, m$,
where 1 means the unit element. The composition rule is given by a
multiplication table MTABLE with m rows and columns. Compute the
truth value that we have a group.

14.   For given n, compute the multiplication table for the symmetric
group S_n. Use the functions PERMUTE1 or PERMUTE2 for enumerating
permutations. Use a global variable MTABLE for the result (no explicit
result here!).

15.   Write a function INVERSE computing the set of inverses of
$g_1, g_2, \ldots, g_r$. The right argument should be the vector
$(g_1, g_2, \ldots, g_n)$. Let the resulting vector be the explicit result.

16.   For a given finite group $G = \{1, 2, \ldots, n\}$ with multiplication table
names MTABLE, consider two measures P and Q and calculate their
convolution. In other words, with

$$P = (p_1, p_2, \ldots, p_n)$$
$$Q = (q_1, q_2, \ldots, q_n)$$
$$R = (r_1, r_2, \ldots, r_n)$$

compute

$$r_z = \sum_{xy=z} p_x p_y = \sum_{x \in G} p_x p_{x^{-1} z}$$

Try to do it without looping, using instead matrix multiplication; compare
with Section 12.1.

17.   Assume that the subgroup N (given as a vector) of a group is normal.
Use the same notation as in Exercise 13. Let r be of the order of N.
     Compute a matrix COSETS with $s = m/r$ rows and r columns. The
rows should be the cosets of N with N itself forming the first row.

18.   Given that the above has been done, write a function that imple-
ments the first isomorphism theorem $G \to F = G/N$. Also compute the
multiplication table for the factor group F.

PART III

DESIGNING MATHEMATICAL EXPERIMENTS

CHAPTER 13

Designing the Experiment

1. CHOICE OF DATA STRUCTURES

After examining the structure and usage of APL, we can return to the topic of Chapter 7 in Part I and give some more detailed advice on how to plan and carry out the experiment.

We have argued in Chapter 7 that mathematical experiments are among the main tools of the working mathematician and have been with us since the birth of mathematics. It must be admitted, however, that experiments on the computer are less familiar and that we have only scant and recent knowledge of how to use them as a general research instrument.

While we can offer no foolproof recipe for planning such experiments, we can draw some conclusions from what we have learned in the previous chapters. In this section we shall single out the crucial relation between mathematics and data structures, a relation that must be made precise by the experimenter.

The objects and relations that we are confronting in mathematics have to be translated into the data structures employed by the programming language. We have already emphasized the difficulty in representing an *infinite mathematical structure* on the machine. We shall return to this later.

But even when the mathematical structure is finite we must give some thought to the choice of data structures. Consider the problem in Exercises 14–16, in Section 4, Chapter 12, dealing with the symmetric group. We enumerated the group elements as $1, 2, 3, \ldots, n!$, formed the multiplication table, and used it in the later computations. The data structure was just a square matrix MTABLE to represent the relations between objects.

Since the group was finite, this data structure seems natural. But this is not the whole story. As soon as n is not quite small, we shall be in trouble computationally, since the size of the memory of the computer will not allow us to store the matrix with its $(n!)^2$ elements requiring $4(n!)^2$ bytes of storage. Even if the storage is sufficient, this data structure is clearly wasteful; we have to do better.

There are several ways out of this particular dilemma. We can, for example, represent a group element by a permutation vector of length n. We do not have to store all these vectors, only instruct the computer how to manipulate them. Here we happen to have a group, so that we would have to tell the system how to find an inverse and how to carry out composition according to the group operation. We would proceed analogously for other mathematical structures. See the discussion at the end of Section 10.5 for how this can be done.

We have chosen this example because it is simple and also illustrates a general principle: try to *use data structures that closely correspond to the definition of the mathematical structure* to be represented; avoid mechanical representations that arbitrarily map the mathematical structure to a data structure.

As a second example, say that we are studying a function $f(x)$ on $\mathbb{R}$ or a sequence $\{f_n\}$ on $\mathbb{N}$. We could approximate them by a finite vector by discretizing and truncating the argument x or n. This may be adequate in a "quick-and-dirty" experiment but not if high accuracy is required. Instead we would use the *defining relations* (unless we have completely unstructured data) and write an APL program that incorporates them. In this example the appropriate data structure is therefore the user-defined function.

We show still another example. Say that we are going to perform experiments on formal languages. These are (finitely generated) infinite algebraic structures, and we have already seen that they should be represented by their defining relations, in this case by their vocabulary and productions. The vocabulary could be represented by a vector, and each production is also easy to store in vector form.

But this is not very convenient for the experimenter if he is going to carry out a sequence of experiments with different languages. It is better to write a general initialization program that interrogates the user about the objects and their relations, using "quad" or "quote quad" to handle the input obtained from interrogating the experimenter and then store the results as numeric or character matrices. Since all the vectors needed to represent productions are unlikely to be of equal length, they will not fit exactly into a rectangular array and some waste of storage space will result. If this is deemed serious enough to warrant further programming, one could use "format" and "execute" to define and name the vectors. Most of the time this would just be cosmetic programming and can be dispensed with. In this example the appropriate data structure is therefore an *interrogating defined function*.

Or say that we are looking at linear integral equations and we evaluate the integrals by some quadrature formula like the trapezoidal rule. We could then represent the functions as vectors on which the kernels act as matrices multiplying vectors. However, we may be more ambitious numerically and represent the functions by splines. This would also lead to vector–matrix representations but with a different interpretation.

The art of designing mathematical experiments cannot be learned through some magic recipe, only by doing them. As the experimenter goes along accumulating experience he will no doubt invent new ways of representing mathematical structures by data structures.

2. LOOKING FOR THE UNEXPECTED

The purpose of an experiment is either to suggest conjectures or to offer empirical evidence in favor of or against a given conjecture. In the second case one should exercise caution in not making the task too narrow. Even when the conjecture is precisely defined by the experimenter, it can be a wise strategy to widen the question, ask additional questions, and in general *prepare for serendipity*. The advantage of this has been illustrated repeatedly in Part I.

This is easier said than done. If we know in advance what we are looking for, perhaps we need not look very far for an answer. What makes mathematical experiments so exciting—and sometimes disappointing—is that they can lead to strange and surprising results and hence motivate an analysis that would not have been undertaken otherwise.

Ideally the program written by the experimenter should have automatic tests built into it, ringing bells if unexpected things happen, signaling oddities and peculiarities. In practice this is seldom possible and we shall have to settle for something less ambitious.

Say that we are looking at a sequence of objects and want to get some feeling for whether it converges. When designing the program it is then a good idea to incorporate a computing module that also looks into speed of convergence. If $\{a_n\}$ is a sequence in a Hilbert space, for example, one could let the extra module compute quantities like

$$\frac{\|a_{n+2} - a_{n+1}\|}{\|a_{n+1} - a_n\|}, \qquad n = 1, 2, \ldots, N$$

or plot

$$\log\|a_{n+1} - a_n\| \qquad \text{against} \quad \log n$$

In other words, we would do some crude data analysis on the speed of convergence, even though we started out by asking only whether the sequence converged or not.

Or, say we are given a real-valued function $f(x)$ specified by an algorithm, with x in some space X; we want to find its maximum value and where it occurs. Let the code also include a module that investigates how flat or how peaked the maximum is. Such questions are often analytically one order of magnitude harder than the original one of just finding the maximum. Computationally, however, they need not require much more CPU time than the original task. Perhaps something practically important will be found, especially in an applied context.

Next we show a third example. We are solving some equation with given parameters; the task is to find a root (or several roots) or to establish the existence of a root empirically. Here one could ask in addition how sensitive is the solution; that is, how much does it vary when we vary the parameters a little? If the solution is very unstable, it may be of little practical interest.

Or say that we want to verify some relation like $x > y$. One should then not just compute the truth value of the relation but also ask how much greater x is than y. This is rather obvious, but if x and y are not just real numbers but more general mathematical objects, for example, Hermitian operators, and the relation is a reasonably complicated one, then the advice is less trivial. If the conjecture is that a certain operator is positive definite, then one could look for its smallest eigenvalue or something similar that expresses "how positive" the operator is.

In Experiment 1 (Chapter 3) we were interested in the speed with which $\text{Var}(\alpha^*)$ tended to zero for the optimal choice of α^*. Any α^* could be expressed as a coefficient vector $\{c_t\}$. We included in the program a module that calculated and displayed the sequence $\{c_t\}$, not just $\text{Var}(\alpha^*)$. In this way we stumbled on the almost parabolic behavior of the sequence. Graphical output is important when we prepare for the unexpected!

We can sum up by suggesting the following: *Widen the scope of the query, make qualitative questions quantitative, include graphical output, and unleash your curiosity when designing the experiment!*

3.   PREPARE FOR CONTINUED EXPERIMENTS

When you run the experiment you may find something that attracts your attention and requires further experimentation. Or if something goes wrong, perhaps the parameters were not well chosen or the number of iterations was not enough. Again, you will repeat the experiment in modified form.

To be prepared for further experiments we had better include code that collects knowledge already gained so that we do not have to recompute exactly the same results. This means setting up tables, or arrays of some form, that are successively filled in as the experiment proceeds. Of course, such tables must be made global variables.

This will take up storage, but you can always delete the tables later, if storage becomes scarce, by the systems command

$$)ERASE\ NAME1\ NAME2\ NAME3\ ...$$

which also works for deleting functions and groups.

It is important psychologically that the experimenter's attention not be drawn away from the mathematics and the experiment itself by trivial typing mistakes. Therefore, it is good practice to prepare for repeated use of *any* long statement by making it into a defined function. Not only does this make it possible to execute the statement again without having to type it in once more—with the resulting typing errors—but it also enables you to debug it in function mode. You cannot do this on an isolated statement outside function mode, except, of course, by the important method of saving formulas as character strings.

As before, you can always)ERASE these functions (or character strings = formulas) when you are quite sure that they will not be used again (do not be too sure about this!) if they take up too much storage.

It is not always obvious which partial results should be saved and stored in tables. If the main experiment is preceded by a long initialization phase interrogating the experimenter, it is usually a good idea to save the results of this phase in global variables. Otherwise the whole initialization will have to be done over again with the deleterious psychological effect that would accompany this waste of the experimenter's time.

More advice, closely related to what has just been said, is to design a master program that controls execution of the main function(s) in such a way that it can be continued without new initialization. For example, if the master program executes some function(s) using a counter T to keep track of the number of iterations, write the master program in such a way that T is not set to one automatically when it is called again. Instead, let T start with its last value and loop through the iterations until a given number (say, given as an argument or by interrogation) of new iterations has been carried out.

In the beginning of the experiment we do not know very much about which, if any, modifications will be needed in continued experiments.

Perhaps we shall have to let the master program call other functions; this often seems to happen. If so, then it is not necessary to rewrite the program, changing the names of these functions. That could take up valuable time. Instead we prepare for this by using the "execute" function (see Section 10.6) and include steps interrogating the user, to which he should respond with function names. The "quote quad" should then be used for the input since we expect character vector inputs.

It must be admitted that sometimes it just is not possible to foresee what will happen and what modification can be expected. Then we could carry out a small pilot experiment, perhaps with parameter values chosen to make the computing almost trivial. Whether trivial or not, the computing may indicate that something is wrong or that we have forgotten to prepare for some contingency. Even a *quite small pilot experiment can be of help in designing the final version of the experiment and its programs*.

4.   INTERPRETING THE RESULTS

The mathematical experiment typically consists of the four phases

(a)   planning and designing algorithms,
(b)   programming and debugging,
(c)   execution, and
(d)   interpretation of the experimental data.

The last phase can be the most sensitive one, and one should prepare for it during the planning phase. Otherwise it may happen that the results do not allow any meaningful interpretation and time has been wasted.

We have emphasized repeatedly that interpretation is difficult when the data are finite, as they always are, but we want to relate them to infinite (or large finite) mathematical structures.

Say that we are interested in the convergence/divergence of the sum $\sum_1^\infty a_\nu$. We compute some of the partial sums $\sum_1^n a_\nu$, say with $+\backslash$, print them out, and try to form an opinion about whether they converge or not. Say that we get the partial sums

0.1428571429	0.2678571429	0.378968254	0.478968254
0.5698773449	0.6532106782	0.7301337551	0.8015623266
0.8682289932	0.9307289932	0.9895525226	1.045108078
1.097739657	1.147739657	1.195358705	

It is not easy to answer the question without hesitating, and, when some mathematicians were shown the data, they gave both positive and negative answers. In this case the data were generated by the command

$$+ \setminus \div 6 + \iota\, 1\, 5$$

so that we have essentially the terms of a harmonic series. It diverges, of course, but the numerical evidence does not point unequivocally in one direction.

What we should have done in a case like this was prepare for the interpretation by writing code that does simple, but it is hoped enlightening, data analysis of the data. In the special case of an infinite series we could try to relate the behavior of the partial sums S_n to n, for example, by doing regressions like

$$S_{n+h} - S_n = cn^{-p} + \text{error}$$

or

$$\log(S_{n+h} - S_n) = cn + \text{error}$$

and so on with other analytical combinations. This would probably help to guard against over optimistic interpretations. The reason why we do not simply take $h = 1$ is to compensate for oscillating or random behavior in the individual terms a_k. We shall go into this more thoroughly in Chapter 14.

In a more general context, it is good practice to include some crude data analysis when preparing the programs. If we have a series of functions, and the question is no longer convergence but uniform convergence, then we could do the data analysis on the sup norms $\|S_{n+h} - S_n\|$. Of course, we would then have to accept a discrete approximation to the sup norm; that is all we can do numerically.

Or, in another example, we want to find out whether a function $f(x)$ is positive for all x in some domain X; then we would probably compute $f(x)$ for a finite set of x values densely located in X. If one of the computed $f(x)$ values is zero or negative, we have the answer to our question. If they are all positive, we cannot be quite sure. It could be useful to have some data analysis that tries to learn about $f'(x)$, and possibly a higher derivative, and use the resulting knowledge to infer what happens in between the x points for which we have computed $f(x)$.

Sometimes replication of the experiment will help; perhaps repeating it unchanged except for some parameter. In particular, probabilistic experiments are notoriously difficult to interpret. This is true especially if we do not replicate. We should then also appeal to other basic ideas in the design of experiments, such as allocation of resources and balancing of factors. This must, of course, be done during the planning phase.

The all important question is: *Do the results indicate the presence of a real mathematical phenomenon or is what we believe we are seeing just the result of numerical coincidences?* As we gain experience of planning mathematical experiments we shall become more proficient in the art of interpreting the experimental data.

CHAPTER 14

An Experiment in Heuristic Asymptotics

1. GUESSING ABOUT ASYMPTOTICS

To illustrate how the design principles of Chapter 13 can appear in practice we shall look at a fairly ambitious experiment. It deals with asymptotics.

Suppose that f is a function defined for a real, positive argument x. We wish to derive asymptotic expressions for $f(x)$ as x tends to infinity. We do not know much about f to start with, except its values at finite number of x values, and we are looking for plausible hypotheses about its asymptotic behavior.

In the spirit of mathematical experiments we would then compute $f(x_1), f(x_2), \ldots , f(x_n)$ for x values large enough to bring out the asymptotic behavior. Looking at the resulting values, perhaps plotting them after various simple transformations of x and $y = f(x)$, we would hope to be able to guess the asymptotic form.

The study we shall describe here was motivated by the belief that we could automate this procedure to some extent, still leaving the important decisions to the experimenter, saving him from some drudgery, and at the same time make the search for hypotheses much more powerful. We are not aiming for a completely automatic algorithm in the spirit of artificial intelligence, where heuristic methods have been considered systematically. Instead we aim for a set of programs that can be called by the analyst to do the number crunching, *leaving the major strategic decisions to the analyst himself*. He should use his analytical insight, which may be useful even when incomplete, and, of course, the final step, the proof of the hypothesis, is entirely up to him.

We shall therefore not approach our task as one belonging to the area of artificial intelligence but as a strictly utilitarian undertaking. What we offer in the next sections is still very preliminary and can and will no doubt be improved a great deal. As a matter of fact, considerable improvements have already been made and are contained in the program library in Part IV. Nevertheless, it seems that in its present limited form it can already be used with some hope of success.

How would the typical patterns look? In asymptotics we are accustomed to statements like

$$f(x) \sim c/x \tag{1}$$

or

$$f(x) \sim ae^{-bx} \tag{2}$$

or

$$f(x) \sim (\cos ax)/\sqrt{x} \tag{3}$$

Primitive functions used are often the simplest rational and transcendental functions from classical analysis. The combinations of the primitive functions are also quite simple most of the time.

But what do we mean by "simple"? Our problem is rather meaningless until we have defined the pattern family within which we are searching. Therefore, the first thing we have to do is delimit the patterns. This will be done in Section 2, where a regular structure is given that will at least serve as a starting point. As we become more experienced we shall have to add to it.

We shall see in Sections 3 and 4 that, if f is monotonic (at least from some x value on), our heuristic is working remarkably well. Unfortunately the same cannot be said about oscillating patterns (see Section 5), for which we need a more flexible pattern family.

Some advice will be given in Section 6 on how to improve the software, especially in the oscillating case. An improved version of the software is given in Part IV.

2. SYNTHESIS OF THE PATTERNS

We shall start from a generator set G consisting of three classes

$$G = G^x \cup G^{\text{assign}} \cup G^{\text{function}} \tag{1}$$

with the following interpretation.

The generators in G^x play the role of variables restricted to certain domains by the bond value attributed to it. They all have $\omega_{\text{in}}(g) = 0$, $\omega_{\text{out}}(g) = 1$. A generator in G^x will be denoted by x, possibly with markers attached (say, subscripts).

The generators in G^{assign} are used to *specify parameters* by setting them equal to a constant. They also have $\omega_{\text{in}}(g) = 0$, $\omega_{\text{out}}(g) = 1$, and the constant to which g sets a parameter must be contained in the out-bond value $\beta_{\text{out}}(g)$.

The generators in G^{function} denote the *primitive functions* that are used as computational modules generating our function patterns. They can have arbitrary (finite) in- and out-arities with bond values, again sets indicating domain (for the in values) and range (for the out values).

In Fig. 1 we show a generator from G^x in (a), two assignment operators in (b) and (c), and three function generators in (d), (e), and (f). It goes without saying that two generators cannot be the same unless they agree in all their attributes, including bond values, markers, etc. Therefore, the generator in (a) is not the same as that in (g), and that in (e) is not the same as that in (h).

So much for the generators. Now the similarities. For any $s \in S$ we shall let $sg = g \; \forall g \in G^x \cup G^{\text{function}}$, while for $g \in G^{\text{assign}}$ we ask that $sG^{\text{assign}} = G^{\text{assign}}$ such that

$$\beta_{\text{out}}(sg) = \beta_{\text{out}}(g) \tag{2}$$

and that for any two g_1, $g_2 \in G^{\text{assign}}$ with $\beta_{\text{out}}(g_1) = \beta_{\text{out}}(g_2)$ there exists a similarity s such that $sg_1 = g_2$. Hence the similarity transformations only change parameter values.

We are now ready to introduce the regularities. The *local regularity* will be given by the bond relation $\rho = \text{INCLUSION}$; that is, inclusion of out-bond values in in-bond values. In order that this be a legitimate bond

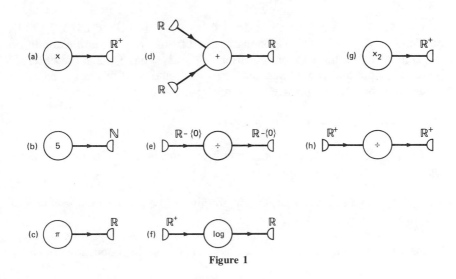

Figure 1

relation, we must verify that it be S invariant (see Grenander, 1976, p. 27), so that, if $\beta_1 \, \rho \, \beta_2$ holds, then $\beta_1' \, \rho \, \beta_2'$ should also hold when one and the same similarity is applied to the generators involved. But this is so since all bond values are preserved by the present similarity group.

The *global regularity* will be chosen by using the connection type $\Sigma =$ POSET; the graph structure for any given configuration should be consistent with a partial ordering of its generators. We do not require that the graphs be connected, although that is the interesting case.

This defines a regular structure $\mathscr{C}(\mathscr{R})$, which will now be decomposed into equivalence classes by an identification rule R. For a given regular configuration c consider

$$\text{Project}(G^{\text{assign}})c = c' \tag{3}$$

obtained by deleting all generators in c that do not belong to G^{assign}. Similarly

$$\text{Annihilate}(G^{\text{assign}})c = c'' \tag{4}$$

which deletes all generators in c that belong to G^{assign}. Write $c = \sigma(c', c'')$, where the connector σ is expressed in a suitable coordinate system; see Grenander (1980).

For two regular configurations c_1 and c_2, we shall identify them $c_1 R c_2$ if and only if

$$
\begin{aligned}
&c_1' = c_2' \\
&c_1'' \text{ computes the same function as } c_2'' \\
&\sigma_1 = \sigma_2
\end{aligned}
\tag{5}
$$

where c_1', c_1'', σ_1' have been obtained from c_1 as described in (3) and (4), and similarly, c_2', c_2'', σ_2 have been obtained from σ_2. The middle condition in (5) means that the subconfigurations c_1'' and c_2'' have the same external bonds (domain and range) and that the functional relations between input values and output values are the same.

Theorem. *R is an identification rule.*

Proof: We have to show that the four conditions in Grenander (1976, Definition 3.1.1) are satisfied.

(i) R is an equivalence; this is obvious since it is defined in terms of the three equalities in (5).

(ii) If for a pair c_1 and c_2 in $\mathscr{C}(\mathscr{R})$ we have $c_1 R c_2$, then the external bonds and bond values of c_1' equal those of c_2'. Similar conditions hold for

c_1'' and c_2''. But c_1' is combined with c_1'' by the same connector that combines c_2' with c_2'', so that the same external bonds are closed or left open. Hence c_1 has the same external bonds and bond values as c_2.

(iii) Assume again that $c_1 R c_2$ and apply the same similarity s to both c_1 and c_2. Since connections are not changed by applying a similarity, it is clear that sc_1 and sc_2 are globally regular. But since s does not change, bond values sc_1 and sc_2 are also locally regular, so that sc_1, $sc_2 \in \mathcal{C}(\mathcal{R})$.

Note that s does not change c_1'', c_2'', so that sc_1'' computes the same function as sc_2''. On the other hand, $c_1' = c_2'$, so that $sc_1' = sc_2'$. This implies that $(sc_1) R (sc_2)$.

(iv) Now consider six regular configurations c_1, c_2, c_{11}, c_{12}, c_{21}, c_{22} with $c_1 = \sigma(c_{11}, c_{12})$, $c_2 = \sigma(c_{21}, c_{22})$, and $c_{11} R c_{21}$, $c_{12} R c_{22}$. Is $c_1 R c_2$?

To show that this is so we observe that the assignment generators appearing in c_{11} are the same as in c_{21}; the same applies for c_{12} and c_{22}. But we combine them with the same connector σ in both cases so that

$$\text{Project}(G^{\text{assign}})c_1 = \text{Project}(G^{\text{assign}})c_2 \tag{6}$$

On the other hand, the subconfiguration

$$\text{Annihilate}(G^{\text{assign}})c_1 \tag{7}$$

computes the same function as the subconfiguration

$$\text{Annihilate}(G^{\text{assign}})c_2 \tag{8}$$

since the analogous statement is true for c_{11} and c_{21} as well as for c_{12} and c_{22}, and the connectors used are the same. The latter fact also implies that the subconfiguration in (6) is connected in the same way to (7) as to (8), so that the third condition in (5) holds. Hence $c_1 R c_2$.   Q.E.D.

Remark. The assignment generators fix the parameters in a formula, while the function generators carry out arithmetic operations. The G^x generators denote the variables in a formula and the markers help us keep distinct variables apart. Hence a similarity means changing the parameter values in a formula but leaving the structure of the formula unchanged. The image with in-arity zero means a completely specified function, while a pattern means a set of functions that are the same except for the parameter values.

The main object of study will be the *pattern*, for any fixed $c \in \mathcal{C}(\mathcal{R})$, of the form

$$\mathcal{P} = \{I | I = s[c]_R \quad \forall s \in S\} \tag{9}$$

The definition in (9) makes sense since we have shown that R is an identification rule where images $[c]_R$ are well defined. Note that any $\mathscr{P}$ as in (9) is S-invariant, so that $\mathscr{P}$ is indeed a pattern; see Grenander (1976, p. 104).

To make the discussion as concrete as possible, let us look at a few examples. In Fig. 2a we have shown a configuration that computes the function $3/\sqrt{x}$. Note the bond values indicated. In Fig. 2b we show a configuration for computing const$/\sqrt{x}$, where the constant is an unspecified real number.

The configuration in Fig. 2c is of course not the same as that in Fig. 2a. Both are, however, R-congruent since they have the same c' and σ with the previous notation, and the two c'' compute the same function.

In Fig. 2d we show a configuration that calculates $e^{2x}\cos 3x$. Note that

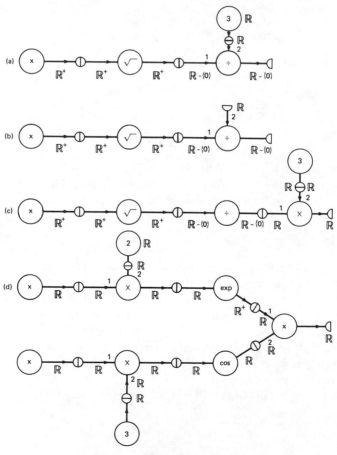

Figure 2

the same G^x generator x appears twice, indicating that it holds the same potential value from $\mathbb{R}$. If we had intended to let the two arguments be independent of each other, we would have used markers; for example, subscripts: x_1 and x_2.

Now when the regular structure has been synthesized, we shall turn to the problem of recognizing patterns in asymptotics. We shall look for patterns; that is, look for the structure of the function rather than for the image itself. Therefore the emphasis will be on the structure, not on the parameter values. The latter cannot be neglected, however, since most of the recognition devices we shall develop also include the determination (approximately) of the parameters.

3. RECOGNITION OF MONOTONIC PATTERNS

We shall distinguish between two main cases: when the function f approaches zero monotonically and when it does not. In the first case, which will be treated in this section, there is no loss of generality in assuming that f is nonnegative, and this will be done below.

Let us imagine that we are confronted with a function f that has been proven to tend to zero decreasingly as $x \rightarrow \infty$. We can calculate it numerically for given values of x but do not know its asymptotics.

To learn about the asymptotics we carry out a mathematical experiment, calculating $f(x)$ for $x = x_1, x_2, \ldots, x_n$ arranged in increasing order. We shall have more to say later about how the xs ought to be chosen. The resulting f values are placed in the n-vector F, the x abscissas into another n-vector X.

The first step is, of course, to plot F against X. While this is helpful in order to develop some feeling for the order of magnitude that characterizes the asymptotic behavior of f, our experience indicates that it is unlikely that plotting alone will lead to a satisfactory answer.

If we are given, for example, the vector F as

1	0.6666666667	0.5454545455	0.48	0.4379562044
	0.4081632653	0.3856749311	0.3679369251	0.3534857624
	0.3414171521	0.3311392768	0.3222468932	0.3144521825
	0.3075444662	0.3013655785	0.2957941917	0.2907354935
	0.2861141852	0.2818696118	0.2779522965	

we can plot it as in Fig. 3. Note that the automatic choice of coordinates makes the vertical axis begin at 0.25. The law of decrease is certainly not apparent. Can the reader guess what the behavior is? Probably not! We shall see later.

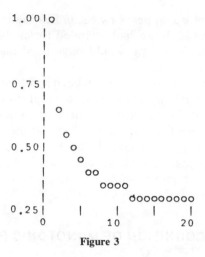

Figure 3

To get any further we must be more precise about how we shall judge the approximation of F compared to some F^*-vector calculated from a particular image. In what follows we shall use an l_p-norm of the form

$$\|F - F^*\| = \left(\sum_{k=1}^{n} w_k k^q |f_k - f_k^*|^p \right)^{1/p} \tag{1}$$

for finite positive p and

$$\|F - F^*\| = \sup_k (w_k k^q |f_k - f_k^*|) \tag{2}$$

for infinite p. In (1) and (2) the $\{w_k\}$ is a *sequence of weights* that will be chosen depending upon how $\{f_k\}$ behaves; therefore (1) and (2) are not norms strictly speaking, only measures of discrepancy.

The *exponent q* expresses how much we want to stress large k values. We shall often let $q = 0$; a value like 0.5 or 1 will occasionally be tried when we get the impression that the approximation should be judged with more stress on large k values.

The *exponent p* will often be chosen as 2. Especially for calculations of F^* this is, of course, convenient since it enables us to use least squares procedures to get the best fit.

But how about $\{w_k\}$? Since *we are interested in asymptotics, not in getting approximations for all x values,* we should let w_k increase as f_k decreases. We could choose some weights like $w_k = 1/f_k$, which would be arbitrary but not unnatural.

We shall indeed do something similar. If f is monotonic, we shall do just this, but, since we want to prepare for nonmonotonic behavior, we

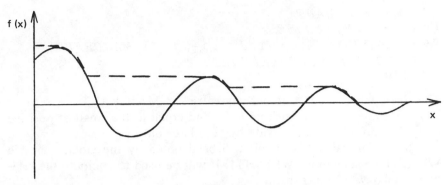

Figure 4

have to be a bit careful. If the behavior is something like that of the full-drawn curve in Fig. 4, we can, of course, not choose w_k as we said. We want positive weights, similar to the *monotonic envelope* shown in the figure as a dashed curve.

More precisely, what we have in mind is the smallest nonincreasing positive majorant of $|F|$. We could calculate it as

$$\text{majorant } (k) = \max_{l \geq k} f_l \qquad (3)$$

and then make sure that zero or negative values are not reached. We therefore write an APL function using $\lceil \ \backslash$ applied to the reversed string, then reverse again:

```
    ∇ Z←MONMAJ U
[1] ⍝COMPUTES SMALLEST NONINCREASING MAJORANT OF U
[2]   Z←ϕ⌈\ϕU
    ∇
```

For completeness we also give the function that computes the largest nondecreasing minorant:

```
    ∇ Z←MONMIN U
[1] ⍝COMPUTES LARGEST NONDECREASING MINORANT OF U
[2]   Z←-MONMAJ-U
```

We can now get the $\{w_k\}$ by applying MONMAJ to $|F|$ as in [2] below and then adjust the last entries in RESULT if they happen to be zero. We shall let MAJORANT be a global variable since it may be needed again:

```
    ∇ RESULT←FWEIGHT F
[1]   N1←ρP
[2]   RESULT←MAJORANT←MONMAJ|F
```

```
[3]    J←MAJORANTι0
[4]    →(J>N1)/DONE
[5]    RESULT[J+0,ιN1-J]←(1+N1-J)ρMAJORANT[J-1]
[6]    DONE:RESULT←÷RESULT
       ∇
```

The statement [3] tests whether any entries in MAJORANT are zero. If any are, they are made positive in [5] and equal to the smallest positive entry. We do not want to divide by zero later on!

The reciprocal of the result will be denoted by the global variable WEIGHT or sometimes WEIGHT1. It will be used to compute the relative error of F^* w.r.t. F as in

```
      ∇ Z←V1 RERROR V2
[1] ⍝COMPUTES RELATIVE ERROR OF V2 W.R.T. V1
[2] ⍝USES WEIGHT-VECTOR
[3] ⍝ASSUMES X-VECTOR GIVEN
[4]   Z←(X NORM V1-V2)÷X NORM V1
      ∇
```

which calls the function NORM, calculating (1) or (2) by going to the label FINITEP or not, depending upon whether p is zero or not:

```
      ∇NORM[□]∇
      ∇ RESULT←U NORM V
[1] ⍝COMPUTES LP-NORM OF V WITH Q-STRESS ON
[2] ⍝LARGE VALUES OF ARGUMENTVALUES IN V
[3]   →(P>0)/FINITEP
[4]   RESULT←⌈/((|U)*Q)×WEIGHT×(|V)
[5]   →0
[6]   FINITEP:RESULT←(+/((|U)*Q)×WEIGHT×(|V)*P)*÷P
      ∇
```

For the user's convenience we supply an initialization program SETUP interrogating the user

```
      ∇ SETUP F
[1]  N←ρF
[2]  'WHAT LP-NORM? P='
[3]  P←□
[4]  'WHAT Q-STRESS OF LARGE ARGUMENTS 3Q='
[5]  Q←□
[6]  (30,(30⌊N))PLOT F VS X
[7]  WEIGHT←FWEIGHT F
[8]  WEIGHT1←Nρ1
      ∇
```

The auxiliary weight vector WEIGHT1 may be modified; its use will be explained later. In [6] we can use one of the standard plotting functions for a typewriter terminal.

Let us now generate patterns, fit them to F, and measure the relative error between F and F^*. Appealing to the discussion in the preceding section, we shall introduce a set of G function generators. For G^x we need just a single generator, say x, since we shall not be concerned with more than one variable in the present study. The $G^{\text{assignment}}$ will be handled indirectly by using least squares approximations for the parameter values that should be assigned.

With a good deal of arbitrariness we select some function generators as in Table 1. The reader can add others that he considers useful for his personalized asymptotics!

We shall now generate patterns by quasi-linearization, assuming linear relationships

$$\psi[f(x)] = c_1\phi_1(x) + c_2\phi_2(x) + \cdots \qquad (4)$$

where $\psi, \phi_1, \phi_2, \ldots$ have been generated from primitives. The particular pattern synthesis that follows is mentioned more as an illustration than as something definitive.

The main reason why we have chosen quasi-linearization is in order to facilitate the least squares computations. If desired, one could certainly employ nonlinear versions of (4), solving the fitting problem by steepest descent algorithms or other optimization techniques. This possibility should be explored.

To do the *weighted least squares,* we write a simple-minded APL program. (Since ⊞ computes not-weighted least squares, we cannot use it directly, although a modified form could be used.) Say that $\psi[f(x)]$ values are given as an U-vector, and that we have a matrix V whose columns are

TABLE 1

Function Generators

Name	Function	In-bond values	Out-bond values
G11	1	—	$\mathbb{R}$
G12	x	$\mathbb{R}$	$\mathbb{R}$
G13	x^2	$\mathbb{R}$	$\mathbb{R}^+$
G21	$1/x$	$\mathbb{R} - \{0\}$	$\mathbb{R} - \{0\}$
G22	e^x	$\mathbb{R}$	$\mathbb{R}^+$
G23	$\ln x$	$\mathbb{R}^+$	$\mathbb{R}$
G41	Multiplication	$\mathbb{R}, \mathbb{R}$	$\mathbb{R}$
G42	Addition	$\mathbb{R}, \mathbb{R}$	$\mathbb{R}$

the vectors of $\phi_1(x)$ values, $\phi_2(x)$ values, and so on. We want to solve the minimum problem

$$\sum_{k=1}^{n} w_k(u_k - c_1 v_{k1} - c_2 v_{k2} - \cdots)^2 = \min \tag{5}$$

which leads to the normal equations

$$V^T W V c^* = V^T W u \tag{6}$$

where

$$W = \text{diag}[w_1, w_2, \ldots, w_n] \tag{7}$$

Hence

$$c^* = (V^T W V)^{-1} V^T W u \tag{8}$$

which is computed by LSQ:

```
     ∇ LSQ V
[1]    N←ρU
[2]  ⍝COMPUTES LEAST SQUARES FIT OF U-VECTOR
         AS LINEAR
[3]  ⍝COMBINATION OF COLUMNS OF V-MATRIX
[4]  ⍝RESULTING COEFF.S STORED IN CSTAR-VECTOR
[5]    DIAGONAL←((ιN)∘.=ιN)×WEIGHT1∘.×Nρ1
[6]    INV←⊟(⍉V)+.×DIAGONAL+.×V
[7]    INNERV←(⍉V)+.×DIAGONAL+.×U
[8]    CSTAR←INV+.×INNERV
[9]    USTAR←V+.×CSTAR
     ∇
```

An alternative version would include a factor k^q in (5). This would allow the user to stress large k values just as in NORM.

Let us do the pattern generation for two terms on the right of Eq. (4). Perhaps a third term ought to be added, but probably not any more. After all, we are not searching for an extremely good fit. *We want the fit to be reasonably accurate, but at the same time the pattern should possess attractive simplicity.*

Say that we have a ψ given by its APL name PSI, together with its inverse ψ^{-1} named PSIINV. Also we have named the two functions ϕ_1 and ϕ_2 as FI1 and FI2.

We can then achieve the best-weighted least squares fit by executing

```
        ∇ FIT
[1]     ⍝FITS GIVEN MONOTONIC PATTERN
[2]     U←⍕PSI,'',⍕F
[3]     V←(N,2)ρ0
[4]     V[;1]←⍕FI1,'',⍕X
[5]     V[;2]←⍕FI2,'',⍕X
[6]     U LSQ V
[7]     FSTAR←⍕PSIINV,'',⍕USTAR
[8]     ⍝PROGRAM STOPS IF USTARVALUES ARE OUTSIDE
            DOMAIN OF PSIINV
[9]     'RELATIVE ERROR IS'
[10]    FSTAR RERROR F
[11]    ι0
        ∇
```

Line [2] computes $\psi[\,f(x)]$, and lines [4] and [5] compute $\phi_1(x)$ and $\phi_2(x)$. In [6] we do the weighted least squares fit, whereafter [7] transforms the fitted data back by applying ψ^{-1}. Finally the relative error of the fit is found in [10].

For ease of reference we arrange most of Table 1 in three lists:

```
              PSIL
        1  2  2  2
        2  1  2  3
              PSINVL
        1  2  2  2
        2  1  3  2
              FI1L
        1  1  1
        1  2  3
              FI2L
        1  1  1  2  2  2
        1  2  3  1  2  3
```

The columns of PSIL give the function generators that will be called for ψ: G12, G21; G22; G23 (similarly for ψ^{-1}, ϕ_1, and ϕ_2). Corresponding to these lists we use three global variables IMAX, JMAX, and KMAX equal to the number of columns in PSIL, FI1L, and FI2L, respectively. For the present choice of generators IMAX = 4, JMAX = 3, and KMAX = 6.

Of course the calling sequence could be continued by repeated application of the given function generators as well as of others. In this pilot study we have not yet had the opportunity to do this.

The master program MONO generates, fits, and evaluates patterns. The argument START gives the initial values of the counters *I, J, K.* Typically one would begin by START ← 1 1 2. The reason for the last value, and for keeping $k > j$, is to avoid singular matrices in LSQ.

Statements [6]–[9] set the names, as character strings, of the function generators used at the moment.

```
      ∇   MONO START
[1]   ⍝GENERATES AND FITS MONOTONIC PATTERNS
[2]    I←START[1]
[3]    J←START[2]
[4]    K←START[3]
[5]   LOOPM:'FOR PATTERN WITH'
[6]    PSI←'G',(⍕PSIL[1;I]),⍕PSIL[2;I]
[7]    FI1←'G',(⍕FI1L[1;J]),⍕FI1L[2;J]
[8]    FI2←'G',(⍕FI2L[1;K]),⍕FI2L[2;K]
[9]    PSIINV←'G',(⍕PSIINVL[1;I]),⍕PSIINVL[2;I]
[10]   'PSI= ',PSI
[11]   'FI1= ',FI1
[12]   'FI2= ',FI2
[13]   FIT
[14]   I,J,K
[15]   K←K+1
[16]   →(K≤KMAX)/LOOPM
[17]   K←1+J←J+1
[18]   →(J≤JMAX)/LOOPM
[19]   I←I+1
[20]   K←1+J←1
[21]   →(I≤IMAX)/LOOPM
      ∇
```

The program will generate configurations not only from 𝒞(ℛ) but *also irregular ones.* It does not test for local regularity. This is signaled by the event that execution stops and DOMAIN ERROR is printed out at the terminal. The user will then have to increment the START-vector.

The program prints out all patterns that are generated and their relative errors: [10], [11], and [12] in MONO and [9] and [10] in FIT. If this is not desired, a simple modification will bypass the printing except for the case when a new relative error is smaller than all the earlier ones encountered so far, or smaller than a certain fraction, say 1.5 or 2.0 of the earlier relative errors.

We have chosen to write FIT in such a way that it operates on function generators given by their names as character strings that can be passed along from MONO or redefined many times by the user. This will facilitate experimentation.

4. SOME EXPERIMENTS

To see how this program works let us look at a few cases. First put $f(x) = 3/\sqrt{x}$, corresponding to the configuration diagrams in Fig. 2a and c. For $X = (1, 2, 4, 7, 9, 10, 11, 13)$ we get the plot in Fig. 5 when executing SETUP F. It is not very informative except for the qualitative behavior.

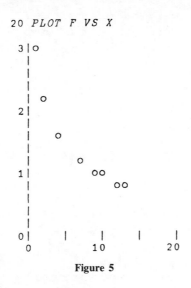

Figure 5

Begin execution by the command

$$MONO \quad 1 \quad 1 \quad 2$$

(remember that K should be larger than J). The terminal responds by printing

```
FOR PATTERN WITH
PSI= G12
FI1= G11
FI2= G12
RELATIVE ERROR IS
0.1959508773
```

```
1  1  2
FOR PATTERN WITH
PSI=  G12
FI1=  G11
FI2=  G13
RELATIVE ERROR IS
0.272414704

1  1  3
FOR PATTERN WITH
PSI=  G12
FI1=  G11
FI2=  G21
RELATIVE ERROR IS
0.06961507911

1  1  4
FOR PATTERN WITH
PSI=  G12
FI1=  G11
FI2=  G22
RELATIVE ERROR IS
0.4102235014

1  1  5
FOR PATTERN WITH
PSI=  G12
FI1=  G11
FI2=  G23
RELATIVE ERROR IS
0.07341900129

1  1  6
FOR PATTERN WITH
PSI=  G12
FI1=  G12
FI2=  G13
RELATIVE ERROR IS
0.7019598955


1  2  3
FOR PATTERN WITH
PSI=  G12
```

```
FI1= G12
FI2= G21
RELATIVE ERROR IS
0.1736356667

1  2  4
FOR PATTERN WITH
PSI= G12
FI1= G12
FI2= G22
RELATIVE ERROR IS
0.9387383446

1  2  5
FOR PATTERN WITH
PSI= G12
FI1= G12
FI2= G23
RELATIVE ERROR IS
0.8491940197

1  2  6
```

The values of I, J, K are printed *after* the description of the pattern to which these values belong. The best fit so far was for the configuration with

$$\psi = g_{12}, \qquad \phi_1 = g_{11}, \qquad \phi_2 = g_{21} \tag{1}$$

with the relative error 0.0696.

As we continue execution and get to the configuration

$$\psi = g_{22}, \qquad \phi_1 = g_{12}, \qquad \phi_2 = g_{23} \tag{2}$$

the execution is halted and we get the error message

```
3  2  5
FOR PATTERN WITH
PSI= G22
FI1= G12
FI2= G23
DOMAIN ERROR
G23[1]  Z←⊛X
       ∧
```

This configuration is not regular.

We restart execution by

$$MONO \ \ 3 \ \ 3 \ \ 4$$

and pattern generation continues. The smallest relative error turns out to be

```
FOR PATTERN WITH
PSI= G23
FI1= G11
FI2= G23
RELATIVE ERROR IS
1.710846957E‾10
```

$$4 \ \ 1 \ \ 6$$

for the configuration

$$\psi = g_{23}, \qquad \phi_1 = g_{11}, \qquad \phi_2 = g_{23} \tag{3}$$

meaning

$$\log f(x) = a + b \log x \tag{4}$$

or

$$f(x) = Ax^B \tag{5}$$

which is exactly the true pattern!

Very good, but this is really too easy since (5) is *exactly* the function given, not just asymptotically equal to it.

To get something more challenging, let us try the function that generated the sequence near the beginning of Section 3. Here $X = (1, 2, \ldots, 20)$, but we shall use only the first ten values of F and X. Executing MONO 1 1 2, we find for the smallest error the pattern given by

```
FOR PATTERN WITH
PSI= G21
FI1= G11
FI2= G23
RELATIVE ERROR IS
0.009864908164
```

$$2 \ \ 1 \ \ 6$$

meaning

$$f(x) = 1/(a + b \log x) \tag{6}$$

What was $f(x)$? We had defined it as

$$f(k) = 1 \Big/ \sum_{l=1}^{k} \frac{1}{l} \tag{7}$$

or in APL code

$$F \leftarrow \div \; + \backslash \div \iota N$$

Of course

$$f(n) \cong 1/(\log n + \gamma) \tag{8}$$

where γ is Euler's constant, so that the heuristic software hits the bull's-eye.

Now let us modify the first example and make the task harder by putting

$$f(x) = (3/\sqrt{x}) + (7/x^{3/2}) \tag{9}$$

For $X = (1, 2, 4, 7, 9, 10, 11, 13)$ the program generates many irregular configurations. Among the regular ones we get the best fit

```
FOR PATTERN WITH
PSI= G12
FI1= G13
FI2= G21
RELATIVE ERROR IS
0.03759032022

1  3  4
```

meaning the pattern

$$f(x) = ax^2 + (b/x) \tag{10}$$

which is quite wrong. It is easy to see why this happened. The x values used are much too small, especially in the beginning of the X-vector.

Instead, let us try $X = (1, 10, 20, 30, \ldots, 100)$. We find the best pattern as

```
FOR PATTERN WITH
PSI= G22
FI1= G11
FI2= G21
RELATIVE ERROR IS
0.01644790361

3  1  4
```

meaning

$$f(x) = \log(a + (b/x)) \tag{11}$$

and the next best one as

```
FOR PATTERN WITH
PSI= G23
FI1= G11
FI2= G23
RELATIVE ERROR IS
0.02401724863

4  1  6
```

meaning

$$f(x) = Ax^B \tag{12}$$

Even for these larger x values we were not led to the correct pattern, although we got close to it.

Let us repeat this experiment, but now stressing large arguments by putting $q = 1$. Then we get, for the above two patterns, and with the changed definition of relative error,

```
        MONO  3  1  4
FOR PATTERN WITH
PSI= G22
FI1= G11
FI2= G21
RELATIVE
0.02049500935
```

and

```
        MONO  4  1  6
FOR PATTERN WITH
PSI= G23
FI1= G11
FI2= G23
RELATIVE ERROR IS
0.01297248208
```

The roles of the two patterns have now changed place with each other, and we have arrived at the correct one.

As another example, we execute MONO on another function that decreases rapidly. In the beginning of execution we get

```
            MONO 1 1 2
FOR PATTERN WITH
PSI=  G12
FI1=  G11
FI2=  G12
RELATIVE ERROR IS
1170.245956

1 1 2
FOR PATTERN WITH
PSI=  G12
FI1=  G11
FI2=  G13
RELATIVE ERROR IS
1607.883366

1 1 3
FOR PATTERN WITH
PSI=  G12
FI1=  G11
FI2=  G21
RELATIVE ERROR IS
239.779869

    1 1 4
```

Note the large relative errors. In cases like this, one should perhaps unstress large arguments by setting q negative, but we have not tried this.

The smallest relative error found was

```
FOR PATTERN WITH
PSI=  G23
FI1=  G11
FI2=  G12
RELATIVE ERROR IS
1.52234435E⁻9
```

$$4\ 1\ 2$$

which means the pattern

$$f(x) = ae^{bx} \tag{13}$$

The true pattern was indeed

$$f(x) = 5e^{-2x}, \qquad x = 1, 2, 4, 7, 9, 10, 11, 13 \tag{14}$$

As a rather different example, we pick another transcendental function that cannot be exactly expressed via the function generators in Table 1, namely the Γ-function. Let us consider $1/\Gamma(x_k)$ for $x_k \downarrow 0$. Put

$$Y \leftarrow \div \iota 2 0$$

and

$$F \leftarrow \div \ ! \ ^-1 + Y$$

Let us consider F as a function of its subscripts. The plot looks as shown in Fig. 6.

Execute MONO 1 1 2. The pattern with the smallest relative error is

```
FOR PATTERN WITH
PSI= G21
FI1= G12
FI2= G23
RELATIVE ERROR IS
0.0173055875
```

or

$$f(x) = 1/(c_1 x + c_2 \log x) \tag{15}$$

which at first glance is puzzling since $\Gamma(z)$ has a first-order pole at $z = 0$ with residue 1 so that one would expect $f(x)$ to behave as $1/x$ close to $x = 0$.

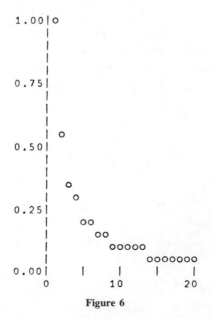

Figure 6

The contradiction is resolved if we print out the c coefficient contained in CSTAR. To get the "correct" CSTAR we have to execute MONO 2 2 5 and interrupt the program *immediately*; after that the value of the relative error is printed out. Otherwise the loop would be reentered and we might get the next update of CSTAR instead.

We get

$$CSTAR$$
$$0.9635768465 \quad 4.97631648E^-10$$

so that c_2 in (15) can be neglected and we are led to the right pattern.

As a last example, consider the function $f(x) = \pi(x)/x$, where $\pi(x)$ is the distribution function of the primes. With $X \leftarrow 100 + 20 \times \iota 20$, where the values are perhaps a bit too small, we execute MONO and get for the best-fitting hypothesis

$$f(x) = 1/(a + b \ln x)$$

which is Legendre's hypothesis! The a value is small and the b value is slightly less than one. This is encouraging.

We conclude that MONO works well if used with caution. It goes without saying that more function generators should be added to Table 1. This will not be difficult.

The CPU time consumed by MONO is quite small and we can well afford to extend the generator space a good deal.

It is annoying to have to restart the program repeatedly when irregular configurations are encountered. To remedy this defect one ought to include a piece of code that tests whether the bond relations are satisfied and, when they are not, makes execution bypass the particular configuration. This has been done in the code presented in Part IV.

5. OSCILLATORY PATTERNS

We now turn to patterns that are not monotonic. In Fig. 7 we have sketched a few.

In Fig. 7a and b the graph turns up or down before settling to a monotonic behavior. Since we are doing asymptotics, both parts a and b should be classified as eventually monotonic, a fact that should be established first, preferably by analytic methods or else numerically. We then apply the pattern recognition algorithms developed in Sections 3 and 4.

In Fig. 7c and d the graphs persist in being nonmonotonic for large x so that other ideas must be advanced for doing the heuristics. It is clear from the very beginning that since the patterns now show much greater flexibil-

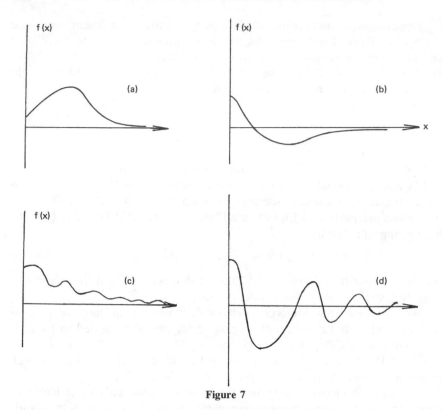

Figure 7

ity and variation, the recognition task will be harder. Wavy patterns especially, as in part c, can be expected to cause trouble.

Our procedure will be based on the idea of first extracting the characteristic behavior of the monotonic decrease of $f(x)$. We can do this by executing

$$MAJ \leftarrow MONMAJ\ F$$

This will give the monotonic majorant of F. Alternatively one could use

$$MIN \leftarrow MONMIN\ F$$

or perhaps use the average (note minus sign)

$$ENVELOPE \leftarrow .5 + MAJ - MIN$$

Sometimes it may be more informative instead to compute the greatest convex minorant of MAJ. Look at Fig. 8. The full-drawn curve is F, the dashed curve is MAJ, and the dashed-dotted curve is the convex minorant of MAJ. Other combinations of this type may be worth exploring.

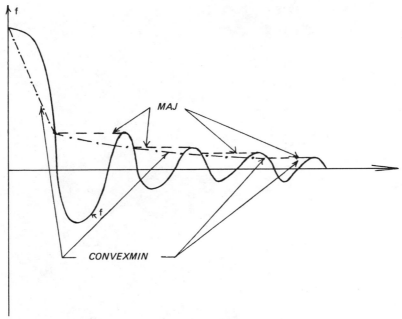

MAJ

CONVEXMIN

Figure 8

We therefore write a program that calculates the greatest convex minorant of a given curve. The x values are placed in the left argument and the $f(x)$ values in the right argument V.

```
      ∇ RESULT←U CONVEXMIN V
[1]   ⍝COMPUTES LARGEST CONVEX MINORANT OF V W.R.T U
[2]   N1←ρV
[3]   RESULT←N1ρV[1]
[4]   I←1
[5]   LOOP:V1←(V[I+ιN1-I]-V[1])÷U[I+ιN1-I]-U[I]
[6]   MIN←⌊/V1
[7]   I1←I+V1ιMIN
[8]   J←I1-I
[9]   RESULT[I+ιJ]←V[I]+MIN×U[I+ιJ]-U[I]
[10]  I←I1
[11]  →(I1<N1)/LOOP
      ∇
```

In the loop [5]–[10] we calculate slopes, statement [5]; find the smallest one, statement [6]; and make the interpolating straight-line segments have increasing slopes, statement [9].

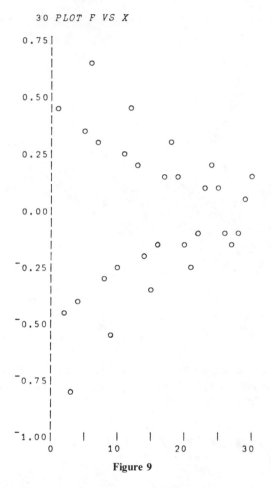

Figure 9

Let f be the pattern $Ae^{-Bx} \cos cx$ plotted for $X = 1, 2, \ldots, 30$ in Fig. 9. It looks more chaotic than it really is. Its monotonic majorant is shown in Fig. 10. It is a good deal more well behaved. Its convex minorant is shown in Fig. 11. This graph gives a fairly good idea of the "law of decrease" of f when the oscillatory behavior is disregarded. Indeed, running MONO with this function, the convex minorant, gives us the best fit

```
FOR PATTERN WITH
PSI= G23
FI1= G11
FI2= G12
RELATIVE ERROR IS
0.04764047185
```

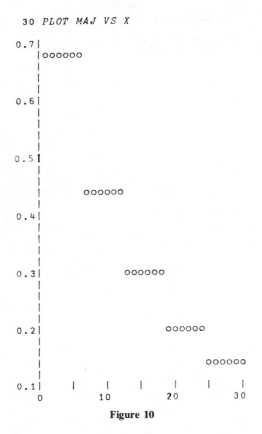

Figure 10

meaning the pattern

$$m(x) = ae^{bx}$$

which is just the right law of decrease. One should not cry triumph immediately, however; we may just have been lucky! More experimentation is needed with these "enveloping" procedures.

Our starting point will be that the oscillating pattern, with or without removing the monotonic envelope, is a simple combination of solutions to Eq. (1) below. More precisely, we shall look at functions f, which after a change of variable satisfy

$$Lf = g \tag{1}$$

where L is a simple differential or difference operator with constant coefficients, and g some primitive function. We therefore have to solve (1), but since we are only concerned with the asymptotic behavior, presumed to be given by the solution of (1), we have to find a function f^* that is as close

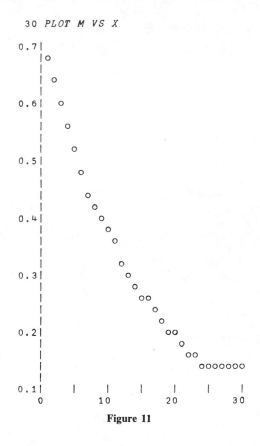

Figure 11

as possible to the given f for large values of the argument. Closeness will be judged in terms of the mean square error for a given weight function and possibly with q stress on large values of the argument.

One could perhaps have started by trying $f(x) \cong$ low-order trigonometric polynomial in the new arguments, but we would then have to fit the frequencies appearing in the trigonometric expression. This is a nonlinear problem, which we avoid by *instead fitting the operator L* to explain the data. This attractive idea will save a good deal of computing time.

Before we go into this, let us point out that it will sometimes lead to a successful search strategy to first decompose f into $f(x) = m(x)n(x)$, where m is some envelope, and then treat $n(x)$ as an oscillating pattern. Then we should not assume that $n(x)$ tends to zero as x tends to infinity. This should be kept in mind when we choose function generators later on.

To find the best autoregression corresponding to difference operators

$$\Delta^p f = c_1 \Delta^{p-1} f + c_2 \Delta^{p-2} f + \cdots + g \tag{2}$$

and in cases where we shall only treat $p = 1$ or 2 here for simplicity, we must first calculate the differences $\Delta^r f$.

Since the x values will not always be equally spaced we have to use divided differences. But which divided differences?

At first glance it seems natural to use the standard divided differences

$$\Delta f = \frac{f(x_{k+1}) - f(x_k)}{x_{k+1} - x_k} \tag{3}$$

and

$$\Delta^2 f = 2 \left(\frac{f(x_{k+2}) - f(x_{k+1})}{x_{k+2} - x_{k+1}} - \frac{f(x_{k+1}) - f(x_k)}{x_{k+1} - x_k} \right) \bigg/ (x_{k+2} - x_k) \tag{4}$$

Software for this was developed and tried with some limited success, but it did not seem to be the best way of dealing with the differences.

We decided after some experimentation to use central divided differences instead; that is, the definitions

$$\Delta f = \frac{f(x_{k+1}) - f(x_{k-1})}{x_{k+1} - x_{k-1}} \tag{5}$$

and

$$\Delta^2 f = 2 \left(\frac{f(x_{k+1}) - f(x_k)}{x_{k+1} - x_k} - \frac{f(x_k) - f(x_{k-1})}{x_k - x_{k-1}} \right) \bigg/ (x_{k+1} - x_{k-1}) \tag{6}$$

If we were just trying to integrate a differential equation numerically, the choice of (5) could be criticized for $p = 1$, but that is not our purpose. We are attempting to recognize a pattern; if we can find an operator L that both is simple and leads to a solution f^* that is close to f, then we can always use the difference equation itself rather than any differential equation to define the structure. We would not necessarily consider it an approximation to a differential equation, although this is sometimes legitimate, at least when the spacings between the x values are small.

The central divided differences in (5) and (6) are computed by CDIVDIFF, whose left argument is the order p, equal to one or two, and whose right argument is a row matrix with two rows. The first row, say U, contains the given f values, and the second, V, has the x values. Statements [13], [17], and [18] compute the central differences in Eqs. (5) and (6). We shall let the vectors $X1M$, $X10$, and $X0M$ be global variables since they will be needed later on.

```
        ∇ RESULT←ORDER CDIVDIFF W ; N1
[1]    ⍝COMPUTES CENTRAL DIVIDED DIFFERENCES OF
[2]    ⍝W[1;] W.R.T. W[2;].RESULTS STORED IN MATRIX
[3]    ⍝WITH 1+ ORDER ROWS:FUNCTION ITSELF IN ROW
[4]    ⍝ONE, FIRST DIFFERENCES IN SECOND ROW,....
[5]    ⍝MATRIX WILL HAVE TWO COLUMNS LESS THAN W
[6]    ⍝ORDER SHALL BE 1 OR 2
[7]     U←W[1;]
[8]     V←W[2;]
[9]     N1←(ρW)[2]
[10]    X1M←(2↓V)-¯2↓V
[11]    RESULT←((ORDER+1),N1-2)ρ0
[12]    RESULT[1;]←¯1↓1↓U
[13]    RESULT[2;]←((2↓U)-¯2↓U)÷X1M
[14]    →(ORDER=1)/0
[15]    X1O←(2↓V)-1↓¯1↓V
[16]    XOM←(1↓¯1↓V)-¯2↓V
[17]    RESULT[3;]←((2↓U)÷X1O×X1M)+((-1↓¯1↓U)
           ÷X1O×XOM)+(¯2↓U)÷XOM×X1M
[18]    RESULT[3;]←2×RESULT[3;]
        ∇
```

In the RESULT matrix we have $n - 2$ columns. The first row has the values $f(x_2), f(x_3), \ldots, f(x_{n-1})$; the second row has the corresponding first differences; and, if $p = 2$, there is also a third row with the second differences.

Look now at the code for the program CGAUTOR given below. It finds the best general autoregression based on central divided differences. The program is a bit more complicated than the earlier ones and deciphering it is less easy. The main work is done in statements [11]–[14]. The diagonal matrix DIAGONAL set up in [7] employs the weights stored in the vector WEIGHT1, which would normally be made equal to WEIGHT computed by FWEIGHT. However, it will sometimes be better to use other weights by a small modification of the code. Then WEIGHT1 will not be the same as WEIGHT.

After the least squares fit is made in [11]–[14], the central differences are computed in [16] or [19] for order 1 or 2, respectively. Then the difference equation is solved for FSTAR in [17] or [20]:

```
        ∇ ORDER CGAUTOR W
[1]    ⍝COMPUTES BEST GENERAL AUTOREGRESSION FOR
[2]    ⍝GENERAL SPACING OF W[2;] OF ORDER GIVEN
[3]    ⍝BY LEFT ARGUMENT.ORDINATES ARE IN W[1;]
```

```
[4]    ⍝ORDER=1 OR 2
[5]    ⍝USES CENTRAL DIFFERENCES
[6]    N1←¯2+(ρW)[2]
[7]    DIAGONAL←((⍳N1)∘.=⍳N1)×WEIGHT1⌈⍳N1]∘.×N1ρ1
[8]    DIFF←ORDER CDIVDIFF W
[9]    DIFF←DIFF[⍉⍳ORDER+1;]
[10]   →(~∧/0=G)/LOOP3
[11]   INV←⊟DIFF[1+⍳ORDER;]+.×DIAGONAL+.
          ×⍉DIFF[1+⍳ORDER;]
[12]   INNERV←DIFF[1+⍳ORDER;]+.×DIAGONAL+.×DIFF[1;]
[13]   CSTAR←INV+.×INNERV
[14]   CSTAR←CSTAR,0
[15]   →⍎'LOOP',⍕ORDER
[16]  LOOP1:CSTAR[1] CORDER1 W[2;]
[17]   FSTAR←W[1;⍳N1] GDIFFEQ N1ρ0
[18]   →0
[19]  LOOP2:CSTAR[1 2] CORDER2 W[2;]
[20]   FSTAR←W[1;⍳N1] GDIFFEQ N1ρ0
[21]   →0
[22]  LOOP3:DIFF←DIFF,[1] G[1+⍳N1]
[23]   INV←⊟DIFF[1+⍳ORDER+1;]+.×DIAGONAL+.
          ×⍉DIFF[1+⍳ORDER+1;]
[24]   INNERV←DIFF[1+⍳ORDER+1;]+.×DIAGONAL+.
          ×DIFF[1;]
[25]   CSTAR←INV+.×INNERV
[26]   →⍎'LOOP',⍕ORDER+3
[27]  LOOP4:CSTAR[1] CORDER1 W[2;]
[28]   FSTAR←W[1;⍳N1] GDIFFEQ CSTAR[2]×G
[29]   →0
[30]  LOOP5:CSTAR[1 2] CORDER2 W[2;]
[31]   CSTAR←W[1;⍳N1] GDIFFEQ CSTAR[3]×G
       ∇
```

Since the heterogeneous equation for $g \not\equiv 0$ requires a different treatment in order to avoid singular matrices, this is tested for in [10]. If $g \not\equiv 0$, we go to LOOP3, after which the treatment is similar to the homogeneous case except that the CSTAR vector has one more component, possibly zero but typically not.

Note that ρCSTAR is equal to ORDER+1 in both cases, but in the homogeneous case CSTAR[ORDER+1] is automatically zero; see statement [14].

The best c coefficients are stored in the vector CSTAR. We had to solve (2) with these c values. This was done in the program GDIFFEQ, to be described below, which assumes that (2) has been expressed, not in the differences, but directly in function values; more about this later.

For the initial value 1, if $p = 1$, and the values (1, 0) and (0, 1), if $p = 2$, we solve the homogeneous equation. In other words *we compute the fundamental solutions*. This is done in statements [8]–[19].

```
      ∇ FSTAR←F GDIFFEQ G;BASIS;FUNDAMENTAL;
        R;I;J;DIAGONAL;INV;INNERV;COEFF
[1]   ⍝COMPUTES BEST LEAST SQUARES
      APPROXIMATION TO F
[2]   ⍝THAT SOLVES LSTAR=G
[3]   ⍝DIFFERENCE OPERATOR L HAS LEADING COEFF
      .S STORED IN
[4]   ⍝VECTOR LO AND REMAINING COEF.S STORED
      IN
[5]   ⍝MATRIX WITH R ROWS AND N-R COLUMNS
[6]   R←(ρL)[1]
[7]   N←ρF
[8]   BASIS←(⍳R)∘.=⍳R
[9]   FUNDAMENTAL←(R,N)ρ0
[10]  FUNDAMENTAL[⍳R;⍳R]←BASIS
[11]  INH←Hρ0
[12]  I←1
[13]  LOOP1:J←R+1
[14]  LOOP2:FUNDAMENTAL[I;J]←-(÷LO[J-R])×
      L[;J-R]+.×FUNDAMENTAL[I;J-⍳R]
[15]  J←J+1
[16]  →(J≤N)/LOOP2
[17]  I←I+1
[18]  →(I≤R)/LOOP1
[19]  J←R+1
[20]  LOOP3:INH[J]←(÷LO[J-R])×G[J]-L[;J-R]+.
      ×INH[J-⍳R]
[21]  J←J+1
[22]  →(J≤N)/LOOP3
[23]  DIAGONAL←((⍳N)∘.=⍳N)×WEIGHT[⍳N]∘.×Nρ1
[24]  INV←⌹FUNDAMENTAL+.×DIAGONAL+.×
      ⍉FUNDAMENTAL
[25]  INNERV←FUNDAMENTAL+.×DIAGONAL+.×F-INH
[26]  COEFF←INV+.×INNERV
[27]  FSTAR←INH++/[1](COEFF∘.×Nρ1)×
      FUNDAMENTAL
      ∇
```

Now consider an arbitrary linear combination of the fundamental solutions with coefficients given by the p-vector COEFF plus the particular solution INH of the inhomogeneous equation. INH is computed in [11] and [19]–[22]. We find the best l_2 fit to f in statements [23]–[27]. The resulting f^* is called FSTAR.

Remark.   One could improve the performance of CGAUTOR by replacing G in GDIFFEG [20] by $(-1 \uparrow \text{CSTAR}) \times G$, where CSTAR has been computed in CGAUTOR.

In statements [16] and [19] of CGAUTOR we also need code to express (2) in terms of function values in the form

$$l_0(k)f(x_k) + l_1(k)f(x_{k-1}) + \cdots = g(x_k) \tag{7}$$

Take, e.g., order one, $p = 1$. Then (2) takes the form

$$\frac{f(x_{k+1}) - f(x_{k-1})}{x_{k+1} - x_{k-1}} = c_1 f(x_k) + g(x_k) \tag{8}$$

which can be rewritten as (7) if we put

$$l_0(k) = \frac{1}{x_{k+1} - x_{k-1}}, \qquad l_1(k) = -c_1, \qquad l_2(k) = -\frac{1}{x_{k+1} - x_{k-1}} \tag{9}$$

Note that the entries of the vector X1M appear in (9); that is why they are stored in a global variable.

The conversion in (9) is effected by

```
        ∇  C CORDER1 X
[1]     ACOMPUTES FIRST ORDER DIFFERENCE OPERATOR
[2]     AIN TERMS OF VECTOR LO, LENGTH TWO LESS
           THAN X
[3]     AAND MATRIX L WITH TWO ROWS
[4]     AUSES CENTRAL DIFFERENCES
[5]     AUSES GLOBAL VAR. X1M
[6]        N1←ρX
[7]        LO←÷X1M
[8]        L←(2,N1-2)ρ0
[9]        L[1;]←-(N1-2)ρC[1]
[10]       L[2;]←-LO
        ∇
```

where Eqs. (9) are realized by statements [7]–[10] in terms of the vector X1M.

Similarly, for $p = 2$ we write

```
  ∇  C CORDER2 X
[1]  ACOMPUTES SECOND ORDER DIFFERENCE OPERATOR
[2]  AIN TERMS OF VECTOR L0, LENGTH TWO LESS THAN X
[3]  AAND MATRIX WITH TWO ROWS. USES CENTRAL∇
[4]  ADIFFERENCES AND GLOBAL VARIABLES X10,X1M,X0M∇
[5]  N1←ρX
[6]  L0←(2÷X10×X1M)-C[1]÷X1M
[7]  L←(2,N1-2)ρ0
[8]  L[1;]←(-2÷X10×X0M)-C[2]
[9]  L[2;]←(2÷X0M×X1M)+C[1]÷X1M
  ∇
```

We can now start to generate hypotheses. We enumerate primitive functions as follows. Adding to the previous collection, we shall use the ones in Table 2. The choice is a bit arbitrary and can no doubt be improved.

To call these primitive functions we introduce the two lists

$$GL = \begin{pmatrix} 1 & 1 & 1 & 2 & 2 & 2 & 2 \\ 0 & 2 & 3 & 1 & 6 & 4 & 5 \end{pmatrix}$$

and

$$FIL = \begin{pmatrix} 1 & 1 & 2 & 2 & 2 & 2 & 2 \\ 2 & 3 & 4 & 1 & 3 & 5 & 7 \end{pmatrix}$$

so that, for example, GL[;1] refers to the function g_{10}, which is identically zero, and FIL[;3] refers to the function $g_{24}(x) = \sqrt{x}$. Other functions can be added by the user.

We define the global variables IMAXO as the number of columns in FIL and JMAXO as the number of columns of GL.

Say that we have selected a ϕ function and a g function. Then we use a pattern of the form

$$Lf[\phi(x)] = \text{const} \times g \tag{10}$$

TABLE 2

Name of function	Function	Name of function	Function
g_{10}	$\equiv 0$	g_{23}	$\ln x$
g_{11}	$\equiv 1$	g_{24}	$\sqrt{x}$
g_{12}	x	g_{25}	$1/\sqrt{x}$
g_{13}	x^2	g_{26}	$1/x^2$
g_{21}	$1/x$	g_{27}	$1/\ln x$

and the ORDER + 1 constants are adjusted for the best fit. In particular, for $g = g_{10}$, we are using a homogeneous difference equation.

To fit for an oscillating pattern in this manner we use FITO:

```
       ∇ FITO ORDER
[1]    ⍝FITS OSCILLATING PATTERN TO GIVEN F AND X
[2]    W←(2,⍴F)⍴F
[3]    W[2;]←⍎FI,' ',⍕X
[4]    G←⍎GFCN,' ',⍕X
[5]    ORDER CGAUTOR W
[6]    'FOR PATTERN WITH'
[7]    'FI = ',FI
[8]    'AND'
[9]    'G = ',GFCN
[10]   'THE RELATIVE ERROR IS'
[11]   F RERROR FSTAR
[12]   ⍳0
       ∇
```

In [3] we calculate $\phi(x)$ by executing the function called FI, and in [4] we get $g(x)$ by executing GFCN. Statement [5] executes CGAUTOR and the remaining statements generate output.

We then call FITO repeatedly by the pattern recognition program OSC:

```
       ∇ OSC[⎕]∇
       ∇ OSC START
[1]    ⍝SEARCHES FOR OSCILLATING PATTERN OF
          ORDER ONE
[2]    ⍝ORDER = ONE
[3]    I1←START[1]
[4]    J1←START[2]
[5]    LOOP;FI←'G',(⍕FIL[1;I1]),⍕FIL[2;I1]
[6]    GFCN←'G',(⍕GL[1;J1]),⍕GL[2;J1]
[7]    I1,J1
[8]    FITO 2
[9]    J1←J1+1
[10]   →(J1≤JMAXO)/LOOPO
[11]   J1←1
[12]   I1←I1+1
[13]   →(I1≤IMAXO)/LOOPO
       ∇
```

It is constructed very much like MONO and requires no further comments.

It is clear that, since parameters are adjusted automatically in CGAUTOR to produce a good fit, the resulting relative errors can be expected to be small, certainly smaller than for MONO.

Also, a good deal more CPU time will be needed for calculating this recognition algorithm than is needed for MONO. This will be noticeable especially when ρF is large.

The interpretation of results requires some thought. If the x values are close together, we are approximating a differential equation and the roots of the characteristic equation can be found by a simple program called ROOTS that need not be described here.

If the x values are further apart from each other, the approximation to a differential equation is not valid. We then have to rely on the difference equation itself, with the coefficients given by CSTAR, to find the pattern.

In FIL we have also included the decreasing functions g_{21}, g_{25}, and g_{27} although they may seem unnatural for asymptotics toward zero. This is done so that we can apply OSC to the envelope of F. It need not tend to zero. A reader can easily change the lists GL and FIL by removing or adding columns and modifying Table 2. If this is done, one should remember to reset the global variables IMAXO and JMAXO.

6. EXPERIMENTS WITH OSCILLATING PATTERNS

Let us start with a simple case, namely,

$$f(x) = \cos \frac{2\pi x}{\text{period}} \, e^{-x/a} \tag{1}$$

where period $= 10$, $a = 4$, and $x = 0.1, 0.2, \ldots, 1.0$. Executing OSC we get, for example, the good fits

```
1 1
FOR PATTERN WITH
FI = G12
AND
G = G10
THE RELATIVE ERROR IS
3.646178347E⁻14

FOR PATTERN WITH
FI = G12
AND
G = G12
THE RELATIVE ERROR IS
5.576990824E⁻15
```

```
1 3
FOR PATTERN WITH
FI = G12
AND
G = G13
THE RELATIVE ERROR IS
5.811429573E⁻15
```

and the much poorer fits

```
2 1
FOR PATTERN WITH
FI = G13
AND
G = G10
THE RELATIVE ERROR IS
0.01890686303
```

```
2 2
FOR PATTERN WITH
FI = G13
AND
G = G12
THE RELATIVE ERROR IS
0.04762275322
```

```
2 3
FOR PATTERN WITH
FI = G13
AND
G = G13
THE RELATIVE ERROR IS
0.04195582175
```

```
2 4
FOR PATTERN WITH
FI = G13
AND
G = G21
THE RELATIVE ERROR IS
0.04709777748
```

Why is this so? After all, we would have expected to find $Lf \cong 0$, i.e., $\phi \equiv x = g_{12}$ and $g \equiv 0 = g_{10}$. For that particular pattern, the error is indeed practically zero, but this also is the case, for example, for $\phi = g_{12}$ and $g = g_{12}$. This puzzling fact is cleared up if we print out CSTAR, which is

$$^-0.1333313581 \quad ^-1.100040942 \quad ^-5.68434188E^-14$$

This shows that the equation is practically the expected (homogeneous) one with CSTAR[3] $\cong 0$. Inspecting the rest of the printout, we see that the recognition algorithm formed the true pattern without hesitation.

Now execute OSC instead for the function

$$f(x) = e^{-\sqrt{x}} - 2e^{-2\sqrt{x}} \tag{2}$$

and $x = 0.8$, 1,6, 2.4, 3.2, 4, 4.8, 5.6, 6.4, 7.2, and 8. The data looks like that in Fig. 12. It is not very revealing as far as pattern structure is concerned. Note that we actually apply OSC to an eventually monotonic function, pretending that we did not know this. We get the output beginning with

```
1  1
FOR PATTERN WITH
FI = G12
AND
G = G10
THE RELATIVE ERROR IS
0.00235031664

1  2
FOR PATTERN WITH
FI = G12
AND
G = G12
THE RELATIVE ERROR IS
0.002445316991

1  3
FOR PATTERN WITH
FI = G12
AND
G = G13
THE RELATIVE ERROR IS
0.00194606587
```

and the pattern $\phi = g_{12}$, $g = g_{13}$ is the best fit, seen by continuing the printing. This means the pattern with $\phi = x$, $g = x^2$, although printing out

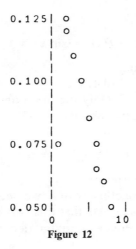

Figure 12

CSTAR, shows that its last component is quite small, so that we have almost a pattern generated by $Lf = 0$, the homogeneous equations. This is wrong.

The reason for this is that the spacings are too large. To see what happens for smaller spacings we repeat the same procedure for $X = 4.1, 4.2, 4.3, \ldots, 5$ and get the best fit for

```
3 1
FOR PATTERN WITH
FI = G24
AND
G = G10
THE RELATIVE ERROR IS
4.64753047E‾8
```

meaning $\phi = \sqrt{x}$, $g \equiv 0$, which is the correct pattern.

Repeated experiments have shown that, as long as we are dealing with F belonging to our $\mathscr{C}(\mathscr{R})$, we can expect to find the true pattern or at least get fairly close. Caution has to be exercised when choosing X, however, and it is usually a good idea to start with a small X set, fairly well concentrated, and then go on with a larger X set selected on the basis of what we have learned. The second X set should include large x values.

If we get a domain error when computing INV, it is an indication that the order of the difference equation should be lowered to 1. This is done by changing [8] in OSC to FITO 1.

Let us now try a couple of patterns that should be harder to recognize.

First consider

$$f(x) = (\sin x)/x \tag{3}$$

which is actually oscillating.

Executing OSC for $X = (0.1, 0.2, \ldots , 1.0)$ we find the best-fitting pattern $\phi = x$ and $g = 1/x^2$. Only the first component of CSTAR is substantially different from zero, about -0.14357, so that we are led to the pattern satisfying the equation

$$D^2f(y) + 0.14357Df(y) = 0, \qquad y = 1/x^2 \tag{4}$$

This is no good for our purpose.

Printing out FSTAR together with F we see, however, that the fit is remarkably close, which suggests that this sort of heuristic software may be of use in studies of numerical approximations: *it may lead to unexpected methods of approximation.*

However, here we want to suggest hypotheses that can be proved or disproved later by analysis, and for this purpose the pattern given by (4) is misleading.

Returning to the pattern (3), we now try instead to isolate the envelope by computing

$$M \leftarrow X \; CONVEXMIN \; MAJORANT.$$

We then apply MONO to $F \leftarrow M$ and find the best fit

```
FOR PATTERN WITH
PSI= G12
FI1= G13
FI2= G21
RELATIVE ERROR IS
0.1268892703
```

with

```
            CSTAR
0.00001635635294  0.7783769656
```

Since CSTAR[1] is practically zero, this leads us to the pattern $\psi(x) = x$, $\phi_2(x) = 1/x$ for M

$$m(x) = \text{const}/x \tag{5}$$

This is the correct pattern for the envelope! Dividing $f(x)$ by $m(x)$, it is easy to find the remaining factor $\sin x$ by using OSC.

This is certainly satisfactory but it was probably just luck that made us get the correct pattern for the envelope. In general, one should not expect more than moderate success from this sort of procedure.

Let us look at still another example that is strictly outside the domain of exact fit and that does not oscillate. Consider the function

$$f(x) = (1 + 20 \ln x)/x \qquad (6)$$

and try OSC for $X = (10.1, 10.2, 10.3, \ldots, 11.0)$; although (6) is actually eventually monotonic, we pretend we are not aware of it. We get by far the best fit for the pattern

```
5 1
FOR PATTERN WITH
FI = G23
AND
G = G10
THE RELATIVE ERROR IS
1.837448972E⁻12
```

or $\phi = \ln x$, $g = 0$. Also

```
        CSTAR
⁻2.000082771 ⁻1.000068482 0
```

so that $f(x) = h(y)$ with the substitution $y = \ln x$ should satisfy approximately

$$h'' + 2h' + h = 0 \qquad (7)$$

The characteristic equation for (7) has the double root $z = {}^-1$ so that one would expect the form

$$h(y) = (A + By)e^{-y} \qquad (8)$$

But (6) implies

$$f(x) \sim 20 \, (\ln x)/x = 20y e^{-y}$$

so that our heuristic led us to the correct pattern in this case.

Repeated experimentation seems to show that our software is only moderately successful in recognizing oscillating patterns; certainly less so then for the monotonic ones. Also, it requires more CPU time.

7. CONCLUSIONS

1. The regular structure used above is fairly flexible and perhaps adequate for simple cases, at least for monotonic patterns. Some additions ought to be made, however, to the generator space G in both cases. For oscillating patterns we also need more powerful recognition devices.

2. The code should be supplemented by tests for checking bond relations. For example, if "log" is applied to a vector V, we should verify that no component is nonpositive, and, if $\div$ is used, that no component is zero. This could perhaps be done by computing the APL expression $\times V$ as suggested by D. E. McClure. Then a bypass should be executed if the test fails.

3. It is not necessary to print out all the patterns that are generated since most of them result in poor fits. Another bypass should therefore be added to the code that takes effect each time that the fit is poor compared to the best one found so far.

4. More experimentation is needed to learn about the performance of the heuristics. This will be helpful in deciding which other generators should be added to G. One soon develops a feeling for when the relative error is small enough to indicate a plausible hypothesis or when it is only a good approximation. The difference is often striking, especially for MONO.

5. One could also extend OSC to handle patterns based on higher-order difference equations. One should be careful, however, *not to make the regular structure too general*. After all, we want not only a good fit but also a reasonably simple pattern. We operate in the spirit of Occam's razor.

The reason why the fit is often remarkably good is that all our primitive functions are analytic and two analytic functions cannot be very close in a small set except when they are close in a larger set. Vaguely speaking, this expresses the principle of analytic continuation.

6. In MONO as well as in OSC one could allow repeated and unbounded application of ψ, ϕ, and g functions to each other, to f, and to x. This would lead to a potentially infinite number of patterns, but again one must be careful not to make $\mathscr{C}(\mathscr{R})$ too large.

7. We have tried the programs on unknown functions (defined by someone else) and again the monotonic patterns were handled well. One difficulty that arises sometimes is that choosing X is not easy. When we do not know approximately where the asymptotic region of f starts we have to experiment with several X sets.

Some of the modifications suggested here have been implemented by T. Ferguson and are contained in code presented in Part IV.

PART IV

MATHEMATICAL SOFTWARE FOR EXPERIMENTS

CHAPTER 15

Organizing a Program Library

1. WHY A LIBRARY?

As the experimenter carries out one experiment after another he will produce code, some of it intended only for immediate use and of little permanent value and some of it of more general interest and therefore deserving more care in its construction. In the latter case it would be a waste not to save it and make it available for future use and perhaps to other experimenters.

One could argue that with the power and versatility of APL it is so easy to write the programs that there is little use for a systematically designed program library. It is true that program libraries are not as essential for the experimenter employing APL as they are in production computing in the conventional languages. Nevertheless, we have found it desirable to save at least the programs for some commonly occurring computing tasks.

But such a library should be built according to criteria other than the standard ones. Indeed, it is typical of the mathematical experiments that they vary so much, and are so seldom like one another, that it is the rule rather than the exception that *existing programs have to be modified at least to some extent to suit the particular experiment.* Perhaps the format of the output has to be changed, perhaps special conditions need particular care, or the mathematical structure is such that the underlying algorithm can be improved.

What is needed then is not so much a finished program, completely fixed and specified in all its details, but an outline of a program that can be used as a starting point to which additions and modifications can be easily made.

In order that this be possible, it is imperative that the documentation included be sufficiently informative to facilitate the changes. Otherwise it might be better to start from scratch and write a completely new program. The documentation must include a description of the algorithm (or a reference to the literature where a description can be found), an explanation of what the names of functions and variables mean, warnings about shortcomings of the code, and sometimes brief descriptions of what some crucial statements in the code are supposed to do.

The library presented in the following chapters has been built with this in mind. The user should approach it in this spirit: consider the programs as building blocks that should be combined and rewritten to fit the particular computing task at hand. Do not treat them as final products ready for immediate use.

To make this possible we have supplied documentation on three levels: *external documentation, internal documentation, and comments in the code.* The external documentation comes in written form and gives an overall view of what is available. The internal documentation is given as describe vectors and should be stored in the work space. Comments are, of course, noncomputing statements in the code itself.

The selection of computing tasks dealt with in the following chapters is certainly subjective and has been motivated by the experiments carried out by the author and his co-workers at Brown University over the past several years. Other mathematical experimenters may have different computing needs; nevertheless, we believe that many of the programs to be presented will be of general use.

All programs have been tested and run; some have been used extensively. In spite of this, there may still be undiscovered bugs in them. The reader is advised to use them with discretion.

2. EXTERNAL DOCUMENTATION

We usually include three types of information in the external documentation.

The *purpose* of the program is described briefly. When this requires formulas, as in ordinary mathematical expressions, it must be done in the written form, employing ordinary characters, so that it can be stored in a work space. It is important to make this specification detailed enough so that it will be reasonably easy for the user to understand what the code is supposed to do and what task we have in mind.

The *algorithm* on which the program is based should be given, either by a complete description, or sometimes, when it is judged to be sufficiently well known, just by its name. Occasionally this can be left out when the choice of algorithm is obvious.

We have tried to use variable names (both in the "purpose" and the "algorithm" section) that are similar to the ones actually used in the code.

References are given to the literature, where more information can be found about the algorithm and the numerical analysis needed to evaluate its properties.

In addition to these three sections we occasionally give an *example*, often one with a trivial choice of parameter values, to help the user in understanding how the program is called and what it does.

We may also include a *warning* where the user's attention is drawn to special cases when the program is highly inefficient or impossible to execute.

3. INTERNAL DOCUMENTATION

In each work space there should be stored a vector DESCRIBE that gives a *concise description of its entire content*. It gives a list of the function names and a brief indication of what tasks they carry out. We do not discuss this DESCRIBE vector in the following text since it will vary depending on what additional programs the user decides to add.

To *each function*, for example, SPECTRAL, there corresponds a *describe vector*, for example, DSPECTRAL. Its name is the name of the function preceded by the letter D.

The format of the information contained in a describe vector is as follows, in eight steps:

Syntax, for example, Z←SS GAUSS PAR, meaning the heading of the function.

Purpose, explaining what the function does, but in a highly abbreviated form; we cannot include formulas, which are given instead in the external documentation when needed.

Index origin, which we usually choose as 1. However, in a few cases it is allowed to be 0, in which case this should be entered here.

Author, the name of the person who was responsible for the final version of the code and whether he is the original author or not.

Last update, to keep track of modifications and when they were made.

Right argument, format and meaning.

Left argument, format and meaning.

Result, format and meaning.

It is important to change these entries appropriately as the programs are modified; otherwise considerable confusion can occur when executing them.

4.   COMMENTS IN THE CODE

Usually, the first few statements in a program are comments describing the computing task, other functions that are called, the role of explicit arguments, and results.

Unless the program is very simple, comments are also given in the main body of the code at statements whose roles are not obvious. In particular, the crucial statements that do the main part of the computing should have comments. It is also a good idea to include comments at statements where the flow of control is not transparent.

Sometimes we may have included too few comments. This is especially so in the older programs before the style of writing code had crystallized in our software group. The user is encouraged to add clarifying comments each time he uses a function and finds the code difficult to read.

5.   PROGRAM STYLE

We have tried to write clear code, very straight, and with only a few long statements, but this has not always been done as consistently as we intended.

The variable names as well as the names of functions have been chosen mnemonically, unless this would result in very long names. We then had to compromise.

Also, we have tried to organize the computing into modular forms, having many short and simple programs rather than a smaller number of long ones.

Whenever there has been a conflict between user efficiency and CPU-time efficiency, we have sacrificed the latter. This means, of course, that such programs should not be used in situations requiring a good deal of number crunching. It may then not even be possible to use the program as a starting point for further modification if the whole idea is unsuitable for production computing. This is sometimes also related to difficulties with the size of the work space, typically when we create large arrays, for example, by the use of the outer product.

CHAPTER 16

Algebra

Most of the programs in this chapter belong to linear algebra. They transform matrices to some standard form, calculate eigenvalues and eigenvectors, or prepare for such operations.

We also give some defined functions designed to do complex arithmetic. These have been written in such a way that they can also be applied to *complex arrays,* not just to single isolated numbers.

A few programs are given for dealing with operations on polynomials. We also include one program especially designed for finding roots of algebraic equations of arbitrary order and with complex coefficients, MULLERM.

Function Name: ADD

EXTERNAL DOCUMENTATION

PURPOSE

To add arbitrary complex arrays: see external documentation of MATMULT.

INTERNAL DOCUMENTATION

```
SYNTAX: Z←A ADD B
PURPOSE: ADDITION OF COMPLEX NUMBERS
INDEX ORIGIN: 1
AUTHORS: B. BENNETT, P. KRETZMER
LAST UPDATE: 4/79
LEFT ARGUMENT: A IS A COMPLEX MATRIX (I.E., A
  SET OF COMPLEX NUMBERS, WHERE THE REAL AND
```

IMAGINARY PARTS ARE SEPARATED ALONG THE
FIRST DIMENSION.)
RIGHT ARGUMENT: B IS ALSO A COMPLEX MATRIX
RESULT: A COMPLEX MATRIX Z

CODE

```
      ∇ Z←A ADD B;AI;BI;S
[1]  ⍝ADDS CORESPONDING ELEMENTS OF TWO
     ⍝COMPLEX ARRAYS
[2]  ⍝AND DOES ALL OTHER ANALOGOUS
     ⍝OPERATIONS OF DYADIC ADD
[3]  ⍝LAST UPDATE: 12-10-78
[4]  AI←⍎'A[2',S←(((⍴⍴A)-1)⍴';'),']'
[5]  A←⍎'A[1',S
[6]  BI←⍎'B[2',S←(((⍴⍴B)-1)⍴';'),']'
[7]  B←⍎'B[1',S
[8]  Z←(A+B),[0.5]AI+BI
      ∇
```

Function Name: BALANCE

EXTERNAL DOCUMENTATION

PURPOSE

To balance a real matrix for the calculation of eigenvalues and eigenvectors.

Diagonal similarity transformations designed to reduce the norm of the
original matrix are used to balance the matrix. Balancing is merely a
device used to reduce errors during the execution of an algorithm. BAL-
ANCE should precede the calling of eigensystem routines.

ALGORITHM

The algorithm presented here is the balance procedure that was developed by B. N. Parlett and C. Reinsch and is discussed in the "Handbook
for Automatic Computation," Vol. II. It balances the norms of corresponding rows and columns, where the norm of a row (column) is the sum
of the absolute values of its entries. Preliminary permutations are carried
out on the matrix before the actual transformation begins. Rows isolating

an eigenvalue are pushed to the bottom of the matrix and columns isolating an eigenvalue are pushed to the left of the matrix. As a result, only the submatrix bounded at the top and to the left by index LOW and at the bottom and to the right by index HI must be balanced.

NOTE

Since the elements of the diagonal matrices are restricted to be exact powers of the radix base, no rounding errors occur during execution. Convergence of the algorithm may be deferred if the original matrix is highly reducible.

REFERENCE

Wilkinson, J. H., and Reinsch, C. (1971). "Handbook for Automatic Computation," Vol. II, pp. 315–326. Springer-Verlag, Berlin, New York.

INTERNAL DOCUMENTATION

```
SYNTAX: Z←BALANCE A
PURPOSE: TO BALANCE ROW AND COLUMN NORMS OF A
 REAL MATRIX BY DIAGONAL SIMILARITY
 TRANSFORMATIONS. BALANCING SHOULD PRECEDE
 THE CALLING OF EIGENSYSTEM PROGRAMS IN ORDER
 TO REDUCE ERROR.
INDEX ORIGIN: 1
AUTHOR: C. HUCKABA
LAST UPDATE: 4/27/79
RIGHT ARGUMENT: A IS AN N×N REAL MATRIX
RESULT: Z IS AN (N+1)×N MATRIX. THE FIRST ROW
 CONTAINS THE VECTOR D WHICH HOLDS INFORMATION
 DETERMINING THE PERMUTATIONS USED AND THE
 SCALING FACTORS. THE REMAINING N ROWS
 CONTAIN THE BALANCED MATRIX. THE PROGRAM
 ALSO COMPUTES AND PRINTS THE VALUES OF THE
 GLOBAL VARIABLES LOW AND HI WHICH ARE USED
 IN THE SUBSEQUENT PROGRAM ELMHES. FOR A MORE
 DETAILED DESCRIPTION OF THE RESULTS, SEE THE
 HANDBOOK FOR AUTOMATIC COMPUTATION, VOL:II,
 P. 319.
```

CODE

```
        ∇BALANCE[□]∇
        ∇ Z←BALANCE A;B;L;K;N;D;J;R;C;V;NOCONV;
          I;G;F;S;COL;ROW
[1]       B←2
[2]       L←1
[3]       K←N←1↑ρA
[4]       D←Nρ0
[5]     ⍝ SEARCH FOR ROWS ISOLATING AN
        ⍝ EIGENVALUE AND PUSH THEM DOWN
[6]       L1:J←K
[7]       L2:R←,(+/|A[J;ιK])-|A[J;J]
[8]       →(R≠0)/L3
[9]       D[K]←J
[10]      →(J=K)/L4
[11]      A[ιK;J,K]←A[ιK;K,J]
[12]      A[K,J;COL]←A[J,K;COL←L+0,ιN-L]
[13]      L4:→(1≤K←K-1)/L1
[14]      L3:→(1≤J←J-1)/L2
[15]      →(K=0)/L16
[16]    ⍝ SEARCH FOR COLUMNS ISOLATING AN
        ⍝ EIGENVALUE AND PUSH THEM LEFT
[17]      L5:J←L
[18]      L6:C←,(+/|A[ιK;J])-|A[J;J]
[19]      →(C≠0)/L7
[20]      D[L]←J
[21]      →(J=L)/L8
[22]      A[ιK;J,L]←A[ιK;L,J]
[23]      A[L,J;COL]←A[J,L;COL←L+0,ιN-L]
[24]      L8:L←L+1
[25]      →L5
[26]      L7:→(K≥J←J+1)/L6
[27]    ⍝ BALANCE THE SUBMATRIX IN ROWS L
        ⍝ THROUGH K
[28]      D[V]←(ρ(V←L+0,ιK-L))ρ1
[29]      L9:NOCONV←0
[30]      I←L
[31]      L10:C←,(+/|A[ROW←L+0,ιK-L;I])-|A[I;I]
[32]      R←,(+/|A[I;ROW])-|A[I;I]
[33]      S←C+R
[34]      C←C×F×F←B⋆⌈(‾1+B⊛R÷C)÷2
```

```
[35]  L14:→(((C+R)÷F)≥0.95×S)/L15
[36]  D[I]←D[I]×F
[37]  NOCONV←1
[38]  A[I;COL]←(A[I;COL←L+0,ιN-L])÷F
[39]  A[ιK;I]←(A[ιK;I])×F
[40]  L15:→(K≥I←I+1)/L10
[41]  →(NOCONV=1)/L9
[42]  L16:'LOW=',(⍕LOW←L),';HI=',⍕HI←K
[43]  Z←D,[1] A
      ∇
```

Function Name: CHOLESKYC

EXTERNAL

PURPOSE

To perform the complex Cholesky decomposition of a hermitian positive definite matrix, i.e., to determine the elements of a lower triangular matrix such that this matrix times its hermitian transpose equals the original matrix. It calls function LTRIC.

ALGORITHM

For $m = 1, 2, k = 1, 2, \ldots, n$, and $i = 1, 2, \ldots, k - 1$,

$$(1) \quad l_{mki} = l_{mii}^{-1} \left(a_{mki} \sum_{j=1}^{i-1} l_{mij} l_{mkj} \right) \qquad \text{calculates elements below main diagonal (done in LTRIC)}$$

$$(2) \quad l_{mkk} = \left(a_{mkk} - \sum_{j=1}^{k-1} l_{mkj}^2 \right)^{1/2} \qquad \text{calculates main diagonal elements}$$

NOTE

The input to this program should be a $2 \times n(n + 1)/2$ complex matrix A, where

$A(1;)$ = Elements in the lower half (diagonal and below) of the real part of the original matrix;

$A(2;)$ = Elements in the lower half of the imaginary part of the original matrix.

INTERNAL DOCUMENTATION

SYNTAX: VL←CHOLESKYC A
PURPOSE: COMPLEX CHOLESKY DECOMPOSITION OF A
 HERMITIAN POSITIVE DEFINITE MATRIX
INDEX ORIGIN: 1
AUTHOR: P. KRETZMER
LAST UPDATE: 5-7-79
RIGHT ARGUMENT: A 2 BY N(N+1)/2 COMPLEX
 MATRIX CONTAINING THE N(N+1)/2 ELEMENTS
 (PROCEEDING ACROSS ROWS) IN THE LOWER HALF
 OF A 2 BY N BY N COMPLEX HERMITIAN POSITIVE
 DEFINITE MATRIX.
RESULT: A 2 BY N(N+1)/2 COMPLEX MATRIX
 CONTAINING THE N(N+1)/2 ELEMENTS IN THE
 LOWER HALF OF A 2 BY N BY N COMPLEX, LOWER
 TRIANGULAR MATRIX SUCH THAT THIS MATRIX
 TIMES ITS HERMITIAN TRANSPOSE IS EQUAL TO A
 IN MATRIX FORM.

CODE FOR CHOLESKYC

```
         ∇ L←CHOLESKYC A;N;I;C;J
[1]   ⍝ COMPLEX CHOLESKY DECOMPOSITION OF A
      ⍝ HERMITIAN POSITIVE DEFINITE MATRIX
[2]   ⍝ CALLS FUNCTIONS: LTRIC, MINUS, MULT,
      ⍝ POWER
[3]   ⍝ LAST UPDATE: 5-7-79
[4]   N←(¯1+((1+8×(ρA)[2])*0.5))÷2
[5]   L←A[;1] POWER 0.5
[6]   I←1
[7]   S2:→0×⍳N<I←1+J←I
[8]   C←L LTRIC ((2,J)↑(0,((((I-1)×I)÷2))↓A)
[9]   →(I≠2)/S3
[10]  L←(2 1 ρL),C,(A[;(I×I+1)÷2] MINUS C
       MULT C× 2 1 ρ 1 ¯1) POWER 0.5
[11]  →S2
[12]  S3:L←L,C,(A[;(I×I+1)÷2] MINUS+/C MULT
       C×⍉((ρC)[2],2)ρ 1 ¯1)POWER 0.5
[13]  →S2
         ∇
```

CODE FOR LTRIC

```
      ∇ L←A LTRIC X;N;I;M;J
[1]   ⍝ CALCULATES ELEMENTS BELOW THE MAIN
      ⍝ DIAGONAL FOR CHOLESKYC
[2]   ⍝ CALLED BY CHOLESKYC; CALLS FUNCTIONS:
[3]   ⍝ MINUS, MULT, DIVBY
[4]    M←((ρA),0)[2]+((ρA),0)[2]=0
[5]   S1:N←(¯1+((1+8×M)*0.5))÷2
[6]    I←2
[7]    →(M≠1)/FIRST
[8]    L←X DIVBY A
[9]    →0
[10]  FIRST:L←X[;1] DIVBY A[;1]
[11]   L←(2 1 ρL),(1 0 DIVBY A[;3]) MULT X[;2]
       MINUS L MULT A[;2]× 1 ¯1
[12]  S2:→0×⍳N<I←I+1
[13]   L←L,(1 0 DIVBY A[;I+I×J÷2]) MULT X[;I]
       MINUS+/(2,J)↑L MULT
       A[;(I×J÷2)+⍳J]×⍉((J←I-1),2)ρ 1 ¯1
[14]   →S2
      ∇
```

Function Name: CHOLESKYR

EXTERNAL DOCUMENTATION

PURPOSE

To perform the Cholesky decomposition of a symmetric positive definite matrix, i.e., to determine the elements of a lower triangular matrix such that this matrix times its transpose equals the original matrix. It calls function LTR1.

ALGORITHM

For $k = 1, 2, \ldots, n$ and $i = 1, 2, \ldots, k - 1$:

(1) $l_{ki} = l_{ii}^{-1} \left(a_{ki} \sum_{j=1}^{i-1} l_{ij} l_{kj} \right)$ calculates elements below main diagonal (done in $LTR1$)

$$(2) \quad l_{kk} = \left(a_{kk} - \sum_{j=1}^{k-1} l_{kj}^2 \right)^{1/2} \quad \text{calculates main diagonal elements}$$

Since A is symmetric, only the lower half of A is needed as input. This vector, which is formed from the matrix A and is required as input to CHOLESKYR, can be computed by calling CONVERT (see external documentation for CONVERT). Note that if a scalar is read into CHOLESKYR, the Cholesky decomposition simply returns its square root.

REFERENCE

Stewart, G. W. (1973). "Introduction to Matrix Computation," pp. 139–142. Academic Press, New York.

INTERNAL DOCUMENTATION

```
SYNTAX:  L←CHOLESKYR A
PURPOSE:  CHOLESKY DECOMPOSITION OF A
 SYMMETRIC POSITIVE MATRIX
INDEX ORIGIN:  1
AUTHOR:  J. ROSS
LAST UPDATE:  6/19/79
RIGHT ARGUMENT:  A IS AN N(N+1)/2 VECTOR
 CONTAINING THE ELEMENTS IN THE LOWER HALF OF
 AN N×N SYMMETRIC POSITIVE DEFINITE MATRIX IN
 THE FOLLOWING ORDER:  (A11,A21,A22,A31,
 A32,A33,...ANN)
RESULT:  L IS AN N(N+1)/2 VECTOR REPRESENTING
 THE LOWER HALF OF AN N×N LOWER TRIANGULAR
 MATRIX SUCH THAT THIS MATRIX TIMES ITS
 TRANSPOSE IS EQUAL TO A IN MATRIX FORM.
```

CODE

```
        ∇ L←CHOLESKYR A;N;I;C;J
[1]   ⍝ CHOLESKY DECOMPOSITION OF A SYMMETRIC
      ⍝ POSITIVE DEFINITE MATRIX
[2]   ⍝ CALLS LTR1
[3]   ⍝ LAST UPDATE: 6/19/79
[4]    N←(¯1+((1+8×⍴A)*0.5))÷2
```

```
[5]     L←A[1]*0.5
[6]     I←1
[7]     S2:→0×ιN<I←1+J←I
[8]     C←L LTR1(J↑(((I-1)×I)÷2)↓A)
[9]     L←L,C,(A[(I×I+1)÷2]-C+.×C)*0.5
[10]    →S2
        ∇
```

CODE FOR LTR1

```
        ∇ L←A LTR1 X;N;I
[1]   ⍝ CALCULATES ELEMENTS BELOW THE MAIN
      ⍝ DIAGONAL FOR CHOLESKYR
[2]   ⍝ LAST UPDATE: 6/19/79
[3]     X←,X
[4]     A←,A
[5]   S1:N←(‾1+((1+8×ρA)*0.5))÷2
[6]     I←1
[7]     L←,X[1]÷A[1]
[8]   S2:→0×ιN<I←I+1
[9]     L←L,(÷A[J+I])×X[I]-A[(J←I×J÷2)+ιJ]+
        .×(J←I-1)↑L
[10]    →S2
        ∇
```

Function Name: CONVERT

EXTERNAL DOCUMENTATION

PURPOSE

To transform a symmetric matrix into a vector (CONVERT) or to reverse this transformation (REVERSE1, REVERSE2).

In the conversion, the elements of the lower half of the matrix (diagonal and below) will be strung out row by row to form the vector. Since the matrix is symmetric, no information will be lost in this procedure. CONVERT would produce the transformation

$$\begin{matrix} a_{11} & a_{12} & a_{13} \\ a_{21} & a_{22} & a_{23} \\ a_{31} & a_{32} & a_{33} \end{matrix} \quad \text{becomes } (a_{11} \quad a_{21} \quad a_{22} \quad a_{31} \quad a_{32} \quad a_{33})$$

NOTE

This program should be called prior to using CHOLESKYR.

To reverse the above conversion, the user may call either REVERSE1 or REVERSE2. The two programs employ slightly different methods to recreate the matrix. Since their relative speeds depend upon the particular problem, the user is advised to experiment. (The utility program TIMER can be called to measure CPU time used in each case.)

INTERNAL DOCUMENTATION

```
SYNTAX:  ∇VA←CONVERT A∇
PURPOSE: TO CONVERT AN N×N SYMMETRIC MATRIX
 INTO A VECTOR
INDEX ORIGIN: 1
AUTHORS: C. HUCKABA AND J. ROSS
LAST UPDATE: 6/19/79
RIGHT ARGUMENT: A SYMMETRIC MATRIX, 'A'
RESULT: THE VECTOR FORM OF 'A' CONTAINING
 THOSE ELEMENTS IN THE LOWER HALF OF 'A'
 STRUNG OUT ROW BY ROW
```

CODE

```
      ∇ VA←CONVERT A
[1]  ⍝ CONVERTS A SYMMETRIC MATRIX INTO A
     ⍝ VECTOR
[2]  ⍝ LAST UPDATE: 6/19/79
[3]   VA←( ,V∘.≥V←ι(1↑ρA))/,A
      ∇
```

Function Name: CUBIC

EXTERNAL DOCUMENTATION

PURPOSE

To find the roots of a third-order equation with real coefficients.

ALGORITHM

See Uspensky (1948, Chap. V).

INTERNAL DOCUMENTATION

SYNTAX: TOTL ← CUBIC PN
PURPOSE: SOLVES CUBIC EQUATION
INDEX ORIGIN: 1
AUTHOR: Y. LEVANON
LAST UPDATE: 1-18-72
RIGHT ARGUMENT: VECTOR OF COEFFICIENTS OF
 EQUATION, STARTING WITH HIGHEST ORDER TERMS.
RESULT: MATRIX WITH THREE ROWS AND TWO
 COLUMNS. ONE ROW FOR EACH ROOT, FIRST COLUMN
 HAS REAL PARTS, SECOND COLUMN HAS IMAGINARY
 PARTS.

CODE

```
      ∇ TOTAL←CUBIC PN;A;AT;B;BT;C;COS2;D;
        DISC;F;K;M;N;P;SR2
[1]   ⍝ ROOTS OF A CUBIC POLYNOMIAL
[2]   ⍝ LAST MODIFIED: 1/18/72
[3]   ⍝ AUTHOR: Y. LEVANON
[4]    P←PN÷PN[1]
[5]    C←P[2]÷3
[6]    A←P[3]-P[2]×C
[7]    B←(2×C*3)+P[4]-C×P[3]
[8]    D←A÷3
[9]    →(0>DISC←(B×B÷4)+D*3)/RL
[10]   AT←((DISC*0.5)-B÷2)*÷3
[11]   BT←-((DISC*0.5)+B÷2)*÷3
[12]   M←(AT+BT)÷2
[13]   N←(|M-BT)×3*÷2
[14]   TOTAL←(3 2)ρ((2×M)-C),0,(-M+C),N,
        (-M+C),-N
[15]   →0
[16]  RL:K←(-4×A÷3)*÷2
[17]   F←(‾2○(3×B÷A×K))÷3
[18]   COS2←(2○F)×K÷2
[19]   SR2←K×(1○F)×0.75*÷2
[20]   TOTAL←(3 2)ρ(-C-2×COS2),0,(-COS2+C+SR2),
        0,(-COS2+C-SR2),0
      ∇
```

Function Name: DIVBY

EXTERNAL DOCUMENTATION

PURPOSE

Divides arbitrary complex arrays componentwise; see external documentation for MATMULT.

INTERNAL DOCUMENTATION

```
SYNTAX: Z← A DIVBY B
PURPOSE: DIVISION OF COMPLEX NUMBERS
INDEX ORIGIN: 1
AUTHORS: B. BENNETT, P. KRETZMER
LAST UPDATE: 4/79
LEFT ARGUMENT: A IS THE DIVIDEND; IT IS A
  COMPLEX MATRIX (I.E., A SET OF COMPLEX
  NUMBERS, WHERE THE REAL AND IMAGINARY PARTS
  ARE SEPARATED ALONG THE FIRST DIMENSION.)
RIGHT ARGUMENT: B IS THE DIVISOR, AND IS ALSO
  A COMPLEX MATRIX
RESULT: A COMPLEX MATRIX
```

CODE

```
        ∇ Z←A DIVBY B;AI;BI;S
[1]   ⍝DIVIDES CORESPONDING ELEMENTS OF TWO
      ⍝COMPLEX ARRAYS
[2]   ⍝AND DOES ALL OTHER ANALOGOUS OPERATIONS
      ⍝OF DYADIC DIVIDE
[3]   ⍝LAST UPDATE: 4/79
[4]    AI←⍎'A[2',S←(((ρρA)-1)ρ';'),']'
[5]    A←⍎'A[1',S
[6]    BI←⍎'B[2',S←(((ρρB)-1)ρ';'),']'
[7]    B←⍎'B[1',S
[8]    Z←((((A×B)+AI×BI)÷(B*2)+BI*2),[0.5]
       ((AI×B)-A×BI)÷(B*2)+BI*2
        ∇
```

Function Name: ELMHES

EXTERNAL DOCUMENTATION

PURPOSE

To reduce a general real matrix to upper-Hessenberg form by a series of similarity transformations.

The reduction is achieved in a stable manner by the use of stabilized elementary matrices. The volume of work involved with many algorithms for finding the eigensystem of a matrix is greatly reduced if the original matrix is first transformed to upper-Hessenberg form. For a discussion of the transformation see "A Survey of Numerical Mathematics."

ALGORITHM

The algorithm presented here is the ELMHES procedure that was developed by R. S. Martin and J. H. Wilkinson and is discussed in the "Handbook for Automatic Computation," Vol. II. In general, the procedure will be more effective if the general matrix has been prepared by BALANCE beforehand. If the matrix has been balanced, then ELMHES must work only on the submatrix that is bounded by LOW and HI. Balancing also makes the matrix more suitable for the pivotal strategy adopted in the algorithm.

NOTE

The reduction is numerically stable in the sense that the derived Hessenberg matrix is usually exactly similar to a matrix that is close to the original. ELMHES is typically used in conjunction with eigensystem routines.

REFERENCES

Wilkinson, J. H., and Reinsch, C. (1971). "Handbook for Automatic Computation," Vol. II, pp. 339–358. Springer-Verlag, Berlin, New York.

Young, D., and Gregory, R., (1973). "A Survey of Numerical Mathematics," Vol. II, pp. 924–930. Addison-Wesley, Reading, Massachusetts.

INTERNAL DOCUMENTATION

SYNTAX: Z←LOHI ELMHES A
PURPOSE: TO REDUCE A REAL MATRIX TO UPPER-
HESSENBERG FORM BY REAL STABILIZED
SIMILARITY TRANSFORMATIONS.
INDEX ORIGIN: 1
AUTHOR: C. HUCKABA
LAST UPDATE: 4/26/79
LEFT ARGUMENT: LOHI IS A TWO ELEMENT VECTOR
CONTAINING THE PARAMETERS LOW AND HI OUTPUT
BY BALANCE. IF A WAS NOT BALANCED, THEN SET
LOHI EQUAL TO (1,N) WHERE N IS THE ORDER
OF A.
RIGHT ARGUMENT: A IS AN N×N REAL MATRIX,
USUALLY PREPARED BY BALANCE.
RESULT: Z IS AN (N+1)×N MATRIX. THE FIRST ROW
CONTAINS THE VECTOR INT WHICH DESCRIBES THE
ROW AND COLUMN INTERCHANGES INVOLVED IN THE
REDUCTION. THE REMAINING N ROWS CONSIST
PARTLY OF THE DERIVED UPPER-HESSENBERG
MATRIX; THE SCALAR QUANTITY USED TO
ELIMINATE AIJ IS STORED IN Z[I+1;J]. FOR A
MORE DETAILED DESCRIPTION OF THE RESULTS,
SEE THE HANDBOOK FOR AUTOMATIC COMPUTATION,
VOL.2, P. 341

CODE

```
        ∇  Z←LOHI ELMHES A;K;L;N;LA;M;I;X;J;
           COL;Y
[1]     K←1↑LOHI
[2]     L←¯1↑LOHI
[3]     N←1↑ρA
[4]     INT←Nρ0
[5]     LA←L-1
[6]     M←K+1
[7]   ⍝ TOP OF TRANSFORMATION
[8]   L1:I←M
[9]     X←0
[10]  J←M
[11]  L2:→((,|A[J;M-1])≤|X)/L3
```

```
[12]    X←,A[J;M-1]
[13]    I←J
[14]  L3:→(L≥J←J+1)/L2
[15]    INT[M]←I
[16]    →(I=M)/L4
[17]  ⍝ INTERCHANGE ROWS AND COLUMNS OF A
[18]    A[I,M;COL]←A[M,I;COL←M+¯1,0,⍳N-M]
[19]    A[⍳L;I,M]←A[⍳L;M,I]
[20]  L4:→(X=0)/L5
[21]    I←M+1
[22]  L6:Y←,A[I;M-1]
[23]    →(Y=0)/L7
[24]    Y←A[I;M-1]←Y÷X
[25]    A[I;COL]←A[I;COL]-Y×A[M;COL←M+0,⍳N-M]
[26]    A[⍳L;M]←A[⍳L;M]+Y×A[⍳L;I]
[27]  L7:→(L≥I←I+1)/L6
[28]  L5:→(LA≥M←M+1)/L1
[29]    Z←INT,[1] A
        ∇
```

Function Names: EVAL and EVAL2

EXTERNAL DOCUMENTATION

PURPOSE

To evaluate complex polynomials at specified points.

EVAL evaluates several complex polynomials at the same point, whereas EVAL2 evaluates one complex polynomial at several points.

ALGORITHM

Both of these programs use Horner's rule to evaluate complex polynomials. For example, for $P(z) = az^2 + bz + c = z(az + b) + c$, one can calculate $P(z_0)$ as follows:

$$az_0 \tag{1}$$
$$az_0 + b \tag{2}$$
$$z_0(az_0 + b) \tag{3}$$
$$z_0(az_0 + b) + c = P(z_0) \tag{4}$$

Since EVAL evaluates several complex polynomials at the same point z_0, the a, b, and c in the above procedure are replaced by a $(2 \times n \times d)$ array of the complex coefficients, in descending order, of the n polynomials. On the other hand, since EVAL2 evaluates a single complex polynomial at several points, the a, b, and c in the above *SCHEME* are replaced by a $(2 \times d)$ array of the complex coefficients, in descending order, of a polynomial $P(z)$, and z_0 is to be interpreted as a vector of complex numbers.

INTERNAL DOCUMENTATION FOR EVAL

SYNTAX:PZ←P EVAL Z
PURPOSE:EVALUATION OF SEVERAL COMPLEX
 POLYNOMIALS AT THE SAME POINT.
INDEX ORIGIN: 1
AUTHOR: *E. SILBERG*
LAST UPDATE: 5/79
LEFT ARGUMENT: (2×N×D) *ARRAY*, N≥1,D≥2, *OF THE*
 COMPLEX COEFFICIENTS,IN DESCENDING ORDER,
 OF N POLYNOMIALS OF DEGREE D-1. (ZEROS ARE
 PERMISSIBLE COEFFICIENTS,FOR EQUALIZING ALL
 DEGREES TO THG MAX. DEGREE.)
RIGHT ARGUMENT: (2×1) *ARRAY, REPRESENTING THE*
 COMPLEX NUMBER AT WHICH ALL OF THE
 POLYNOMIALS ARE TO BE EVALUATED.
RESULT:(2×N) MATRIX, OF THE(COMPLEX) VALUES
 OF THE N POLYNOMIALS, AT THE POINT Z.

INTERNAL DOCUMENTATION FOR EVAL2

SYNTAX:PZ←P EVAL2 Z
PURPOSE:EVALUATION OF A COMPLEX POLYNOMIAL
 AT SEVERAL POINTS.
INDEX ORIGIN: 1
AUTHOR: *E. SILBERG*
LAST UPDATE: 5/79
LEFT ARGUMENT: (2×D) *MATRIX*, D≥2 *OF THE*
 COMPLEX COEFFICIENTS(IN DESCENDING ORDER) OF
 THE POLYNOMIAL P.
RIGHT ARGUMENT: (2×M) *ARRAY*, M≥1, *OF THE M*

COMPLEX NUMBERS Z[;1],...,Z[;M] AT WHICH
P IS TO BE EVALUATED.
RESULT: (2×M) ARRAY, PZ:
PZ[;K] = P(Z[;K]), K=1,2,...,M.

CODE FOR EVAL

```
      ∇ PZ←P EVAL Z;D;I
[1]  ⍝ EVALUATES SEVERAL COMPLEX POLYNOMIALS
     ⍝ AT SAME POINT.
[2]  ⍝ CALLS MULT.
[3]   PZ←P[;;1]
[4]   I←2+D-D←⁻1↑⍴P
[5]  LOOP:PZ←((2,(⍴P)[2])⍴P[;;I])+Z MULT PZ
[6]   →(D≥I←I+1)/LOOP
      ∇
```

CODE FOR EVAL2

```
      ∇ PZ←P EVAL2 Z;D;I
[1]  ⍝ EVALUATES A COMPLEX POLYNOMIAL AT
         SEVERAL POINTS.
[2]  ⍝ CALLS ADD, MULT.
[3]   PZ←P[;1]
[4]   I←2+D-D←⁻1↑⍴P
[5]  LOOP:PZ←P[;I] ADD Z MULT PZ
[6]   →(D≥I←I+1)/LOOP
      ∇
```

Function Name: HSHLDR

EXTERNAL DOCUMENTATION

PURPOSE

To produce the Householder reduction of a real symmetric matrix to a symmetric tridiagonal matrix by a series of similarity transformations.

For a discussion of the Householder transformation see "A Survey of Numerical Mathematics."

ALGORITHM

The algorithm presented here is essentially the TRED3 procedure that was developed by R. S. Martin, C. Reinsch, and J. H. Wilkinson and is discussed in the "Handbook for Automatic Computation," Vol. II. HSHLDR differs from TRED3 in that it operates on the entire matrix instead of only the lower triangle. This modification allows more efficient use of APL matrix operations. The procedure can be used for arbitrary real symmetric matrices and requires no tolerance limit for termination.

A tolerance level for skipping transformations has been set in the procedure. It insures that a transformation is skipped whenever there is a danger that it might not be accurately orthogonal. The maximum error in any eigenvalue as a result of the use of this criterion is $2 \times (\text{tolerance} * .5)$. A discussion of the error incurred in the numerical procedure is given on pp. 222–223 of the "Handbook for Automatic Computation," Vol. II.

REFERENCES

Wilkinson, J. H., and Reinsch, C. (1971). "Handbook for Automatic Computation," Vol. II, pp. 202–211. Springer-Verlag, Berlin, New York.
Young, D., and Gregory, R. (1973). "A Survey of Numerical Mathematics," Vol. II, pp. 901–909. Addison-Wesley, Reading, Massachusetts.

INTERNAL DOCUMENTATION

```
SYNTAX: T←HSHLDR A
PURPOSE: TO TRIDIAGONALIZE A REAL SYMMETRIC
  MATRIX VIA HOUSEHOLDER SIMILARITY
  TRANSFORMATIONS
INDEX ORIGIN: 1
AUTHOR: C. HUCKABA
LAST UPDATE: 1/22/79
RIGHT ARGUMENT: A IS AN N×N REAL SYMMETRIC
  MATRIX
RESULT: T IS AN (N+2)×N MATRIX. THE FIRST ROW
  CONTAINS THE DIAGONAL ELEMENTS OF THE
  RESULTANT TRIDIAGONAL MATRIX. THE SECOND ROW
  CONTAINS THE OFF-DIAGONAL ELEMENTS. THE
  FIRST ELEMENT OF THE SECOND ROW HAS
  ARBITRARILY BEEN SET EQUAL TO ZERO. THE
```

REMAINING N ROWS CONTAIN THE PRODUCT OF THE
HOUSEHOLDER TRANSFORMATION MATRICES, THAT
IS, THEY CONTAIN AN N×N ORTHOGONAL MATRIX P
SUCH THAT P'AP IS THE TRIDIAGONAL MATRIX
REPRESENTED BY THE FIRST TWO ROWS OF T.

CODE

```
        ∇ T←HSHLDR A;TOL;M;V;N;I;L;SIG;U;H;P;K;
          Q;B;D;E
[1]     TOL←1E¯17×(1-⎕CT)÷⎕CT
[2]     M←V∘.=V←ιN←1↑ρA
[3]     D←E←Nρ0
[4]     I←1
[5]     L1:L←(N-I)+1
[6]     ⍝ CHECK TOLERANCE LEVEL FOR SKIPPING
          TRANSFORMATION
[7]     →((SIG←+/((L-2)↑,A[L;])*2)≤TOL)/L2
[8]     SIG←SIG+A[L;L-1]*2
[9]     SIGN←(A[L;L-1]≥0)-(A[L;L-1]<0)
[10]    ⍝ TOP OF TRANSFORMATION
[11]    U←((L-2)↑,A[L;]),(,A[L;L-1]+SIGN×SIG*
          0.5);Iρ0
[12]    H←0.5×U+.×U
[13]    P←(A+.×U)÷H
[14]    K←(U+.×P)÷2×H
[15]    Q←P-K×U
[16]    A←A-(⍉B)+B←U∘.×Q
[17]    D[L-1]←A[L-1;L-1]
[18]    E[L]←A[L;L-1]
[19]    M←M-((M+.×U)∘.×U)÷H
[20]    →L3
[21]    L2:D[L-1]←A[L-1;L-1]
[22]    E[L]←A[L;L-1]
[23]    L3:→((N-2)≥I←I+1)/L1
[24]    D[1]←A[1;1]
[25]    D[N]←A[N;N]
[26]    E[2]←A[2;1]
[27]    T←D,[1] E,[1] M
        ∇
```

Function Name: INVERSE

EXTERNAL DOCUMENTATION

PURPOSE

To compute the inverse of a complex lower triangular matrix by the Gaussian elimination method.

Because of the triangular form of the matrix, the number of steps required to complete the procedure is substantially reduced.

In the following algorithm, A can represent either the real or the imaginary part of the complex matrix. A has been formed (a) by appending an $N \times N$ identity matrix onto the right-hand side of the real part of the matrix or (b) by appending an $N \times N$ matrix of zeros onto the right-hand side of the imaginary part of the matrix. The algorithm is then performed identically on both the real and imaginary matrices.

ALGORITHM

For $j = 1$ to $2N$, set

$$A(1j) = A(1j)/A(11) \text{End}$$

For $i = 2$ to N, $x = 1$ to i, and $j = 1$ to $2N$, set

$$A(ij) = A(ij) - A(ix)A(xj) \text{End}$$

For $j = 1$ to $2N$, set

$$A(ij) = A(ij)/A(ii) \text{End}$$

(Where $A(ij) = $ the ij element of A.)

After the looping is completed, the inverse is formed by dropping the first N columns of each matrix and then combining the remaining matrices into one complex lower triangular matrix.

INTERNAL DOCUMENTATION

```
SYNTAX:  ∇C←INVERSE B
PURPOSE:  COMPUTES INVERSE OF A COMPLEX, LOWER
  TRIANGULAR MATRIX: SEE EXTERNAL DOC FOR
  COMPLEX PACKAGE FOR DEFINITION OF COMPLEX
  MATRIX.
INDEX ORIGIN:  1
AUTHOR:  P. KRETZMER
```

```
LAST UPDATE: 5-4-79
RIGHT ARGUMENT: A 2 BY N BY N COMPLEX, LOWER
 TRIANGULAR MATRIX.
RESULT: A 2 BY N BY N COMPLEX, LOWER
 TRIANGULAR MATRIX THAT IS THE INVERSE OF B.
```

CODE

```
       ∇ C←INVERSE B;N;I;X;E;D
[1]   ⍝COMPUTES INVERSE OF A COMPLEX, LOWER
      ⍝TRIANGULAR MATRIX
[2]   ⍝CALLS ∇Z←A MINUS B,∇Z←A MULT B, ∇Z←A
      ⍝DIVBY B
[3]   ⍝LAST UPDATE: 5-4-79
[4]    N←(ρB)[2]
[5]    E←B[1;;],(N,N)ρ1,Nρ0
[6]    D←B[2;;],(N,N)ρ0
[7]    B←(2,N,2×N)ρ(,E),,D
[8]    B[;1;]←B[;1;] DIVBY B[;1;1]
[9]    I←2
[10] LOOP1:X←1
[11] LOOP2:B[;I;]←B[;I;] MINUS B[;I;X] MULT
      B[;X;]
[12]   X←X+1
[13]   →(X<I)/LOOP2
[14]   B[;I;]←B[;I;] DIVBY B[;I;I]
[15]   I←I+1
[16]   →(I≤N)/LOOP1
[17]   E←(0,N)↓B[1;;]
[18]   D←(0,N)↓B[2;;]
[19]   C←(2,N,N)ρ(,E),,D
       ∇
```

Function Name: JACOBI

EXTERNAL DOCUMENTATION

PURPOSE

To compute the eigenvalues and (optionally) the eigenvectors of a real symmetric matrix by the Jacobi method.

For a discussion of the Jacobi method, see "The Algebraic Eigenvalue Problem."

ALGORITHM

The algorithm presented here was developed by H. Rutishauser and is discussed in the "Handbook for Automatic Computation," Vol. II. It essentially uses row-wise scanning of the upper triangle in determining the individual Jacobi rotations. The procedure is foolproof for all real symmetric matrices and requires no tolerance limit for termination.

The total error incurred in the diagonalization of a symmetric matrix is for every eigenvalue at most

$$E = 18.2n^{1.5} \, r|A0|*$$

where $|A0|^2$ equals the sum of the squares of the elements of the original matrix, r is the number of sweeps required to perform the diagonalization, and * is the smallest positive number such that in the computer $1+*$ is not $= 1$. This error estimate is influenced by the measures taken in the present algorithm in only one way. The contributions of the round-off errors become insignificant as soon as the off-diagonal elements of the iterated matrix Ak become small as compared to the diagonal elements. Thus in the implementation of Rutishauser's algorithm the round-off error can be expected to be substantially smaller than the above bound.

REFERENCES

Wilkinson, J. H., and Reinsch, C. (1971). "Handbook for Automatic Computation," Vol. II, pp. 202–211. Springer-Verlag, Berlin, New York.
Wilkinson, J. H., (1965). "The Algebraic Eigenvalue Problem." Oxford Univ. Press (Clarendon), London and New York.

INTERNAL DOCUMENTATION

```
SYNTAX: Z←E JACOBI S
PURPOSE: COMPUTATION OF THE EIGENVALUES AND
 (OPTIONALLY) THE EIGENVECTORS OF A REAL
 SYMMETRIC MATRIX USING THE JACOBI METHOD.
INDEX ORIGIN: 1
AUTHOR: C. HUCKABA
LAST UPDATE: 11/9/78
```

LEFT ARGUMENT: E=0 IF EIGENVALUES ONLY ARE
DESIRED; E=1 IF BOTH EIGENVALUES AND
EIGENVECTORS ARE DESIRED.
RIGHT ARGUMENT: S IS AN N×N REAL SYMMETRIC
MATRIX
RESULT: IF E=0, Z IS AN N-DIMENSIONAL VECTOR
CONTAINING THE EIGENVALUES OF S. IF E=1, Z
IS AN (N+1)×N MATRIX WITH THE FIRST ELEMENT
IN EACH OF THE N COLUMNS BEING AN EIGENVALUE
WHOSE CORRESPONDING, NORMALIZED EIGENVECTOR
IS REPRESENTED BY THE REMAINING N COLUMN
ELEMENTS BELOW IT.

CODE

```
        ∇ Z←E JACOBI S;M;N;V;B;D;Z;ROT;SWP;SUM;
          THRSH;P;Q;G;H;THETA;TAN;COS;SIN;TAU;
          GS;HS;GV;HV;I;J;K;L
[1]     →(E=0)/L1
[2]     V←(M∘.=M←ιN←(ρS)[1])
[3]     L1:B←D←(,M∘.=M←ιN←(ρS)[1])/,S
[4]     Z←Nρ0
[5]     ROT←SWP←0
[6]     ⍝ TOP OF SWEEP
[7]     L9:→(0=SUM←+/|(,M∘.<M)/,S)/L2
[8]     THRSH←(SWP≤3)×0.2×SUM÷N*2
[9]     P←1
[10]    Q←2
[11]    ⍝ TOP OF ROTATION
[12]    L8:G←100×|S[P;Q]
[13]    →((SWP≤3)∧(((|D[P])+G)≠(|D[P]))∧
          ((((|D[Q])+G)≠(|D[Q]))))/L3
[14]    S[P;Q]←0
[15]    →L4
[16]    L3:→((|S[P;Q])≤THRSH)/L4
[17]    H←D[Q]-D[P]
[18]    →((((|H)+G)≠(|H)))/L5
[19]    TAN←S[P;Q]÷H
[20]    →L6
```

```
[21]  ⍝ COMPUTATION OF TAN OF ROTATION ANGLE
[22]  L5:THETA←0.5×H÷S[P;Q]
[23]   TAN←1÷(|THETA)+(1+THETA*2)*0.5
[24]   →(THETA≥0)/L6
[25]   TAN←-TAN
[26]  L6:COS←1÷(1+TAN*2)*0.5
[27]   SIN←TAN×COS
[28]   TAU←SIN÷(1+COS)
[29]   H←TAN×S[P;Q]
[30]   Z[P]←Z[P]-H
[31]   Z[Q]←Z[Q]+H
[32]   D[P]←D[P]-H
[33]   D[Q]←D[Q]+H
[34]   S[P;Q]←0
[35]  ⍝ UPDATE S
[36]   S[I;P]←GS-SIN×((HS←S[I;Q])+(GS←S[I←⍳
       (P-1);P])×TAU)
[37]   S[I;Q]←HS+SIN×(GS-HS×TAU)
[38]   S[P;J]←GS-SIN×((HS←S[J;Q])+(GS←S[P;J←P+
       ⍳Q-P+1])×TAU)
[39]   S[J;Q]←HS+SIN×(GS-HS×TAU)
[40]   S[P;K]←GS-SIN×((HS←S[Q;K])+(GS←S[P;K←Q+
       ⍳N-Q])×TAU)
[41]   S[Q;K]←HS+SIN×(GS-HS×TAU)
[42]   →(E=0)/L7
[43]  ⍝ UPDATE V
[44]   V[L;P]←GV-SIN×((HV←V[L;Q])+(GV←V[L←⍳N;
       P])×TAU)
[45]   V[L;Q]←HV+SIN×(GV-HV×TAU)
[46]  L7:ROT←ROT+1
[47]  L4:→(N≥Q←Q+1)/L8
[48]   Q←P+2
[49]   →(N>P←P+1)/L8
[50]   D←B←B+Z
[51]   Z←N⍴0
[52]   →(50≥SWP←SWP+1)/L9
[53]  L2:'SWP=',(⍕SWP),';ROT=',⍕ROT
[54]   →(E=1)/L10
[55]   Z←D
[56]   →0
[57]  L10:Z←D,[1] V
       ∇
```

Function Name: MATMULT

EXTERNAL DOCUMENTATION

PURPOSE

To multiply two complex matrices.

A complex matrix is defined to be a set of complex numbers organized into a matrix in which the real and imaginary parts are separated along the first dimension. Thus a complex matrix can consist of a 2-element vector (one complex number), a two-dimensional matrix with two rows, a three-dimensional matrix with two planes, etc. For example, $a + bi$, $c + di$, $e + fi$, and $g + hi$ can be represented as

$$\begin{array}{cccc} a & c & e & g \\ b & d & f & h \end{array} \quad \text{or} \quad \begin{array}{cccc} a & c & b & d \\ e & g & f & h \end{array}$$

For this program, the form must be a three-dimensional matrix with two planes.

ALGORITHM

MATMULT uses a simple loop of multiplication (by the program MULT) and addition to determine the desired product.

INTERNAL DOCUMENTATION

SYNTAX: Z←A MATMULT B
PURPOSE: MULTIPLICATION OF TWO THREE-
  DIMENSIONAL COMPLEX MATRICES
INDEX ORIGIN: 1
AUTHORS: B. BENNETT, P. KRETZMER
LAST UPDATE: 5/79
LEFT ARGUMENT: A THREE-DIMENSIONAL COMPLEX
  MATRIX (I.E. A SET OF COMPLEX NUMBERS WHERE
  THE REAL AND IMAGINARY PARTS ARE SEPARATED
  ALONG THE FIRST DIMENSION)
RIGHT ARGUMENT: B IS ALSO A THREE-DIMENSIONAL
  COMPLEX MATRIX
RESULT: A COMPLEX MATRIX, THE PRODUCT OF A
  AND B

CODE

```
      ∇  Z←A MATMULT B;R;S
[1]  ⍝MULTIPLICATION OF TWO COMPLEX MATRICES
[2]  ⍝CALLS: ∇Z←A MULT B∇
[3]  ⍝LAST UPDATE: 2-20-79
[4]    Z←(2,(ρA)[2],(ρB)[3])ρ0
[5]    R←1
[6]  NR:S←1
[7]  NC:Z[;R;S]←+/A[;R;] MULT B[;;S]
[8]    →((S←S+1)≤(ρB)[3])/NC
[9]    →((R←R+1)≤(ρA)[2])/NR
      ∇
```

Function Name: MAGNITUDE

EXTERNAL DOCUMENTATION

PURPOSE

Computes absolute values componentwise of arbitrary complex array:
See external documentation for MATMULT.

INTERNAL DOCUMENTATION

```
SYNTAX: Z←MAGNITUDE A
PURPOSE: TO COMPUTE THE MAGNITUDE OF A
  COMPLEX NUMBER
INDEX ORIGIN: 1
AUTHORS: B. BENNETT, P. KRETZMER
LAST UPDATE: 4/79
RIGHT ARGUMENT: A IS A COMPLEX MATRIX (I.E.,
  A SET OF COMPLEX NUMBERS, WHERE THE REAL AND
  IMAGINARY PARTS ARE SEPARATED ALONG THE
  FIRST DIMENSION.)
RESULT: A SCALAR (OR VECTOR IF MORE THAN ONE
  MAGNITUDE IS BEING COMPUTED)
```

CODE

```
    ∇ Z←MAGNITUDE A;S;AI
[1] ⍝GIVES MAGNITUDE OF EACH COMPLEX NUMBER
    ⍝OF A COMPLEX MATRIX
[2] ⍝LAST UPDATE: 12-10-78
[3]  AI←⍎'A[2',S←(((ρρA)-1)ρ';'),']'
[4]  A←⍎'A[1',S
[5]  Z←((A*2)+AI*2)*0.5
    ∇
```

Function Name: MULLERM

EXTERNAL DOCUMENTATION

PURPOSE

To find n roots, both real and complex, for an nth degree polynomial equation

$$f(x) = a_0 x^n + a_1 x^{n-1} + \cdots + a_n = 0$$

by using Muller's method.

ALGORITHM

Each root is found by an iterative procedure, each iteration proceeding by the identification of the nearer root of an approximating quadratic using a variation of the quadratic formula.

(1) Define $h = x_3 - x_2$, $L = h/(x_2 - x_1)$, $D = 1 + L$.
(2) Initialize $x_1 = -1$, $x_2 = 1$, $x_3 = 0$:

$$f(x_1) = a_n - a_{n-1} + a_{n-2}$$
$$f(x_2) = a_n + a_{n-1} + a_{n-2}$$
$$f(x_3) = a_n$$

Thus $L = -0.5$, $h = -1$, and the first approximating quadratic is

$$a_n + a_{n-1}x + a_{n-2}x^2$$

(3) For each iteration, compute

$$G = f(x_1)L^2 - f(x_2)D^2 + f(x_3)(L + D)$$

and

$$L_{\text{new}} = -2f(x_3)D/\{G + [G^2 - 4f(x_3)DL(f(x_1)L - f(x_2)D + f(x_3)]^{1/2}\}$$

Then $h_{\text{new}} = L_{\text{new}}h$ and $x_{\text{new}} = x_3 + h_{\text{new}}$. x(new) is a zero of the approximating quadratic in this iteration. The sign in the denominator of L(new) is taken so as to provide the denominator of greater magnitude. This makes h(new) the smaller of two possibilities, i.e., we find the nearer root.

Two special cases:

(a)   If $|f(x_{\text{new}})/f(x_3)| > 10$, L_{new} is reassigned to $L_{\text{new}}/2$. h(new) and x(new) are adjusted appropriately.

(b)   If $f(x_1) = f(x_2) - f(x_3)$, the denominator of L_{new} is zero. Choose L(new) equal to one in this case.

(4)   Continue iterations until $|x_3 - x_2| < $ or $= 10*(-7)$. At this point (equality of $x_{(3)}$ and $x_{(2)}$), a root has been found.

(5)   Use synthetic division to reduce the degree of the polynomial by one. If the new polynomial is still of order ≥ 2, go to TRY to find the next root. If the new polynomial is linear, we solve the linear equation and execution terminates.

NOTES

(1)   The result is accurate to within the tolerance specified in the program, $10*(-7)$.

(2)   MULLERM performs no evaluations of the derivatives of polynomials. The process is particularly economical timewise for solving large-order polynomial equations.

EXAMPLE

$$x^3 + (-2 - 2i)x^2 + 3ix + (1 - i)$$

Coefficient matrix:

$$\begin{matrix} 1 & -2 & 0 & 1 \\ 0 & -2 & 3 & -1 \end{matrix}$$

Result:

$$\begin{matrix} 1 & 0 & 1 \\ 0 & 1 & 1 \end{matrix} \quad \text{(roots are thus } 1, i, 1 + i\text{)}$$

REFERENCE

Muller, D. E. (1956). A method for solving algebraic equations using an automatic computer, *Math. Comp.* **10**, 208–215.

INTERNAL DOCUMENTATION

SYNTAX: RES←MULLERM V
PURPOSE: SOLUTION OF AN NTH DEGREE POLYNOMIAL
 EQUATION FOR BOTH REAL AND COMPLEX ROOTS,
 USING MULLER'S METHOD
INDEX ORIGIN: 1
AUTHORS: B. BENNETT, P. KRETZMER, MODIFIED BY
 U. GRENANDER
LAST UPDATE: 2/79
RIGHT ARGUMENT: A 2×(N+1) COMPLEX MATRIX OF
 THE POLYNOMIAL COEFFICIENTS; I.E. V[1;]
 CONTAINS THE REAL COMPONENTS V[2;] CONTAINS
 THE CORRESPONDING IMAGINARY COMPONENTS
RESULT: A 2×(N+1) COMPLEX MATRIX OF THE ROOTS
 OF THE POLYNOMIAL
WARNING: THE TOLERANCE 1E-7 IN STATEMENTS
 [10] AND [23] CAN BE CHANGED BY THE USER
 WHEN NEEDED.

CODE

```
        ∇ RES←MULLERM V
[1]   ⍝FINDS N ROOTS, BOTH REAL AND COMPLEX,
      ⍝OF AN NTH ORDER POLYNOMIAL
[2]   ⍝USES MULLER'S METHOD
[3]   ⍝LAST UPDATE: 2/79
[4]   RES←⍳0
[5]   BIG:V←⌽V
[6]   X← 2 3 ρ ¯1 1 0 0 0 0
[7]   H← ¯1 0
[8]   F←⍉ 3 2 ρ(V[;1] ADD V[;3] MINUS V[;2]),
       (V[;1] ADD V[;2] ADD V[;3]),V[;1]
[9]   D← 1 0 ADD L← ¯0.5 0
[10]  LOOP:→((+/(((1E¯7>|F[;1]-F[;2])∧1E¯7>|
       F[;2]-F[;3]))=2)/INC1
[11]  G←+/F MULT⍉ 3 2 ρ(L POWER 2),(-(D POWER
       2)),(L ADD D)
[12]  J←(G POWER 2) MINUS 4 0 MULT F[;3] MULT
       D MULT L MULT+/F MULT⍉ 3 2 ρL,(-D), 1 0
```

```
[13]   D←1 0 ADD L←(¯2 0 MULT F[;3] MULT D)
       DIVBY,P[;¯1↑⍋MAGNITUDE P←G
       ADD(2 2 ρ ¯1 1 0 0) MULT J POWER÷2]
[14]   →INC
[15]   INC1:D← 1 0 ADD L← 1 0
[16]   INC:H←H MULT L
[17]   X← 2 ¯3 ↑X,X[;3] ADD H
[18]   F← 2 ¯3 ↑F,+/V MULT X[;3] POWER 0,
       ι(¯1↑ρV)-1
[19]   →(10≥(MAGNITUDE F[;3])÷0.1+MAGNITUDE
       F[;2])/CK
[20]   D← 1 0 ADD L←L DIVBY 2 0
[21]   X[;3]←X[;3] MINUS H←H DIVBY 2 0
[22]   F[;3]←+/V MULT X[;3] POWER 0,ι(¯1↑ρV)-1
[23] CK:→((MAGNITUDE X[;3] MINUS X[;2])>1E¯7)
       /LOOP
[24]   Z←X[;3]
[25]   RES←RES,Z
[26]   V←⌽V
[27]   N←(¯1↑ρV)-1
[28]   I←1
[29] COEF:I←I+1
[30]   V[;I]←V[;I] ADD Z MULT V[;I-1]
[31]   →(I<N)/COEF
[32]   V← 0 ¯1 ↓V
[33]   →(2<¯1↑ρV)/BIG
[34]   Z←-V[;2] DIVBY V[;1]
[35]   RES←⍉((((ρRES)÷2)+1),2)ρRES,Z
       ∇
```

Function Name: MULT

EXTERNAL DOCUMENTATION

PURPOSE

Multiplies arbitrary complex arrays *componentwise*. See external documentation for MATMULT.

INTERNAL DOCUMENTATION

SYNTAX: Z←A MULT B
PURPOSE: MULTIPLICATION OF COMPLEX NUMBERS
INDEX ORIGIN: 1
AUTHORS: B. BENNETT, P. KRETZMER
LAST UPDATE: 4/79
LEFT ARGUMENT: A IS A COMPLEX MATRIX (I.E.,
 A SET OF COMPLEX NUMBERS, WHERE THE REAL AND
 IMAGINARY PARTS ARE SEPARATED ALONG THE
 FIRST DIMENSION.)
RIGHT ARGUMENT: B IS ALSO A COMPLEX MATRIX
RESULT: A COMPLEX MATRIX

CODE

```
      ∇ Z←A MULT B;AI;BI;S
[1]  ⍝MULTIPLIES CORRESPONDING ELEMENTS OF TWO
     ⍝COMPLEX ARRAYS
[2]  ⍝AND DOES ALL OTHER ANALOGOUS OPERATIONS
     ⍝OF  DYADIC MULTIPLY
[3]  ⍝LAST UPDATE: 12-10-78
[4]   AI←⍎'A[2',S←(((⍴⍴A)-1.)⍴';'),']'
[5]   A←⍎'A[1',S
[6]   BI←⍎'B[2',S←(((⍴⍴B)-1)⍴';'),']'
[7]   B←⍎'B[1',S
[8]   Z←((A×B)-AI×BI),[0.5](A×BI)+AI×B
      ∇
```

Function Name: MINUS

EXTERNAL DOCUMENTATION

PURPOSE

Subtracts arbitrary complex arrays componentwise; see external documentation for MATMULT.

INTERNAL DOCUMENTATION

SYNTAX: Z←A MINUS B
PURPOSE: SUBTRACTION OF COMPLEX NUMBERS
INDEX ORIGIN: 1
AUTHORS: B. BENNETT, P. KRETZMER
LAST UPDATE: 4/79
LEFT ARGUMENT: A IS A COMPLEX MATRIX (I.E.,
A SET OF COMPLEX NUMBERS, WHERE THE REAL AND
IMAGINARY PARTS ARE SEPARATED ALONG THE
FIRST DIMENSION.)
RIGHT ARGUMENT: B IS ALSO A COMPLEX MATRIX
RESULT: A COMPLEX MATRIX

CODE

```
      ∇ Z←A MINUS B;AI;BI;S
[1]  ⍝SUBTRACTS CORRESPONDING ELEMENTS OF TWO
     ⍝COMPLEX ARRAYS
[2]  ⍝AND DOES ALL OTHER ANALOGOUS OPERATIONS
     ⍝OF DYADIC SUBTRACT
[3]  ⍝LAST UPDATE: 12-10-78
[4]   AI←⍎'A[2',S←(((ρρA)-1)ρ';'),']'
[5]   A←⍎'A[1',S
[6]   BI←⍎'B[2',S←(((ρρB)-1)ρ';'),']'
[7]   B←⍎'B[1',S
[8]   Z←(A-B),[0.5] AI-BI
      ∇
```

Function Name: PDIVBY

EXTERNAL DOCUMENTATION

PURPOSE

To find the quotient and remainder when dividing one polynomial by another.

ALGORITHM

If degree(divisor) > degree(dividend), then quotient = 0 and remainder = divisor. End PDIVBY. Else, $I = 1$.

Do for $I = 1$ to $1 +$ degree(dividend) $-$ degree(divisor): quotient(I) $=$ dividend(1)/divisor(1); dividend $=$ dividend $-$ (quotient(I) $\times$ divisor); $I = I + 1$. End.

Remainder $=$ dividend

End PDIVBY.

INTERNAL DOCUMENTATION

SYNTAX: Z←P1 PDIVBY P2
PURPOSE: TO FIND THE QUOTIENT AND REMAINDER
 WHEN DIVIDING ONE POLYNOMIAL BY ANOTHER.
INDEX ORIGIN: 1
AUTHOR: A LATTO
LAST UPDATE: 8/18/80
RIGHT ARGUMENT: THE VECTOR OF COEFFICIENTS OF
 THE DIVISOR IN DESCENDING ORDER WITH RESPECT
 TO POWERS OF X.
LEFT ARGUMENT: THE VECTOR OF COEFFICIENTS OF
 THE DIVIDEND IN DESCENDING ORDER WITH
 RESPECT TO POWERS OF X.
RESULT: A 2×N ARRAY WHERE N=1+DEGREE
 (DIVIDEND). THE FIRST ROW IS THE VECTOR OF
 COEFFICIENTS OF THE QUOTIENT ORDERED IN THE
 MANNER DISCUSSED ABOVE. THE SECOND ROW IS
 THE VECTOR OF COEFFICIENTS OF THE REMAINDER
 ORDERED AS ABOVE.

CODE

```
      ∇ Z←P1 PDIVBY P2;DEGP2;LEAD;SHIFT
[1]   →(((ρP1)-(P1≠0)ι1)≥(ρP2)-(P2≠0)ι1)/
      DIVIDE
[2]   Z←(2,ρP2)ρ((ρP2)ρ0),P2
[3]   →0
[4]   DIVIDE:Z←(ρP1)ρ0
[5]   LEAD←P2[DEGP2←(0≠P2←(-ρP1)↑P2)ι1]
[6]   P2←(SHIFT←(DEGP2-(P1≠0)ι1))⌽P2
[7]   LOOP:P1←P1-P2×Z[(ρP1)-SHIFT]←P1[DEGP2-
      SHIFT]÷LEAD
[8]   P2←¯1⌽P2
```

```
[9]    →(0≤SHIFT←SHIFT-1)/LOOP
[10]   Z←(2,ρP1)ρZ,P1
       ▽
```

Function Name: POLY

EXTERNAL DOCUMENTATION

PURPOSE

To find the product of two polynomials.
If

$$Y = a_0x^n + a_1x^{n-1} + \cdots + a_n \quad \text{and} \quad Z = b_0x^m + b_1x^{m-1} + \cdots + b_m$$

then

$$YZ = ZY = a_0b_0x^{n+m} + (a_0b_1 + a_1b_0)x^{n+m-1}$$
$$+ (a_0b_2 + a_1b_1 + a_2b_0)x^{n+m-2} + \cdots + a_nb_m$$

ALGORITHM

A is an array of $N = n + 1$ coefficients of the nth order polynomial while B is an array of $M = m + 1$ coefficients of the mth order polynomial. The inner product of A and B is computed, and an N by $N - 1$ matrix of zeros is catenated to the result, thus forming an N by $N + M - 1$ matrix. The first row is left unshifted, the second row shifted rightward by one, . . . , and the Nth row shifted rightward by $N - 1$. The resulting $N + M - 1$ coefficients are then obtained by summing across the rows of the shifted matrix. This algorithm is due to D. McClure.

EXAMPLE

$$A = 3 \quad 5 \quad 6 \quad 2 \quad 3: \quad 3x^4 + 5x^3 + 6x^2 + 2x + 3$$
$$B = 4 \quad 3 \quad 5 \quad 1: \quad 4x^3 + 3x^2 + 5x + 1$$

A POLY B:

$$12 \quad 29 \quad 54 \quad 54 \quad 53 \quad 25 \quad 17 \quad 3:$$
$$12x^7 + 29x^6 + 54x^5 + 54x^4 + 53x^3 + 25x^2 + 17x + 3$$

INTERNAL DOCUMENTATION

```
SYNTAX:  ∇Z←A POLY B∇
PURPOSE:  COMPUTATION OF THE COEFFICIENTS OF
  THE PRODUCT OF TWO POLYNOMIALS.
INDEX ORIGIN:  1
AUTHOR:  P. KRETZMER
LAST UPDATE:  2-25-79
RIGHT ARGUMENT:  B←VECTOR OF THE M
  COEFFICIENTS OF A POLYNOMIAL IN DESCENDING
  ORDER; I.E. THE VECTOR, B, OF COEFFICIENTS
  IS ARRANGED WITH THOSE CORRESPONDING TO THE
  HIGHEST POWER APPEARING FIRST.
LEFT ARGUMENT:  A←VECTOR OF THE N COEFFICIENTS
  OF A POLYNOMIAL IN DESCENDING ORDER.
RESULT:  Z←VECTOR OF THE N+M-1 COEFFICIENTS,
  IN DESCENDING ORDER, OF THE RESULTING
  POLYNOMIAL.
```

CODE

```
      ∇ Z←A POLY B;N
 [1] ⍝COMPUTATION OF THE COEFFICIENTS OF
     ⍝PRODUCT OF TWO POLYNOMIALS
 [2] ⍝LAST UPDATE: 2-25-79
 [3]  N←⍴A
 [4]  Z←+/[1](-⁻1+⍳N)⌽(A∘.×B),(N,N-1)⍴0
      ∇
```

Function Name: POLYC

EXTERNAL DOCUMENTATION

PURPOSE

To multiply two complex polynomials.

ALGORITHM

POLYC uses a simple loop to compute the complex coefficients, in descending order, of the product of two complex polynomials P and Q, both of whose coefficients are also given in descending order. The loop computes the coefficients successively from highest to lowest by rotation, multiplication, and summation:

$$P = p_{m,1} \ p_{m-1,1} \ \cdots \ p_{01} \qquad Q = q_{n,1} \ q_{n-1,1} \ \cdots \ q_{01}$$
$$ = p_{m,2} \ p_{m-1,2} \ \cdots \ p_{02}, \qquad = q_{n,2} \ q_{n-1,2} \ \cdots \ q_{02}$$

$$P1 = 0 \ \ 0 \ \ 0 \ \cdots \ 0 \ \ 0 \ \ p_{m,1} \ p_{m-1,1} \ \cdots \ p_{01}$$
$$ = 0 \ \ 0 \ \ 0 \ \cdots \ 0 \ \ 0 \ \ p_{m,2} \ p_{m-1,2} \ \cdots \ p_{02}$$

$$Q1 = q_{11} \ \ q_{21} \ \cdots \ q_{n-1,1} \ \ q_{n,1} \ \cdots \ 0 \ \ \ 0 \ \ 0$$
$$ = q_{12} \ \ q_{22} \ \cdots \ q_{n-1,2} \ \ q_{n,2} \ \ \ 0 \ \cdots \ 0 \ \ 0$$

INTERNAL DOCUMENTATION

SYNTAX:R←P POLYC Q
PURPOSE:MULTIPLICATION OF TWO POLYNOMIALS
 WITH COMPLEX COEFFICIENTS.
INDEX ORIGIN: 1
AUTHOR: E. SILBERG
LAST UPDATE: 5/79
LEFT ARGUMENT: 2 BY M+1 ARRAY, M≥1, OF THE
 COMPLEX COEFFICIENTS(IN DESCENDING ORDER) OF
 THE FIRST POLYNOMIAL, P.
RIGHT ARGUMENT: 2 BY N+1 ARRAY, OF THE
 COEFFICIENTS OF THE SECOND POLYNOMIAL, Q.
RESULT:2 BY M+N+1 ARRAY, OF THE COMPLEX
 COEFFICIENTS(IN DESCENDING ORDER) OF THE
 PRODUCT OF P AND Q.

CODE

```
        ∇ R←P POLYC Q;M;N;P1;Q1;R;J;S
[1]   ⍝ PRODUCT OF TWO COMPLEX POLYNOMIALS.
[2]   ⍝ CALLS MULT.
[3]     P1←((2,N←1↓ρQ)ρ0),P
[4]     Q1←(⌽Q),(2,M←1↓ρP)ρ0
[5]     R←(2,S←⁻1+M+N)ρ0
[6]     J←1
[7]   LOOP:R[;J]←+/P1 MULT(-J)⌽Q1
[8]     →(S≥J←J+1)/LOOP
        ∇
```

Function Name: POLFACT

EXTERNAL DOCUMENTATION

PURPOSE

To factor a mod-2 polynomial into the product of two distinct mod-2 polynomials. Recall that a mod-2 polynomial is a polynomial whose coefficients are mod 2.

ALGORITHM

This program uses Berlekamp's method of factoring mod-p polynomials. Let $f(x)$ of degree d be the mod-2 polynomial to be factored. If we can find a mod-2 polynomial $g(x)$ with degree >1 and $<d$ such that $f(x)$ divides $g(x)$ $(g(x) - 1)$, then by Fermat:

$$f(x) = \text{g.c.d.} \ (f(x), g(x)) \ \text{g.c.d.} \ (f(x), g(x) - 1) \qquad (1)$$
$$(\text{g.c.d.} = \text{greatest common denominator})$$

Thus if we can find a $g(x)$ as described above we can obtain a factor of $f(x)$ by finding g.c.d.$(f(x), g(x))$ and then divide this factor into $f(x)$ to obtain the desired factorization.

Let $g(x) = b_0 + b_1 x + \cdots + b_{d-1} x^{d-1}$. To see if $f(x)$ divides $g(x)(g(x) - 1)$ look at

$$g(x)^2 = b_0 + b_1 x^2 + \cdots + b_{d-1} x^{2(d-1)} \qquad (2)$$

Divide $f(x)$ into x^{2i} $(i = 0, 1, 2, \ldots, d \ - 1)$ to get

$$x^{2i} = f(x)g(x) + r_i(x) \qquad (3)$$

where

$$r_i(x) = r_{i,0} + r_{i,1} x + \cdots + r_{i,d-1} x^{d-1}$$

($r(x)$ is the remainder polynomial). Upon substituting (3) into (2) and subtracting we have

$$g(x)(g(x) - 1) = b_0 r_0(x) + b_1 r_1(x) + \cdots + b_{d-1} r_{d-1}(x)$$
$$- (b_0 + b_1 x + \cdots + b_{d-1} x^{d-1}) + (\text{multiple of } f(x)) \quad (4)$$

Since the polynomial on the right side of (4) (excluding the multiple of $f(x)$) is of degree $<d - 1$, $f(x)$ divides it if and only if it is $= 0$ (call this polynomial $g(x)$). This condition translates into the matrix equation

$$(b_0, \ldots, b_{d-1})(Q - I) = (0, \ldots, 0) \qquad (5)$$

with

$$Q = \begin{pmatrix} r_{0,0}, & \cdots & , r_{0,d-1} \\ \vdots & \vdots & \vdots \\ r_{d-1,0}, & \cdots & , r_{d-1,d-1} \end{pmatrix}$$

I = the appropriate size identity matrix

The d unknowns $(b_0, \ldots, b_{d-1})$ are the coefficients of $g(x)$. Once these coefficients are determined and hence $g(x)$ determined we find g.c.d.$(f(x), g(x))$ by applying Euclid's algorithm for the g.c.d. of two natural numbers to polynomials. Call functions: DIVIDE, ECHELON, REMAIN.

WARNING

In Eq. (5) we are really finding vectors in the nullspace of $(Q - I)$. If the polynomial to be factored is irreducible or the result of a mod-2 polynomial raised to some power, the dimension of this nullspace will be 1 and the only solution to (5) is the trivial one $b = (a, 0, \ldots, 0)$. In this case the program outputs the message 'Irreducible or Power Thereof'. Thus if $f(x) =$ mod-2 polynomial to some power this program will not factor $f(x)$.

NOTE

Given that $f(x)$ is factorable, it can be factored further by using each of the outputted factors as input, thus factoring the factors.

REFERENCES

Childs, L. (1979). *In* ''A Concrete Introduction to Higher Algebra,'' Euclid's algorithm applied to polynomials, p. 132, and Berlekamp's Method, pp. 185–192. Springer-Verlag, Berlin, New York.

INTERNAL DOCUMENTATION

```
SYNTAX:  Z←POLFACT N
PURPOSE:  TO FACTOR A MOD 2 POLYNOMIAL INTO
 THE PRODUCT OF TWO DISTINCT MOD 2
 POLYNOMIALS.
INDEX ORIGIN:  1
AUTHOR:  B. BUKIETT, MODIFIED BY R. COHEN
```

LAST UPDATE: 10/80
RIGHT ARGUMENT: THE VECTOR OF COEFFICIENTS OF
THE POLYNOMIAL TO BE FACTORED (COEFFICIENTS
ARE IN DESCENDING ORDER WITH RESPECT TO THE
POWERS OF X).
RESULT: THE TWO DISTINCT FACTORS OF THE MOD 2
POLYNOMIAL IN THE SAME VECTOR FORM DESCRIBED
ABOVE.
NOTE: IF THE POLYNOMIAL TO BE FACTORED IS
IRREDUCIBLE OR THE RESULT OF A MOD 2
POLYNOMIAL RAISED TO SOME POWER, THE
MESSAGE,'IRREDUCIBLE OR POWER THEREOF'IS THE
ONLY OUTPUT.

CODE FOR POLFAC

```
       ∇ Z←POLFACT N;A;I;J;K;L;M;S;ANS;ANSW;
         NUM;ZAP;C;H;P;Q
 [1]    ANS←(A←(C←ρNUM←N)-1)ρ1
 [2]    Q←I-I←(2ρA)ρ1,Aρ0
 [3]    J←1
 [4]    BEGIN:Q[(J+1)÷2;J]←1
 [5]    →((J←J+2)<C)/BEGIN
 [6]    HARD:P←1,(J-1)ρ0
 [7]    Q[(J+1)÷2;ιρP]←P←φP←P REMAIN N
 [8]    →((J←J+2)≤¯1+2×A)/HARD
 [9]    Q←ECHELON Q←|Q-I
 [10]   J←1
 [11]   BACK:→((+/Q[;J])≠1)/ON
 [12]   ANS[J]←0
 [13]   ON:→((J←J+1)≤A)/BACK
 [14]   ANSW←ANS
 [15]   →((H←¯1+2*K←+/ANS)=1)/MSG
 [16]   L←Kρ2
 [17]   UP:M←LTH
 [18]   ANSW[S←ANS/ιA]←φM
 [19]   →((+/(2|ANSW+.×Q))=0)/EUCLID
 [20]   →((H←H-2)≥2)/UP
 [21]   EUCLID:ANSW←((ANSWι1)-1)↓ANSW←φANSW
 [22]   LI:ZAP←N REMAIN ANSW
 [23]   N←ANSW
 [24]   ANSW←ZAP
```

```
[25]    →((+/ZAP)≠0)/LI
[26]    □←N
[27]    Z←NUM DIVIDE N
[28]    →0
[29] MSG:'IRREDUCIBLE OR POWER THEREOF'
        ∇
```

CODE FOR DIVIDE

```
        ∇ Z←N DIVIDE ANSWER;LENG;QUOTIENT;DROP
[1]  ⋒CALLED BY POLFACT TO DIVIDE F(X) BY THE
     ⋒FIRST FACTOR FOUND IN ORDER TO FIND THE
[2]  ⋒SECOND FACTOR. THE LEFT ARGUMENT IS
[3]  ⋒THE VECTOR OF COEFFICIENTS OF F(X) AND
     ⋒THE RIGHT ARGUMENT
[4]  ⋒IS THE VECTOR OF COEFFICIENTS OF THE
     ⋒FIRST FACTOR.
[5]  ⋒(COEFFICIENTS ARE IN DESCENDING ORDER)
[6]     QUOTIENT←ρLENG←1+(ρN)-ρANSWER
[7]  TOP:QUOTIENT←QUOTIENT,1
[8]     N←(|(((ρANSWER)↑N)-ANSWER),(⁻1×((ρN)-
        ρANSWER))↑N
[9]     DROP←(Nι1)-2
[10]    N←((Nι1)-1)↓N
[11]    QUOTIENT←QUOTIENT,DROPρ0
[12]    →((+/N)≠0)/TOP
[13]    Z←1↓(LENG+1)ρQUOTIENT
        ∇
```

CODE FOR ECHELON

```
        ∇ Z←ECHELON Q;G;B;K
[1]  ⋒CALLED BY POLFACT TO REDUCE |Q-I| TO A
     ⋒LOWER TRIANGULAR MATRIX.
[2]  ⋒THE RIGHT ARGUMENT IS |Q-I| (| |
     ⋒DENOTES ABSOLUTE VALUE).
[3]     B←1
[4]  BACK:→(((G←(((B-1)↓Q[B;])ι1)+B-1)=C)/NEXT
[5]     →(G=B)/GOON
[6]     H←Q[;B]
```

```
[7]    Q[;B]←Q[;G]
[8]    Q[;G]←H
[9]    GOON:K←1
[10]   UP:→((K=B)∨Q[B;K]≠1)/DOWN
[11]   Q[;K]←|Q[;K]-Q[;B]
[12]   DOWN:→((K←K+1)≤A)/UP
[13]   NEXT:→((B←B+1)≤A)/BACK
[14]   Z←Q
       ∇
```

CODE FOR REMAIN

```
       ∇ Z←P REMAIN N;D;E;F
[1]    ACALLED BY POLFACT TO FIND THE ROWS OF
       ATHE REMAINDER MATRIX
[2]    AWHICH CORRESPOND TO POWERS OF X ≥DEGREE
       AOF F(X). THE LEFT
[3]    AARGUMENT IS THE VECTOR OF COEFFICIENTS
       AOF X TO SOME POWER ≥DEGREE OF F(X).
[4]    ATHE RIGHT ARGUMENT IS THE VECTOR OF
       ACOEFFICIENTS OF F(X).
[5]    A(COEFFICIENTS ARE IN DESCENDING ORDER)
[6]    TOP:D←(ρN)↑P
[7]    E←|D-N
[8]    P←E,(ρE)↓P
[9]    F←Pι1
[10]   P←(F-1)↓P
[11]   →((ρP)≥(ρN))/TOP
[12]   Z←P
       ∇
```

Function Name: POWER

EXTERNAL DOCUMENTATION

PURPOSE

Computes powers *componentwise* of arbitrary complex arrays; see external documentation for MATMULT.

INTERNAL DOCUMENTATION

SYNTAX: Z←A POWER B
PURPOSE: EXPONENTIATION OF COMPLEX NUMBERS
INDEX ORIGIN: 1
AUTHORS: A. ERDAL, U. GRENANDER
LAST UPDATE: 11/80
LEFT ARGUMENT: A IS A COMPLEX MATRIX (I.E., A
  SET OF COMPLEX NUMBERS, WHERE THE REAL AND
  IMAGINARY PARTS ARE SEPARATED ALONG THE
  FIRST DIMENSION.)
RIGHT ARGUMENT: B IS A SCALAR OR VECTOR
  CONTAINING THE POWER(S) TO WHICH THE COMPLEX
  NUMBERS ARE BEING RAISED.

RESULT: A COMPLEX MATRIX

CODE

```
        ∇ Z←A POWER N;RE;IM;TH;S;MAG;X;Y;ZERO
[1]     RE←⍕'A[1',S←(((ρρA)-1)ρ';'),']'
[2]     IM←⍕'A[2',S
[3]     MAG←((RE*2)+IM*2)*0.5
[4]     MAG←MAG+ZERO←MAG=0
[5]     TH←¯1○IM÷MAG
[6]     MAG←MAG*N
[7]     TH←N×(TH×RE≥0)+(RE<0)×(○1)-TH
[8]     X←(1-ZERO)×MAG×2○TH
[9]     Y←(1-ZERO)×MAG×1○TH
[10]    Z←(ρA)ρ0
[11]    Z←X,[0.5] Y
        ∇
```

Function Name: QUARTIC

EXTERNAL DOCUMENTATION

PURPOSE

To find roots of a fourth-order equation with real coefficients.

ALGORITHM

See Uspensky (1948, Chap. V).

NOTE

This function calls the function cubic.

INTERNAL DOCUMENTATION

```
SYNTAX: TOTL ← QUARTIC PN
PURPOSE: SOLVES QUARTIC EQUATION
INDEX ORIGIN: 1
AUTHOR: Y. LEVANON
LAST UPDATE: 1-18-72
RIGHT ARGUMENT: VECTOR OF COEFFICIENTS OF
 EQUATION STARTING WITH HIGHEST ORDER TERMS
RESULT: MATRIX WITH FOUR ROWS AND TWO
 COLUMNS. ONE ROW FOR EACH ROOT. REAL PARTS
 IN FIRST COLUMN, IMAGINARY PARTS IN SECOND
 COLUMN.
```

CODE

```
        ∇ TOTL←QUARTIC PN;D;DIS0;D2;E;E2;KK;KM;
          KS;M;N;P;PC;R;RR;RV;R2;TETA;VCTR;Y
[1]   ⍝ ROOTS OF A QUARTIC POLYNOMIAL.
[2]   ⍝ LAST MODIFIED: 1/18/72
[3]   ⍝ AUTHOR: Y. LEVANON
[4]     P←PN÷PN[1]
[5]     RR← 0 0
[6]     PC←1,(-P[3]),((P[2]×P[4])-4×P[5]),
          (4×P[3]×P[5])-(P[5]×P[2]*2)+P[4]*2
[7]     Y←(CUBIC PC)[1;1]
[8]     KS←(0.75×P[2]*2)-2×P[3]
[9]     →(0≠R2←(P[2]×P[2]÷4)+Y-P[3])/LE
[10]    R←0
[11]    →(0>DIS0←(Y*2)-4×P[5])/LD
[12]    →(0>E2←KS-2×DIS0*0.5)/LA
[13]    E←(E2*0.5),0
[14]    D←((KS+2×DIS0*0.5)*0.5),0
[15]    →FIN
[16] LA:→(0>D2←KS+2×DIS0*0.5)/LB
[17]    D←(D2*0.5),0
[18]    →LC
[19] LB:D←0,(-D2)*0.5
```

```
[20]   LC:E←0,(-E2)*0.5
[21]    →FIN
[22]   LD:TETA←(‾3○((-4×DIS0)*0.5)÷KS)÷2
[23]    RV←((KS*2)-4×DIS0)*0.25
[24]   LQ:M←RV×(2○TETA)
[25]    N←RV×(1○TETA)
[26]    D←M,N
[27]    E←M,-N
[28]    →FIN
[29]   LE:R←(|R2)*0.5
[30]    KM←KS-R2
[31]    KK←((P[2]×P[3])-2×P[4])+(P[2]*3)÷4)÷R
[32]    →(0>R2)/LP
[33]    →(0>D2←KM+KK)/LG
[34]    D←(D2*0.5),0
[35]   LK:→(0>E2←KM-KK)/LF
[36]    E←(E2*0.5),
[37]    →FIN
[38]   LG:D←0,(-D2)*0.5
[39]    →LK
[40]   LF:E←0,(-E2)*0.5
[41]    →FIN
[42]   LP:RV←((KM*2)+KK*2)*0.25
[43]    TETA←(‾3○(-KK÷KM))÷2
[44]    RR←(-R),R
[45]    →LQ
[46]   FIN:VCTR←RR+R,0
[47]    TOTL←(4 2)ρ(((VCTR+D),(VCTR-D),
        (-VCTR-E),-VCTR+E)÷2)-8ρ(P[2]÷4),0
        ∇
```

Function Name: REVERSE1

EXTERNAL DOCUMENTATION

PURPOSE

To reverse the effect of CONVERT; see external documentation of CONVERT.

INTERNAL DOCUMENTATION

SYNTAX: *Z←REVERSE*1 *N*
PURPOSE: *TO REVERSE THE CONVERSION OF A*
  SYMMETRIC MATRIX INTO A VECTOR
INDEX ORIGIN: 1
AUTHOR: *J. ROSS*
LAST UPDATE: 7/9/79
RIGHT ARGUMENT: *N, THE FIRST COMPONENT OF THE*
  SHAPE OF THE DESIRED SYMMETRIC MATRIX
RESULT: *Z, AN N×N SYMMETRIC MATRIX. THE*
  ELEMENTS OF THE LOWER HALF (DIAGONAL AND
  BELOW) OF Z STRUNG OUT ROW BY ROW WOULD FORM
  THE VECTOR ιV, WHERE V=(((2N+1)*2)-1)÷8
NOTE: *THIS PROGRAM CAN BE USED TO CONVERT AN*
  ARBITRARY VECTOR (OF APPROPRIATE SHAPE) TO A
  SYMMETRIC MATRIX WITH THE COMMAND Z←AV
  *[REVERSE*1 *N], WHERE N IS AS ABOVE AND AV IS*
  THE VECTOR.

CODE

```
     ∇ Z←REVERSE1 N;M;V
[1]  Z←((V∘.≠V)×M)+⍉M←(V∘.≥V)×(0.5×V×V-1)∘.
     +V←ιN
     ∇
```

Function Name: REVERSE2

EXTERNAL DOCUMENTATION

PURPOSE

To invert the effect of CONVERT; see external documentation of
CONVERT.

INTERNAL DOCUMENTATION

SYNTAX: *∇Z←REVERSE*2 *A∇*
PURPOSE: *TO REVERSE THE CONVERSION OF A*
  SYMMETRIC MATRIX INTO A VECTOR

```
INDEX ORIGIN: 1
AUTHORS: C. HUCKABA AND J. ROSS
LAST UPDATE: 6/20/79
RIGHT ARGUMENT: A VECTOR WHICH CONTAINS THE
 LOWER HALF OF THE SYMMETRIC MATRIX STRUNG
 OUT ROW BY ROW, THE RESULT OF 'CONVERT' THE
 NUMBER OF ELEMENTS IN THIS VECTOR MUST BE A
 MEMBER OF THE ARITHMETIC SERIES: 1,3,6,10,
 15.....
RESULT: AN N×N SYMMETRIC MATRIX
```

CODE

```
     ∇ Z←REVERSE2 A;F;N
[1] ⍝ REVERSES THE CONVERSION OF A SYMMETRIC
    ⍝ MATRIX TO A VECTOR
[2] ⍝ LAST UPDATE: 6/20/79
[3]  F←1+(÷¯2)×V∘.=V←⍳N←(¯1+(1+8×⍴A)*÷2)÷2
[4]  Z←F×(⍉M)+M←(N,N)⍴( ,V∘.≥V)\A
     ∇
```

Function Names: SYNDIV and SYNDIV1

EXTERNAL DOCUMENTATION

PURPOSE

To perform synthetic division of a polynomial

$$a_0 x^n + a_1 x^{n-1} + \cdots + a_n$$

by a root for the general case of complex coefficients and a complex root (SYNDIV) and for the special case of real coefficients and a real root (SYNDIV1).

The user must provide V, the complex matrix or the vector of coefficients, $a_0, a_1, \ldots, a_n$, in this order. He must also provide RT, the complex or real root. Both programs successively update the coefficients according to $V(I)$ gets $V(I) + \text{RT} \times V(I - 1)$.

The coefficients $a'_0, a'_1, \ldots, a'_{n-1}$ of the resulting polynomial

$$a'_0 x^{n-1} + a'_1 x^{n-2} + \cdots + a'_{n-1}$$

are then placed in Z, a complex matrix or vector. The remainder of the division is also printed.

EXAMPLE

$$
\begin{array}{cccc}
V & 1 & -2 & 0 & 1 \\
 & 0 & -2 & 3 & -1;
\end{array}
\qquad
\begin{array}{ccc}
RT & 1 & 5
\end{array}
$$

$$
\begin{array}{lcccc}
V \ SYNDIV \ RT & 1 & -1 & -16 & \text{remainder: } -20 \\
 & 0 & 3 & 1 & -80
\end{array}
$$

$(x^3 + (-2 - 2i)x^2 + 3ix + (1 - i))/(x - (1 + 5i))$
$= x^2 + (-1 + 3i)x + (-16 + i),$ remainder: $-20 - 80i$

INTERNAL DOCUMENTATION FOR SYNDIV

```
SYNTAX: ∇Z←V SYNDIV RT∇
PURPOSE: PERFORMS SYNTHETIC DIVISION OF A
 COMPLEX POLYNOMIAL BY A COMPLEX ROOT.
INDEX ORIGIN: 1
AUTHORS: P. KRETZMER, B. BENNETT
LAST UPDATE: 2-12-79
RIGHT ARGUMENT: RT: TWO ELEMENT VECTOR
 REPRESENTING THE COMPLEX ROOT.
LEFT ARGUMENT: V: COMPLEX MATRIX OF THE
 POLYNOMIAL COEFFICIENTS.
RESULT: Z: COMPLEX MATRIX OF THE COEFFICIENTS
 OF THE RESULTING POLYNOMIAL; FOLLOWED BY THE
 REMAINDER FROM THE DIVISION
NOTE: SEE ALSO EXTERNAL DOCUMENTATION FOR
 COMPLEX ARITHMETIC PACKAGE.
```

INTERNAL DOCUMENTATION FOR SYNDIV1

```
SYNTAX: Z←V SYNDIV1 RT
PURPOSE: PERFORMS SYNTHETIC DIVISION OF A
 REAL POLYNOMIAL BY A REAL ROOT.
INDEX ORIGIN: 1
AUTHORS: P. KRETZMER, B. BENNETT
LAST UPDATE: 6/29/79
LEFT ARGUMENT: V, A VECTOR OF THE POLYNOMIAL
 COEFFICIENTS
RIGHT ARGUMENT: RT, THE SCALAR ROOT
```

RESULT: Z, A VECTOR OF THE COEFFICIENTS OF THE RESULTING POLYNOMIAL; FOLLOWED BY THE REMAINDER FROM THE DIVISION

CODE FOR SYNDIV

```
      ∇ Z←V SYNDIV RT;N;I
[1]  ⍝SYNTHETIC DIVISION OF A COMPLEX
     ⍝POLYNOMIAL BY A COMPLEX ROOT
[2]  ⍝CALLS COMPLEX FUNCTIONS, ADD AND MULT
[3]  ⍝LAST UPDATE: 6/18/79
[4]   N←(¯1↑⍴V)-1
[5]   I←2
[6]  COEF:V[;I]←V[;I] ADD RT MULT V[;I-1]
[7]   →((I←I+1)≤N+1)/COEF
[8]   Z←(⍕(0 ¯1 ↓V)),(2 13 ⍴' REMAINDER:',
      14⍴' '),⍕(2 ¯1 ↑V)
      ∇
```

CODE FOR SYNDIV1

```
      ∇ Z←V SYNDIV1 RT;N;I
[1]   ⍝SYNTHETIC DIVISION OF A REAL POLYNOMIAL
      ⍝BY A REAL ROOT
[2]   ⍝LAST UPDATE: 6/18/79
[3]    N←(⍴V)-1
[4]    I←2
[5]   COEF:V[I]←V[I]+RT×V[I-1]
[6]    →((I←I+1)≤N+1)/COEF
[7]    Z←(⍕(¯1↓V)),' REMAINDER: ',⍕(¯1↑V)
      ∇
```

Function Name: TQL

EXTERNAL DOCUMENTATION

PURPOSE

To compute the eigenvalues and eigenvectors of a symmetric tridiagonal matrix by the QL algorithm.

If the tridiagonal matrix has arisen from the Householder reduction of a full symmetric matrix using HSHLDR, then TQL may be used to find the eigenvectors of the full matrix without first finding the eigenvectors of the tridiagonal matrix. For a discussion of the QL algorithm see the "Handbook for Automatic Computation," Vol. II.

ALGORITHM

The algorithm presented here is the $tg12$ procedure that was developed by H. Bowdler, R. S. Martin, C. Reinsch, and J. H. Wilkinson and is discussed in the "Handbook for Automatic Computation," Vol. II. It produces eigenvectors always orthogonal to working accuracy and almost orthonormal to working accuracy. The procedure can be used for arbitrary, real symmetric, tridiagonal matrices. The number of iterations required to find any one eigenvalue has been restricted to 30. According to the "Handbook for Automatic Computation," Vol. II, the computed eigenvalues have errors that are small compared with the standard operator norm of the tridiagonal matrix. The combination of HSHLDR and TQL is one of the most efficient methods when the complete eigensystem of a full symmetric matrix is required.

REFERENCE

Wilkinson, J. H., and Reinsch, C. (1971). "Handbook for Automatic Computation," Vol. II, pp. 227–240. Springer-Verlag, Berlin, New York.

INTERNAL DOCUMENTATION

```
SYNTAX: ∇Z←TQL T
PURPOSE: TO FIND THE EIGENVALUES AND
  EIGENVECTORS OF AN N×N SYMMETRIC TRIDIAGONAL
  MATRIX USING THE QL ALGORITHM. IF THE
  TRIDIAGONAL MATRIX HAS ARISEN FROM THE
  HOUSEHOLDER REDUCTION OF A FULL SYMMETRIC
  MATRIX USING HSHLDR, THEN TQL MAY BE USED TO
  FIND THE EIGENVECTORS OF THE FULL MATRIX.
INDEX ORIGIN: 1
AUTHOR: C. HUCKABA
LAST UPDATE: 6/22/79
RIGHT ARGUMENT: IF THE EIGENVALUES AND
  EIGENVECTORS OF A FULL SYMMETRIC MATRIX ARE
```

REQUIRED THEN T IS THE RESULT OF HSHLDR. T
IS AN (N+2)×N MATRIX WHOSE FIRST ROW
CONTAINS THE DIAGONAL ELEMENTS OF THE
TRIDIAGONAL MATRIX AND WHOSE SECOND ROW
CONTAINS A ZERO AND THE OFF-DIAGONAL
ELEMENTS. THE REMAINING N ROWS CONTAIN THE
PRODUCT OF THE HOUSEHOLDER TRANSFORMATION
MATRICES, THAT IS, THEY CONTAIN AN N×N
ORTHOGONAL MATRIX P SUCH THAT P'AP IS THE
TRIDIAGONAL MATRIX REPRESENTED BY THE FIRST
TWO ROWS OF T.IF THE EIGENVALUES AND
EIGENVECTORS OF A GIVEN SYMMETRIC
TRIDIAGONAL MATRIX ARE REQUIRED THEN AN
ARGUMENT OF THE ABOVE FORM SHOULD BE CREATED
WITH THE LAST N ROWS BEING AN N×N IDENTITY
MATRIX.
RESULT: FIRST, A LISTING OF THE NUMBER OF
ITERATIONS WHICH WERE REQUIRED THEN Z, AN
(N+1)×N MATRIX WITH THE FIRST ELEMENT IN
EACH OF THE N COLUMNS BEING AN EIGENVALUE
WHOSE CORRESPONDING NORMALIZED EIGENVECTOR
IS REPRESENTED BY THE REMAINING N COLUMN
ELEMENTS BELOW IT. THE EIGENVALUES ARE
LISTED IN ASCENDING ORDER. WHEN TQL IS USED
IN CONJUNCTION WITH HSHLDR, THE EIGENVECTORS
ARE THOSE OF THE ORIGINAL FULL SYMMETRIC
MATRIX.

CODE

```
        ∇  Z←TQL  T;EPS;D;E;V;N;NUMIT;B˙;F;L;J;H;
        M;G;P;R;SIGN;C;S;I;PERM
[1]  ⍝ FINDS THE EIGENVALUES AND EIGENVECTORS
     ⍝ OF A SYMMETRIC
[2]  ⍝ TRIDIAGONAL MATRIX BY USING THE QL
     ⍝ ALGORITHM
[3]  ⍝ LAST UPDATE: 6/22/79
[4]    EPS←⎕CT÷1-⎕CT
[5]    D←T[1;]
```

```
[6]    E←T[2;]
[7]    V←T[2+ι(N←¯1↑ρT);]
[8]    E←1⌽E
[9]    NUMIT←ιE[N]←B←F←0
[10]   L←1
[11] L1:J←0
[12]   H←EPS×(|D[L])+|E[L]
[13]   B←B⌈H
[14]   M←L
[15] L2:→((|E[M])≤B)/L3
[16]   →(N≥M←M+1)/L2
[17] L3:→(M=L)/L4
[18] L5:→(J=30)/LG
[19]   J←J+1
[20] ⍝ FORM SHIFT
[21]   G←D[L]
[22]   P←(D[L+1]-G)÷2×E[L]
[23]   R←((P*2)+1)*0.5
[24]   SIGN←(P≥0)-(P<0)
[25]   D[L]←E[L]÷P+SIGN×R
[26]   H←G-D[L]
[27]   D[L+ιN-L]←D[L+ιN-L]-H
[28]   F←F+H
[29] ⍝ QL TRANSFORMATION
[30]   P←D[M]
[31]   C←1
[32]   S←0
[33]   I←M-1
[34] L7:G←C×E[I]
[35]   H←C×P
[36]   →((|P)<|E[I])/L8
[37]   C←E[I]÷P
[38]   R←((C*2)+1)*0.5
[39]   E[I+1]←S×P×R
[40]   S←C÷R
[41]   C←1÷R
[42]   →L9
[43] L8:C←P÷E[I]
[44]   R←((C*2)+1)*0.5
[45]   E[I+1]←S×E[I]×R
[46]   S←1÷R
```

```
[47]    C←C÷R
[48]  L9:P←(C×D[I])-S×G
[49]    D[I+1]←H+S×(C×G)+S×D[I]
[50]    H←V[K←ιN;I+1]
[51]    V[K;I+1]←(S×V[K;I])+C×H
[52]    V[K;I]→(C×V[K;I])-S×H
[53]    →(L≤I←I-1)/L7
[54]    E[L]←S×P
[55]    D[L]←C×P
[56]    →((|E[L])>B)/L5
[57]  L4:D[L]←D[L]+F
[58]    NUMIT←NUMIT,J
[59]    →(N≥L←L+1)/L1
[60]    'NUMIT=',(▼NUMIT)
[61]  D←D[PERM←▲D]
[62]    V←V[;PERM]
[63]    Z←D,[1] V
[64]    →0
[65]  L6:'MAX NO. OF ITERATIONS EXCEEDED'
        ∇
```

CHAPTER 17

Analysis

We have collected some programs for computing tasks in analysis that have occurred often in our experiments. They include defined functions for Fourier analysis, approximation theory, and orthogonalization, and functions for solving differential and integral equations as well as for root finding programs.

We also give simple programs for numerical integration and differentiation. A fairly slow program is included for computing conformal mappings.

Function Name: BSPL

EXTERNAL DOCUMENTATION

PURPOSE

To evaluate, at a point x, all B-splines of order $1, 2, \ldots, k$ that are possible and not zero at x.

DEFINITIONS

Let $t = t_i$ be a nondecreasing sequence of numbers called knots or nodes. The ith (normalized) B-spline of order k for the knot sequence t, is denoted by $B_{i,k,t}$ and is defined by the rule

$$B_{i,k,t}(x) = (t_{i+k} - t_i)(t_i, \ldots, t_{i+k})(. - x)_+^{k-1}$$
$$\text{for all real numbers } x$$

The above placeholder notation is used to indicate that the kth divided difference of the function $(t - x)_+^{k-1}$ of the two variables t and x is to be taken by fixing x and considering $(t - x)_+^{k-1}$ as a function of t alone. The resulting number depends on the particular value of x chosen; therefore $B_{i,k,t}$ is a function of x as well as of t.

ALGORITHM

Rather than incorporating the definition of B-splines into coding, a recurrence relation is used. For a given knot sequence t,

$$B_{i,k}(x) = 0 \qquad \text{for} \quad x \text{ not in } (t_i, t_{i+k})$$

where k is the order of the B-splines.

Beginning with

$$B_{i,1}(x) = \begin{cases} t_i < x < t_{i+1} \\ 0 \qquad \text{otherwise} \end{cases}$$

the value of B at x is then calculated by

$$B_{i,k}(x) = \frac{x - t_i}{t_{i+k-1} - t_i} B_{i,k-1}(x) + \frac{t_{i+k} - x}{t_{i+k} - t_{i+1}} B_{i+1,k-1}(x)$$

For some x such that x is in t_i, t_{i+1}, the values of all B-splines not automatically zero at x fit into a triangular array as follows (note: each $B_{i,k}$ is at value x):

$$
\begin{array}{ccccccc}
& & & 0 & B_{i,1} & 0 & \\
& & 0 & B_{i-1,2} & B_{i,2} & 0 & \\
& 0 & B_{i-2,3} & B_{i-1,3} & B_{i,3} & 0 & \\
& & & \vdots & & & \\
0 & B_{i-k+1,k} & B_{i-k+2,k} & \cdots & B_{i-1,k} & B_{i,k} & 0
\end{array}
$$

The bordering zeros are to emphasize the fact that all other B-splines (of order k or less) not mentioned in the table vanish at x.

WARNING

Double, triple, etc., knots are acceptable as long as they are not in the interval one to the right of the interval containing x.

REFERENCE

Deboor, C. (1978). "A Practical Guide to Splines," Springer-Verlag, New York.

INTERNAL DOCUMENTATION

```
SYNTAX: X BSPL T
PURPOSE: TO EVALUATE, AT A POINT X, ALL
  B-SPLINES OF ORDER 1,2,...,K THAT ARE
  POSSIBLY NOT ZERO AT X.
```

```
INDEX ORIGIN: 1
AUTHOR: P. FORCHIONE
LAST UPDATE: 8/7/80
RIGHT ARGUMENT: THE SEQUENCE OF KNOTS OR
 NODES USED IN FORMING THE B-SPLINES.
LEFT ARGUMENT: THE POINT X AT WHICH B-SPLINES
 ARE TO BE EVALUATED.
RESULTS: 1)  THE TRIANGULAR ARRAY:
```

$$0 \quad B_{I,1} \quad 0$$

$$0 \quad B_{I-1,2} \quad , \quad B_{I,2} \quad 0$$

$$0 \quad B_{I-2,3} \quad , \quad B_{I-1,3} \quad , \quad B_{I,3} \quad , \quad 0$$

$$\ldots\ldots\ldots\ldots\ldots\ldots\ldots\ldots\ldots\ldots\ldots\ldots\ldots\ldots$$

$$0 \quad B_{I-K+1,K} \quad , \quad B_{I-K+2,K} \quad , \quad \ldots, \quad B_{I-1,K} \quad , \quad B_{I,K} \quad , \quad 0$$

```
WHERE I=INDEX OF THE KNOT INTERVAL X FALLS IN
AND K=THE HIGHEST ORDER B-SPLINES USED.
        2) THE VALUE AT X OF B
```
$$_{I,K}$$
```
NOTE. THE PROGRAM FINDS THE LARGEST VALUE OF
K WHICH WILL GENERATE THE FULL TRIANGULAR
ARRAY ABOVE. EXAMPLE 1) X=5 T=1,2,3,4,5,6.
THEN I=5 AND K=1. EXAMPLE 2) X=3 T=1,2,3,4,
5,6,7. THEN I=3 AND K=3. IF, IN EXAMPLE 2)
THE USER DESIRED B-SPLINES OF UP TO ORDER
4, A DOUBLE KNOT COULD BE PLACED AT 1,2 OR
3. IF IN EXAMPLE 1 THE USER DESIRED
B-SPLINES OF ORDER > 1 NOTHING COULD BE
DONE BECAUSE DOUBLE KNOTS ARE NOT ALLOWED
IN THE INTERVAL ONE TO THE RIGHT OF THE
INTERVAL CONTAINING X (THE ALGORITHM BREAKS
DOWN IN THIS CASE). HOWEVER, GIVEN X=5
T=1,2,3,4,5,6,7 THEN I=5 AND K=2 AND THE
USER COULD PLACE MULTIPLE KNOTS AT 7 TO
OBTAIN B-SPLINES OF ORDER >2.
```

CODE

```
       ∇ X BSPL T;I;K;X;J;R;M;B
[1]    T←T[⍋T]
```

```
[2]     K←⌊/I,(ρT)-I←+/X≥T
[3]     →E1×ι(X<1↑T)∨(X>¯1↑T)
[4]     →(K=1+0×B← 0 1 0)/OUT
[5]     →(T[I+1]=T[I+2])/E2
[6]     □←B
[7]     J←2
[8]     R←(I-K)↓ιI
[9]     LP:□←B←1⌽0,0,((¯1↓B)×(X-T[M])÷T[M+J-1]
        -T[M])+(1↓B)×(T[M+J]-X)÷T[M+J]-T[1+M←
        (-J)↑R]
[10]    →LP×ιK≥J←J+1
[11]    ' '
[12]    OUT:'B(',(⍕I,K,X),') = ',⍕Z←B[K+1]
[13]    →0
[14]    E1:'X-VALUE IS NOT IN THE INTERVAL
        [T(1),T(N)]. CHOOSE A VALID X-VALUE'
[15]    'OR SPECIFY A NEW KNOT SEQUENCE.'
[16]    →0
[17]    E2:'DOUBLE KNOTS ARE NOT ALLOWED IN THE
        INTERVAL ONE TO THE RIGHT'
[18]    'OF THE INTERVAL CONTAINING X. CHOOSE A
        DIFFERENT X-VALUE OR A'
[19]    'DIFFERENT KNOT SEQUENCE.'
        ∇
```

Function Name: CONFORMAL

EXTERNAL DOCUMENTATION

PURPOSE

To compute an approximation of a conformal mapping from a given plane region R to a disk by the Rayleigh–Ritz method. (See Warschawski, pp. 233–236, for a discussion of this method.)

The problem is as follows: Let R be the interior of a plane region bounded by a closed rectifiable Jordan curve C. Let z_0 be a particular point in R. Find a function $w = f(z)$ that maps R conformally onto a disk, such that $f(z_0) = 0$ and $f'(z_0) = 1$.

It can be shown that there is a unique solution f to this problem and that f is fully determined by the solution of the following minimization problem:

$$\int_C |F(z)|^2 \, ds = \text{minimum, subject to } F(z_0) = 1$$

Specifically, the conformal mapping function $f(z)$ is related to $F(z)$ as follows: $F(z) = (f'(z)) * .5$, or, equivalently,

$$f(z) = \int_{z_0}^z F^2(p) \, dp$$

where the lower limit of integration is determined by the given condition $f(z_0) = 0$.

Now suppose that one restricts $F(z)$ to the set S_m of polynomials of degree $<$ or $= m$. It can be shown that the solution of the minimization problem over S_m is given by

$$F(z) = \sum_{j=0}^m b_j P_j(z)$$

where the $P_j(z)$ are orthonormal polynomials (i.e.,

$$\int_C P_j(z)\overline{P_k(z)} \, ds = \begin{cases} 1 & \text{if } j = k \\ 0 & \text{if } j \text{ is not } = k \end{cases}$$

and where

$$b_j = \frac{P_j(z_0)}{\sum_{j=0}^m P_j(z_0)\overline{P_j(z_0)}} = \frac{P_j(z_0)}{\sum_{j=0}^m |P_j(z_0)|^2}$$

Thus, the minimization problem, and therefore, the conformal mapping problem, is reduced to the determination of the orthonormal polynomials

$$P_j(z), \quad j = 0, 1, 2, \ldots, m$$

For numerical purposes, the user must define the curve C by n points $z_1, z_2, \ldots, z_n$ (counterclockwise), which will be used to construct an approximation of $f(z)$.

The program CONFORMAL calls GRAMS, which constructs the orthonormal polynomials, using the z_ks to approximate the curve C in the integrals. (See the discussion of the program GRAMS.)

In computing the solution

$$F(z) = \sum_{j=0}^m b_j P_j(z)$$

CONFORMAL calculates the coefficients b_j by calling the program EVAL to evaluate all of the P_js at the point z_0. The solution is then reexpressed in terms of ordinary powers:

$$F(z) = \sum_{j=0}^{m} a_j z^j$$

Next, $G(z) = F^2(z)$ is computed by calling the program POLYC for multiplication of two complex polynomials:

$$G(z) = \sum_{j=0}^{2m} q_j z^j$$

Now the conformal mapping $f(z)$ can be obtained by a single integration:

$$f(z) = \int_{z_0}^{z} G(r)\, dr = \int_{z_0}^{z} \left(\sum_{j=0}^{2m} g_j r^j \right) dr$$

$$= \sum_{j=0}^{2m} \left(\int_{z_0}^{z} g_j r^j\, dr \right) = \sum_{j=0}^{2m} \frac{g_j}{(j+1)} r^{j+1} \quad \text{from } z_0 \text{ to } z$$

$$= \sum_{j=0}^{2m} \frac{g_j}{(j+1)} z^{j+1} - \sum_{j=0}^{2m} \frac{g_j}{(j+1)} z_0^{j+1}$$

$f(z)$ is the approximate conformal mapping from the region R to a disk. The explicit result of the program CONFORMAL is a 2 by $2m + 2$ array of the complex coefficients, in descending order, of f. After using CONFORMAL, one may evaluate f at any point z, inside or on the curve C, by using the program EVAL2, which evaluates a complex polynomial at one or several points. (See the discussion of EVAL2.)

WARNING

The program CONFORMAL should be used only for small-scale experimental work, since CPU time can become very large, especially if one or both of m ($=$ degree of F) and n ($=$ no. of boundary points) are large. For a particular curve C, the optimal combination of m and n may not be obvious, and some experimentation may be needed.

	1	2	3	4
$m = $ degree of $F(z)$	3	6	3	6
$n = $ no. of boundary points	8	8	32	32
Radius along x axis of image circle	3.5176	3.8068	3.3397	3.3575
Radius along y axis of image circle	3.2229	3.1757	3.3168	3.3627
CPU time (sec)	8.05	18.42	14.61	30.13
Approx. % error	10	18	1	0.15

For the ellipse (x,y) = (4 cos t, 3 sin t), CPU times and levels of accuracy for some choices of m and n are shown in the accompanying tabulation.

REFERENCES

1. On the Rayleigh–Ritz method and other methods:

Warschawski, S. E. (1956). Recent results in numerical methods of conformal mapping, "Proceedings of Symposia in Applied Mathematics," (John H. Curtiss, ed.), McGraw-Hill, New York: Vol. 6, pp. 219–250.

2. Background and other methods:

Andersen, C. *et al.* (1962). *In* "Conformal Mapping, Selected Numerical Methods" (Christian Gram, ed.), pp. 114–261. Regnecentralen, Copenhagen.

INTERNAL DOCUMENTATION

```
SYNTAX:F←M CONFORMAL Z
PURPOSE:CONFORMAL MAPPING FROM A PLANE REGION
 TO A DISC BY THE RAYLEIGH-RITZ METHOD.
INDEX ORIGIN: 1
AUTHOR: E. SILBERG
LAST UPDATE: 5/79
RIGHT ARGUMENT: (2×(N+1) MATRIX, N≥3)
 Z[;1] = ZO = POINT TO BE MAPPED TO THE
 ORIGIN.
 Z[;K]; K=2,3,...,N+1 IS A SET OF N POINTS,
 DEFINING THE BOUNDARY CURVE (COUNTERCLOCK-
 WISE).
LEFT ARGUMENT: M = (INTEGER≥1)
 =DEGREE OF THE HIGHEST-ORDER POLYNOMIAL IN
 THE ORTHOGONALIZATION PROCEDURE. (THE DEGREE
 OF THE CONFORMAL MAPPING FUNCTION WILL BE
 1+(2×M).)
RESULT:F IS AN ARRAY OF DIMENSION 2 BY(2×M)+2
 OF THE COMPLEX COEFFICIENTS, IN DESCENDING
 ORDER, OF THE POLYNOMIAL CONFORMAL MAPPING
 FUNCTION F.
```

CODE

```
     ∇ F←M CONFORMAL Z;Z0;CR;E;B;A;G;R
[1]  ⍝ CONFORMAL MAPPING FROM A PLANE REGION
     ⍝ TO A DISC.
[2]  ⍝ CALLS EVAL,EVAL2, GRAMS,MAGNITUDE,
     ⍝ DIVBY,MATMULT,POLYC.
[3]   E←(CR←⌽M GRAMS Z← 0 1 ↓Z) EVAL Z0←Z[;1]
[4]   B←((2,(ρE)[2])ρE[1;],-E[2;]) DIVBY(+/
      (MAGNITUDE E)*2),0
[5]   G←A POLYC A←(2,(ρA)[3])ρA←((2,1,(ρB)[2])
      ρB) MATMULT CR
[6]   F←(G DIVBY(2,R)ρ(⌽⍳R),(R←1↓ρG)ρ0),0
[7]   F[;R+1]←-F EVAL2 Z0
     ∇
```

Function Names: COSFILON and SINFILON

EXTERNAL DOCUMENTATION

PURPOSE

To approximate integrals of the form

$$\int_a^b f(x)\ \cos(tx)\ dx$$

and

$$\int_a^b f(x)\ \sin(tx)\ dx \qquad \text{by Filon's method}$$

The name of the user-defined function f should be supplied as one of the parameters of the program. In addition, the user should include the lower and upper bounds of integration, the number n (one-half the number of subintervals), and the t values for which the integral is to be evaluated.

ALGORITHM

(1) Divide the interval (a,b) into $2n$ equal parts, each of length

$$h = (b - a)/2n$$

(2) Perform the following computations:

(a)   For COSFILON, compute $C1$ and C, where

$C1$ = the sum of all even ordinates of the curve $y = f(x)\cos(tx)$ (i.e., $y_2 + y_4 + \cdots + y_{2n}$);
C = the sum of all the odd ordinates of the curve $y = f(x)\cos(tx)$ between a and b, inclusive, less half the first and the last ordinates.

(b)   For SINFILON, compute $S1$ and S, where

$S1$ = the sum of all even ordinates of the curve $y = f(x)\sin(tx)$;
S = the sum of all the odd ordinates of the curve $y = f(x)\sin(tx)$ between a and b, inclusive, less half the first and the last ordinates.

(3)   Compute the constants A, B, G for each $p = th$, where

$$A(p) = 1/p + (\sin 2p)/2p^2 - (2\sin^2 p)/p^3$$
$$B(p) = 2((1 + \cos^2 p)/p^2 - (\sin 2p)/p^3)$$
$$G(p) = 4((\sin p)/p^3 - (\cos p)/p^2)$$

making special provision for p small, as noted in the references.

(4)   Then apply the formula to find the desired approximation:

(a)   COSFILON:

$$z = h(A(p)(f(b)\sin tb - f(a)\sin ta) + B(p)C + G(p)C1)$$

(b)   SINFILON:

$$z = h(A(p)(f(b)\cos tb - f(a)\cos ta) + B(p)S + G(p)S1)$$

WARNING

For t large and n small, the approximation will be somewhat inaccurate, as the error is of the 4th order. Also, no t value is allowed to be zero.

REFERENCES

Tranter, C. J. (1971). "Integral Transforms in Mathematical Physics," pp. 61–71. Chapman and Hall, London.
Abramowitz, M., and Stegun, I. A. (eds.) (1964). "Handbook of Mathematical Functions," National Bureau of Standards Applied Mathematics Series, Vol. 55, pp. 890–91. Washington, D.C.

INTERNAL DOCUMENTATION FOR COSFILON

```
SYNTAX: Z←FN COSFILON VAL
PURPOSE: TO APPROXIMATE THE INTEGRAL OF F(X)
 COS(TX) BY FILON'S METHOD
INDEX ORIGIN: 1
AUTHOR: G. PODOROWSKY
LAST UPDATE: 11/26/79
LEFT ARGUMENT: FN--A CHARACTER STRING
 FN←NAME OF USER-DEFINED FUNCTION WHICH
 ACCEPTS INPUT IN VECTOR FORM
 (X[1],X[2],...,X[2N]) AND RETURNS AS AN
 EXPLICIT RESULT THE VALUES F(X[1]),F(X[2]),
 ...,F(X[2N]) IN VECTOR FORM.
RIGHT ARGUMENT: VAL--A VECTOR
 VAL[1 2]=LOWER AND UPPER BOUNDS OF
 INTERGRATION,RESPECTIVELY
 VAL[3]=N, WHERE THE NUMBER OF SUBINTERVALS
 EQUALS 2×N
VAL[4...M+4]=THE M CONSTANTS (T) FOR WHICH
 INTEGRAL IS BEING COMPUTED
RESULT: Z--A VECTOR OF LENGTH M CONTAINING
 THE INTEGRAL APPROXIMATIONS, ONE FOR EACH
 CONSTANT
```

CODE FOR COSFILON

```
      ∇ Z←FN COSFILON VAL;A;B;C;C1;G;H;I;J;M;N;
       P;P2;P3;T;V1;V2;X
[1]   ⍝ FILON'S METHOD FOR APPROXIMATION OF
      ⍝ INTEGRAL OF F(X)COS(TX)
[2]   ⍝ LAST UPDATE: 11/26/79
[3]    T←3↓VAL
[4]    X←VAL[1]+(0,⍳2×N)×H←(VAL[2]-VAL[1]÷2×N←
       VAL[3]
[5]    V1←((⍴T),⍴X)⍴⍎FN,' X'
[6]    V2←2○T∘.×X
[7]    C1←+/[2]  V1[;I]×V2[;I←(0=2|I)/I←⍳2×N]
[8]    C←(+/[2]  V1[;J]×V2[;J←(1=2|J)/J←⍳2×N+1])
       -0.5×((V1[;1+2×N]×V2[;1+2×N])+V1[;1]
       ×V2[;1])
```

```
[9]    A←(÷P)+((1○2×P)÷(2×P2←P×P))-((2×(1○P)×
       (1○P))÷P3←(P←T×H)*3)
[10]   B←2×(((1+(2○P)×(2○P))÷P2)-(1○2×P)÷P3)
[11]   G←4×(((1○P)÷P3)-(2○P)÷P2)
[12]   →(0≠ρM←(0.001≥P)/ιρP)/PSMALL
[13]   COMPUTE:Z←H×(A×(V1[;1+2×N]×1○T×X[1+2×N])
       -V1[;1]×1○T×X[1])+(B×C)+G×C1
[14]   →0
[15]   PSMALL:A[M]←((P[M]*3)×2÷45)-((P[M]*5)
       ×2÷315)+((P[M]*7)×2÷4725)
[16]   B[M]←(2÷3)+((P[M]*2)×2÷15)-((P[M]*4)
       ×4÷105)+((P[M]*6)×2÷567)
[17]   G[M]←(4÷3)-((P[M]*2)×2÷15)+((P[M]*4)
       ÷210)-((P[M]*6)÷11340)
[18]   →COMPUTE
       ∇
```

INTERNAL DOCUMENTATION FOR SINFILON

```
SYNTAX: Z←FN SINFILON VAL
PURPOSE: TO APPROXIMATE THE INTEGRAL OF
 F(X)SIN(TX) BY FILON'S METHOD
INDEX ORIGIN: 1
AUTHOR: G. PODOROWSKY
LAST UPDATE: 11/26/79
LEFT ARGUMENT: FN--A CHARACTER STRING
 FN←NAME OF USER-DEFINED FUNCTION WHICH
 ACCEPTS INPUT IN VECTOR FORM
 (X[1],X[2],...,X[2N]) AND RETURNS AS AN
 EXPLICIT RESULT THE VALUES F(X[1]),F([2]),
 ...,F(X[2N]) IN VECTOR FORM.
RIGHT ARGUMENT: VAL--A VECTOR
 VAL[1 2]=LOWER AND UPPER BOUNDS OF
 INTEGRATION, RESPECTIVELY
 VAL[3]=N, WHERE THE NUMBER OF SUBINTERVALS
 EQUALS 2×N
 VAL[4...M+4]=THE M CONSTANTS (T) FOR WHICH
 INTᴸGRAL IS BEING COMPUTED
RESULT: Z--A VECTOR OF LENGTH M CONTAINING
 THE INTEGRAL APPROXIMATIONS, ONE FOR EACH
 CONSTANT.
```

CODE FOR SINFILON

```
        ∇ Z←FN SINFILON VAL;A;B;S;S1;G;H;I;J;M;
          N;P;P2;P3;T;V1;V2;X
[1]     ⍝ FILON'S METHOD FOR APPROXIMATION
        ⍝ OF INTEGRAL OF F(X)SIN(TX)
[2]     ⍝ LAST UPDATE: 11/26/79
[3]       T←3↓VAL
[4]       X←VAL[1]+(0,⍳2×N)×H←(VAL[2]-VAL[1])÷2
          ×N←VAL[3]
[5]       V1←((ρT),ρX)ρ⍎FN,' X'
[6]       V2←10T∘.×X
[7]       S1←+/[2] V1[;I]×V2[;I←(0=2|I)/I←⍳2×N]
[8]       S←(+/[2] V1[;J]×V2[;J←(1=2|J)/J←⍳2×N+
          1])-0.5×((V1[;1+2×N]×V2[;1+2×N])+V1[;1]
          ×V2[;1])
[9]       A←(÷P)+((1○2×P)÷(2×P2←P×P))-((2×(1○P)
          ×(1○P))÷P3←(P←T×H)*3)
[10]      B←2×(((1+(2○P)×(2○P))÷P2)-(1○2×P)÷P3)
[11]      G←4×(((1○P)÷P3)-(2○P)÷P2)
[12]    →(0≠ρM←(0.001≥P)/⍳ρP)/PSMALL
[13]    COMPUTE:Z←H×(A×(V1[;1+2×N]×2○T×X[1+2×N])
          -V1[;1]×2○T×X[1])+(B×S)+G×S1
[14]    →0
[15]    PSMALL:A[M]←((P[M]*3)×2÷45)-((P[M]*5)
          ×2÷315)+((P[M]*7)×2÷4725)
[16]      B[M]←(2÷3)+((P[M]*2)×2÷15)-((P[M]*4)×4÷
          105)+((P[M]*6)×2÷567)
[17]      G[M]←(4÷3)-((P[M]*2)×2÷15)+((P[M]*4)÷
          210)-((P[M]*6)÷11340
[18]    →COMPUTE
        ∇
```

Function Name: DETERMINANT

EXTERNAL DOCUMENTATION

PURPOSE

To evaluate determinant of real square matrix.

WARNING

This is a straightforward algorithm, unsuitable for large matrices.

INTERNAL DOCUMENTATION

SYNTAX: Z←DETERMINANT A
PURPOSE: COMPUTES THE DETERMINANT OF THE
 REAL SQUARE MATRIX A.
INDEX ORIGIN: 1
AUTHOR: T. FERGUSON
LAST UPDATE: 6/80
RIGHT ARGUMENT: SQUARE MATRIX A
RESULT: VALUE OF DETERMINANT OF A.

CODE

```
      ∇ Z←DETERMINANT A;B;P;I
[1]   ⍝EVALUATES A DETERMINANT
[2]   I←⎕IO
[3]   Z←1
[4]   L:P←(|A[;I])⍳⌈/|A[;I]
[5]   →(P=I)/LL
[6]   A[I,P;]←A[P,I;]
[7]   Z←-Z
[8]   LL:Z←Z×B←A[I;I]
[9]   →(0 1 ∨.=Z,1↑⍴A)/0
[10]  A← 1 1 ↓A-(A[;I]÷B)∘.×A[I;]
[11]  →L
      ∇
```

Function Name: GRAMS

EXTERNAL DOCUMENTATION

PURPOSE

To compute coefficients of polynomials of degree 0–N that are orthonormal about the closed curve defined by the right argument Z. The result is a 2 by $N + 1$ by $N + 1$ complex, lower triangular matrix C (see external document for complex arithmetic package for definition of a complex matrix) whose ith row contains the coefficients in ascending order of the polynomial of degree $i - 1$.

THEORY

The result matrix C must satisfy the following relation if the polynomials are to be orthonormal about the closed curve:

$$CPC^H = I$$

$$C = \begin{bmatrix} c_{00} & 0 & 0 & \cdots & 0 \\ c_{10} & c_{11} & 0 & \cdots & 0 \\ c_{20} & c_{21} & c_{22} & \cdots & 0 \\ & & & \vdots & \\ c_{n0} & c_{n1} & c_{n2} & \cdots & c_{nn} \end{bmatrix}$$

and

$$P = \begin{bmatrix} \int ds & \int z\, ds & \int z^2\, ds & \cdots & \int z^n\, ds \\ \int z\, ds & \int z\bar{z}\, ds & \int z\bar{z}^2\, ds & \cdots & \int z\bar{z}^n\, ds \\ \int z^2\, ds & \int z^2\bar{z}\, ds & \int z^2\bar{z}^2\, ds & \cdots & \int z^2\bar{z}^n\, ds \\ & & \vdots & & \\ \int z^n\, ds & \int z^n\bar{z}\, ds & \int z^n\bar{z}^2\, ds & \cdots & \int z^n\bar{z}^n\, ds \end{bmatrix}$$

The integrals in P are taken about the closed curve in the complex plane defined by the m consecutive points in Z, the right argument. Thus

$$P = C^{-1}C^{H-1} = BB^H \qquad \text{where} \quad C = B^{-1}$$

ALGORITHM

1. Compute P using the program INTEGRAL.
2. Find B by calling CHOLESKYC and performing a complex Cholesky decomposition of P.
3. Take the inverse of the complex, lower triangular matrix B using INVERSE to arrive at C, the result.

NOTES

N, the degree of the highest order orthonormal polynomial desired, must be at least 1. GRAMS is called by CONFORMAL, which computes a conformal mapping function from a closed curve to a circle by the Rayleigh–Ritz method (see external documentation for CONFORMAL for further details).

EXAMPLE

$$N \quad 2$$

$$Z \quad \begin{matrix} 4 & 2 & 0 & -2 & -4 & -2 & 0 & 2 \\ 0 & 2.6 & 3 & 2.6 & 0 & -2.6 & -3 & -2.6 \end{matrix}$$

2 GRAMS *Z*

0.217	0	0	0	0	0
0	0.062	0	0	0	0
−0.032	0	0.018	0	0	0

Interpretation:

$z_1 = 4, \quad z_2 = 2 + 2.6i, \quad z_3 = 3i, \quad \ldots, \quad z_8 = 2 - 2.6i$
$P_0 = 0.217, \quad P_1 = 0.062z, \quad P_2 = -0.032 + 0.018z^2$

INTERNAL DOCUMENTATION

SYNTAX: ∇*C*←*N GRAMS Z*
PURPOSE: COMPUTES COEFFICIENTS OF POLYNOMIALS
 OF DEGREE 0 *TO N ORTHONORMAL ABOUT THE*
 CLOSED CURVE DEFINED BY Z.
INDEX ORIGIN: 1
AUTHOR: P. KRETZMER
LAST UPDATE: 5-11-79
RIGHT ARGUMENT: A 2 BY M COMPLEX MATRIX
 CONSISTING OF M CONSECUTIVE POINTS WHICH
 DEFINE A CLOSED CURVE IN THE COMPLEX PLANE.
LEFT ARGUMENT: A SCALAR EQUAL TO THE DEGREE
 OF THE HIGHEST ORDER ORTHONORMAL POLYNOMIAL
 DESIRED.
RESULT: A 2 BY N+1 BY N+1 COMPLEX, LOWER
 TRIANGULAR MATRIX WHOSE ITH ROW CONTAINS
 THE COEFFICIENTS IN ASCENDING ORDER OF THE
 POLYNOMIAL OF DEGREE I-1.

CODE

```
     ∇ C←N GRAMS Z;Q;P;B;BR;BI
[1] ⍝ COMPUTES COEFFICIENTS OF POLYNOMIALS
    ⍝ OF ORDER 0 TO N
[2] ⍝ ORTHONORMAL ABOUT CLOSED CURVE DEFINED
    ⍝ BY Z
```

```
[3]  ⍝ CALLS FUNCTIONS: INTEGRAL, CHOLESKYC,
     ⍝ INVERSE
[4]  ⍝ LAST UPDATE: 5-11-79
[5]  P←N INTEGRAL Z
[6]  B←CHOLESKYC(2,(Q×Q+1)÷2)ρ((2×Q*2)ρ,(⍳Q)
     ∘.≥⍳Q←N+1)/,P
[7]  BR←((⍳Q)∘.≥⍳Q)×(Q,Q)↑(((⍳Q)×(⍳Q)-1)
     ÷2)⌽(Q,(Q×Q+1)÷2)ρ,B[1;]
[8]  BI←((⍳Q)∘.≥⍳Q)×(Q,Q)↑(((⍳Q)×(⍳Q)-1)
     ÷2)⌽(Q,(Q×Q+1)÷2)ρ,B[2;]
[9]  C←INVERSE(2,Q,Q)ρ(,BR),,BI
     ∇
```

Function Name: INTEGRAL

EXTERNAL DOCUMENTATION

PURPOSE

To compute integrals about a closed curve defined (counterclockwise) by the complex matrix $Z = z_1, z_2, \ldots, z_m$.

Since this program is called by GRAMS in its orthogonalization procedure, INTEGRAL is used to construct the matrix

$$\begin{bmatrix} \int_e ds & \cdots & \int_e \bar{z}^n \, ds \\ \int_e z \, ds & \cdots & \int_e z\bar{z}^n \, ds \\ & \vdots & \\ \int_e z^n \, ds & \cdots & \int_e z^n\bar{z}^n \, ds \end{bmatrix}$$

The matrix P is the summary of all inner products of powers, where the (hermitian) inner product is defined as

$$\langle f, g \rangle = \int_e f \bar{g} \, ds$$

where f and g are complex functions and s is arclength along the curve C.

ALGORITHM

An approximation of P can be constructed as follows:

(1) Assume that the approximate increment in arclength at the point z_i is

$$(\tfrac{1}{2})(|z_{i+1} - z_i| + |z_i - z_{i-1}|)$$

(2) Let the complete set of arclengths at the points $z_1, \ldots, z_m$ be denoted by $l_1, l_2, \ldots, l_m$, i.e.,

$$l_i = (\tfrac{1}{2})(|z_{i+1} - z_i| + |z_i - z_{i-1}|)$$

In doing the integrals, $l_{(i)}$ plays the role of ds at the point $z_{(i)}$ on the curve C.

(3) Now the matrix P can be formed by matrix multiplications:

$$\begin{bmatrix} 1 & 1 & \cdots & \\ z_1 & z_2 & \cdots & z_m \\ z_1^2 & z_2^2 & \cdots & z_m^2 \\ & & \vdots & \\ z_1^n & z_2^n & \cdots & z_m^n \end{bmatrix} \begin{bmatrix} l_1 & & & 0 \\ & l_2 & & \\ & & l_3 & \\ & & & \ddots \\ 0 & & & l_m \end{bmatrix} \begin{bmatrix} 1 & \bar{z}_1 & \cdots & \bar{z}_1^n \\ 1 & \bar{z}_2 & \cdots & \bar{z}_2^n \\ & & \vdots & \\ 1 & \bar{z} & \cdots & \bar{z}^n \end{bmatrix}$$

The $(k + 1)(r + 1)$th entry is $\sum_{i=1} z_i^k \bar{z}_i^r l_i$, which is approximately $\int_e z^k \bar{z}^r \, ds$.

INTERNAL DOCUMENTATION

```
SYNTAX: ∇P←N INTEGRAL Z
PURPOSE: COMPUTES LINE INTEGRALS OF POWERS
 OF Z AND Z CONJUGATE ABOUT THE CLOSED CURVE
 DEFINED BY THE RIGHT ARGUMENT OF INPUT.
INDEX ORIGIN: 1
AUTHOR: P. KRETZMER
LAST UPDATE: 5-11-79
RIGHT ARGUMENT: A 2 BY M COMPLEX MATRIX
 CONSISTING OF M CONSECUTIVE POINTS
 WHICH DEFINE A CLOSED CURVE IN THE
 COMPLEX PLANE.
LEFT ARGUMENT: A SCALAR EQUAL TO THE HIGHEST
 POWER OF Z AND Z CONJUGATE FOR WHICH LINE
 INTEGRALS ARE COMPUTED.
RESULT: A 2 BY N+1 BY N+1 COMPLEX HERMITIAN
 MATRIX (SEE EXTERNAL DOC FOR COMPLEX PACKAGE
 FOR DEFINITION) WHOSE ELEMENTS ARE THE LINE
 INTEGRALS AS SPECIFIED IN ∇GRAMS EXTERNAL
 DOC.
```

CODE

```
      ∇ P←N INTEGRAL Z;M;Z1;DS;I;LSR;LSI,LS;
        D;RS;ZT
[1]   A COMPUTES INTEGRALS ABOUT CLOSED
      A CURVE DEFINED BY Z
[2]   A SEE GRAMS EXTERNAL DOC FOR MORE
      A DETAILS
[3]   A CALLED BY GRAMS; CALLS FUNCTIONS:
      A MAGNITUDE, MINUS, MATMULT
[4]   A LAST UPDATE: 5-11-79
[5]   M←(ρZ)[2]
[6]   Z1←MAGNITUDE(Z,Z[;1]) MINUS Z[;M],Z
[7]   DS←0.5×(1↓Z1)+⁻1↓Z1
[8]   D←(M,M)ρ,⍉DS,[1](M,M)ρ0
[9]   D←(2,M,M)ρ(,D),(M*2)ρ0
[10]  I←2
[11]  LS←(2,M)ρ(Mρ1),Mρ0
[12]  LS←LS,[1] ZT←Z
[13] LOOP:LS←LS,[1] ZT←ZT MULT Z
[14]  →(N≥I←I+1)/LOOP
[15]  LSR←LS[⁻1+2×⍳N+1;]
[16] LSI←LS[2×⍳N+1;]
[17]  LS←(2,(N+1),M)ρ(,LSR),,LSI
[18]  RS← 1 3 2 ⍉LS
[19]  RS[2;;]←-RS[2;;]
[20]  P←LS MATMULT D MATMULT RS
      ∇
```

Function Name: INTEQ

EXTERNAL DOCUMENTATION

PURPOSE

To approximate the eigenvalues and eigenvectors of a real symmetric and continuous kernal.

We set up the integral equation as

$$\int_0^1 k(x, y)v(y) \, dy = cv(x)$$

where k is real, continuous, and symmetric.

ALGORITHM

Discretize the system using $N + 1$ equally spaced points on $0 \le x$, $y \le 1$, where $N = 1/\text{step-size}$. Choose a rule of quadrature with an associated set of weights $P_0, \ldots, P_N$ to be applied along the y axis of the discretization matrix. The discretized system is then

$$
\begin{bmatrix}
P_0 k(0, 0) & P_1 k(0, 1/N) & \cdots & P_N k(0, 1) \\
P_0 k(1/N, 0) & P_1 k(1/N, 1/N) & \cdots & P_N k(1/N, 1) \\
& \vdots & & \\
P_0 k(1, 0) & P_1 k(1, 1/N) & \cdots & P_N(1, 1)
\end{bmatrix}
\begin{bmatrix}
v(0) \\
v(1/N) \\
\vdots \\
v(1)
\end{bmatrix}
=
\begin{bmatrix}
c(v(0)) \\
c(v(1/N)) \\
\vdots \\
c(v(1))
\end{bmatrix}
$$

We now need to call an eigenvalue routine to solve our system, but this requires that our matrix be symmetric. Thus perform the transformation (preserves eigenvalues)

$$Z = DMD^{-1}$$

where M is the k-matrix and D is a diagonal matrix whose (i,i) element is $(P_i)^{1/2}$.

Now we can apply the routines HSHLDR and TQL to the symmetric matrix Z to obtain the eigenvalues C and eigenvectors E. The eigenvectors are not those of the original system because of the transformation, so perform the reversing transformation

$$V = D^{-1}E$$

to obtain the eigenvectors of the original system.

ERROR BOUNDS FOR SIMPSON AND TRAPEZOIDAL QUADRATURE

We want to find bounds on our approximation error:

$|c' - c|$,   where $c' = $ actual eigenvalue, $c = $ computed eigenvalue

TRAPEZOIDAL RULE

We must first specify the Lipschitz condition:

$$|k(x, y) - k(x_i, y_j)| < L(|x - x_i| + |y - y_i|)$$
$$|x - x_i| < 1/2n \quad \text{and} \quad |y - y_j| < 1/2n$$

where $n = $ number of subintervals and (x_i, y_j) is any mesh point. L is the Lipschitz constant. Then the conclusion is

$$|c' - c| \le (\tfrac{1}{4} + \sqrt{\tfrac{1}{12}})(L)/(n - 1)$$

SIMPSON'S RULE

Let $k(x, y)$ have a continuous second derivative such that $|k_{xx}(x, y)| \leq M$; then

$$|c' - c| \leq (1 + 5/3)(5/18)(M)/(n - 1)^2$$

NOTE

The function calls HSHLDR and TQL.

REFERENCE

Wielandt, H. (1956). Error bounds for eigenvalues of symmetric integral equations, *in* "Proceedings of Symposia in Applied Mathematics," (John H. Curtiss, ed.), Vol. 6, pp. 261–282. McGraw-Hill, New York.

INTERNAL DOCUMENTATION

```
SYNTAX: R←HW INTEQ K
PURPOSE: TO APPROXIMATE THE EIGENVECTORS
 AND EIGENVALUES OF A REAL SYMMETRIC AND
 CONTINUOUS KERNAL.
INDEX ORIGIN: 1
AUTHOR: R. COHEN
LAST UPDATE: 12/2/80
RIGHT ARGUMENT: CHARACTER STRING: THE NAME
 OF THE APL FUNCTION WHICH GENERATES THE
 KERNAL. THIS MUST BE A FUNCTION OF ONE
 ARGUMENT WHICH RETURNS AS AN EXPLICIT
 RESULT K(X,Y). A 2 ROW MATRIX WHOSE FIRST
 ROW CONTAINS X VALUES AND WHOSE SECOND
 ROW CONTAINS Y VALUES WILL BE FED INTO
 THIS ARGUMENT. (THE ARGUMENT MUST BE A
 RIGHT ARGUMENT)
LEFT ARGUMENT: A FIVE ELEMENT VECTOR WHOSE
 FIRST COMPONENT IS THE STEPSIZE TO BE USED
 IN THE QUADRATURE RULE. IF THE USER
 DESIRES TRAPEZOIDAL QUADRATURE THE RE-
 MAINING ELEMENTS MUST BE 1 .5 1 1. IF
 SIMPSON QUADRATURE IS DESIRED THE REMAINING
 ELEMENTS MUST BE 3 1 4 2.
```

RESULT: AN N×N MATRIX WHERE N=1+1÷STEPSIZE.
THE FIRST ELEMENT IN EACH COLUMN IS AN
EIGENVALUE AND THE REMAINING ELEMENTS
CONSTITUTE ITS ASSOCIATED EIGENVECTOR.
NOTE: THE LIMITS OF INTEGRATION USED ARE
0 TO 1.

CODE

```
      ∇ R←HW INTEQ K;MAT;Z;POINTS;VEC;MT;
        D;S;DINV
[1]   MAT←( ,⍉MAT) , ,MAT←(Z,Z)ρ0,ι¯1+Z←1+⌊÷HW[1]
[2]   POINTS←HW[1]×(2,(÷2)×Z×Z+1)ρ((2×Z*2)
      ρVEC←,(ιZ)∘.≤ιZ)/MAT
[3]   MAT←MT+MAT×MAT≠MT←⍉MAT←(Z,Z)ρ(VEC)\⍋K,
      'POINTS'
[4]   D←(-0,ιZ-1)⌽[1](Z,Z)ρ((HW[3]*0.5),
      ((Z-2)ρ(¯2↑HW)*0.5),HW[3]*0.5),
      (Z×Z-1)ρ0
[5]   MT←(DINV)+.×(MAT←TQL HSHLDR(S←÷HW[2]×Z-1)
      ×D+.×MAT+.×DINV←⌹D)[1↓ι Z+1;]
[6]   R←MAT[1;],[1] MT÷S×(Z-1)*0.5
      ∇
```

Function Name: KUTTA

EXTERNAL DOCUMENTATION

PURPOSE

To numerically solve an rth-order differential equation:

$$d^r x_1/dt^r = q(t, x_1, dx_1/dt, d^2x_1/dt^2, \ldots, d^{r-1}x_1/dt^{r-1})$$

The equation should be reduced to the system of first-order equations:

$$\dot{x}_1 = x_2$$
$$\dot{x}_2 = x_3$$
$$\vdots$$
$$\dot{x}_{r-1} = x_r$$
$$\dot{x}_r = q(t, x_1, x_2, \ldots, x_r)$$

with initial conditions $x_1(t_0) = x_{10}, \ldots, x_r(t_0) = x_{r0}$.

ALGORITHM

This system of r first-order initial-value problems is then numerically solved using the classical Runge–Kutta formula. Consider the general system:

$$\dot{x}_1 = f_1(t, x_1, x_2, \ldots, x_r)$$
$$\vdots$$
$$\dot{x}_r = f_r(t, x_1, x_2, \ldots, x_r)$$

Then the classical Runge–Kutta formula is given by

$$x_{n+1} = x_n + 1/6(k1_n + 2k2_n + 2k3_n + k4_n)$$

where

$$k1_n = hF(t_n, x_n)$$
$$k2_n = hF(t_n + 0.5h, x_n + 0.5k1_n)$$
$$k3_n = hF(t_n + 0.5h, x_n + 0.5k2_n)$$
$$k4_n = hF(t_n + h, x_n + k3_n)$$

The algorithm uses the Runge–Kutta formula above beginning with the user supplied initial data:

$$t_0, x_0, h, t_f$$

The user-supplied function F is called by the program. h is the step-size and $t(f) = t(0) + nh$ is the final point for which computations are performed. There are therefore n iterations. For iteration i,

$$t_0 + (i - 1)h = t_{i-1}, x_{i-1} \qquad \text{becomes} \quad x_i$$

The program prints the initial data and the results of each iteration:

$$t_0, x_0$$
$$t_0 + h = t_1, x_1$$
$$t_0 + 2h = t_2, x_2$$
$$\vdots$$
$$t_0 + nh = t_f, x_f$$

The local formula error $e(n)$, the error arising in one iteration from the use of the approximate formula, is proportional to $h * 5$. Thus halving the step-size reduces this error by the factor $\frac{1}{32}$. The accumulated formula error is at most a constant times $h * 4$.

Although reducing step-size decreases accumulated formula error by the same factor raised to the fourth power, making h too small (thus increasing the required number of iterations) increases the accumulated

round-off error. The user should keep this trade-off in mind in choosing a
step-size that will minimize total error.

EXAMPLE

Z gets F X
Z gets $(X(1) + X(2) + X(3))$, $(4 \times X(2)) - 2 \times X(3)$

'F' KUTTA 0   1   0   0.5   2

0	1	0
0.5	2.38	2.02
1	6.80	6.49
1.5	19.15	18.74
2	52.92	52.35

REFERENCE

Boyce, W., and DiPrima, R. (1965). "Elementary Differential Equations
and Boundary Value Problems," 2nd ed. Wiley, New York.

INTERNAL DOCUMENTATION

```
SYNTAX: ∇RES←NAME KUTTA DATA∇
PURPOSE: SOLVES A SYSTEM OF R FIRST ORDER
 INITIAL VALUE PROBLEMS NUMERICALLY USING
 CLASSICAL RUNGE-KUTTA
INDEX ORIGIN: 1
AUTHORS: B. BENNETT, P. KRETZMER
LAST UPDATE: 10-31-78
RIGHT ARGUMENT: DATA: VECTOR
 DATA[1]←INITIAL VALUE OF INDEPENDENT VARIABLE
 DATA[2],...,DATA[R+1]←INITIAL VALUES OF
 DEPENDENT VARIABLES X[1],...,X[R] RESPEC-
 TIVELY
 DATA[R+2]←STEP SIZE
 DATA[R+3]←FINAL VALUE OF THE INDEPENDENT
 VARIABLE
LEFT ARGUMENT: NAME: CHARACTER STRING
 NAME←NAME OF USER DEFINED FUNCTION WHICH
 ACCEPTS INPUT IN VECTOR FORM (T,X[1],...,
 X[R]) AND RETURNS THE VALUES F[1],...,F[R]
 IN VECTOR FORM SUCH THAT F[1] IS THE RIGHT
```

HAND SIDE OF THE EQUATION IN THE SYSTEM:
D(X[I])=F[I]
RESULT: RES: MATRIX
RES[;1]←TIME STEPS
RES[;2],...,RES[;R+1]←VALUES OF THE DEPENDENT
  VARIABLES X[1],...,X[R]
  RES[1;]←INITIAL VALUES OF T,X[1],...,X[R]
  RESPECTIVELY
WARNING: IF STEP SIZE IS INCONSISTENT WITH
  THE FINAL VALUE OF THE INDEPENDENT VALUE
  SPECIFIED, THE NUMBER OF ITERATIONS IS
  CHOSEN SUCH THAT THE LAST ITERATION WILL
  CAUSE THE FINAL VALUE TO BE EXECUTED.

CODE

```
        ∇ RES←NAME KUTTA DATA;K1;K2;K3;K4;
          K;I;V;R;N;VAR;H
[1]     ⍝NUMERICAL SOLUTION OF A SYSTEM OF
        ⍝R IVP'S BY CLASSICAL RUNGE-KUTTA
[2]     ⍝USER CONSTRUCTED EXTERNAL FUNCTION;
        ⍝SPECIFY ITS NAME AS A CHARACTER
        ⍝STRING (LEFT ARGUMENT)
[3]     ⍝LAST UPDATE: 10-31-78
[4]      R←-3+ρDATA
[5]      N←⌈(DATA[R+3]-DATA[1])÷H←DATA[R+2]
[6]      RES←((N+1),R+1)ρ0
[7]      RES[1;]←VAR←DATA[⍳R+1]
[8]      I←2
[9]      V←1+⍳R
[10]    LOOP:K1←H×⍎NAME,' VAR'
[11]     K2←H×⍎NAME,' (VAR[1]+H÷2),VAR[V]
          +K1÷2'
[12]     K3←H×⍎NAME,' (VAR[1]+H÷2),VAR[V]
          +K2÷2'
[13]     K4←H×⍎NAME,' (VAR[1]+H),VAR[V]+K3'
[14]     K←(K1+(2×K2+K3)+K4)÷6
[15]     RES[I;]←VAR←VAR+H,K
[16]     →((I←I+1)≤N+1)/LOOP
        ∇
```

Function Names: MRFFT and MRFFT1

EXTERNAL DOCUMENTATION

PURPOSE

To compute the discrete Fourier transform of a real- or complex-valued string.

Specifically, let x_j, for $j = 0, 1, \ldots, N - 1$ be a sequence of N complex values. Let W_N denote the Nth root of unity:

$$W_N = \exp(2\pi/N)$$

The discrete Fourier transform of (x_j) is the sequence a_k given by

$$a_k = \sum_{j=0}^{N-1} x_j W_N^{jk} \qquad (1)$$

for $k = 0, 1, \ldots, N - 1$.

ALGORITHM

MRFFT and the alternative version MRFFT1 are Mixed Radix Fast Fourier Transforms.

Naïve implementation of the discrete Fourier transform is equivalent in computational complexity to computing the product of a complex $N \times N$ matrix and a complex N vector. When N has an integer factorization

$$N = r_1 r_2, \ldots, r_m$$

then the computational complexity can be substantially reduced and is equivalent in MRFFT to successive matrix–vector multiplications of order r_k for $k = 1$ to m. The main idea on which the algorithm is based is described in the fundamental reference by Cooley and Tukey (see below).

NOTES

The programs work in index origin 0 or 1. The input string X is a real N-vector or a $2 \times N$ (complex) array and the explicit result A is a $2 \times N$ (complex) array. If the user's index origin is zero, then the APL indices correspond exactly to the indices j and k in Eq. (1) above. If the user's index origin is one, then the APL indices are greater by one than the indices in (1).

The left argument of MRFFT and of MRFFT1 is an APL vector with integer elements whose product is N. As a general rule, the factors should be small in order to minimize execution time. However, our experience with the program indicates that factors in the range of 6–10 give faster execution than ones in the range 2–5. (When the smallest factors are used, the amount of interpretation increases even though the number of operations—multiplies and adds—decreases.)

The program FACTOR can be used to find the prime factorization of N.

REFERENCES

Cooley, J. W., and Tukey J. W. (1965). An algorithm for the machine calculation of complex Fourier series, *Math. Comp.* **19,** 297–301.
Singleton, R. C. (1969). An algorithm for computing the mixed radix fast Fourier transform, *IEEE Trans. Audio and Electroacoust.* **AU-17,** 93–104.

INTERNAL DOCUMENTATION

```
SYNTAX: ∇A←PR MRFFT X∇
PURPOSE: MIXED RADIX FAST FOURIER TRANSFORM
 OF A REAL OR COMPLEX-VALUED STRING WITH
 N ELEMENTS
INDEX ORIGIN: 0 OR 1
AUTHOR: D. E. MCCLURE
LAST UPDATE: 10/12/80
RIGHT ARGUMENT: X: A REAL-VALUED VECTOR WITH
 N ELEMENTS OR A COMPLEX-VALUED VECTOR WITH
 N ELEMENTS (2×N MATRIX WHOSE ROWS
 CORRESPOND TO THE REAL AND IMAGINARY PARTS
 OF THE VECTOR)
LEFT ARGUMENT: PR: EITHER THE SCALER N OR A
 VECTOR OF POSITIVE INTEGERS WHOSE PRODUCT
 IS N
RESULT: A: COMPLEX VECTOR (2×N MATRIX)
 WHICH IS THE DISCRETE FOURIER TRANSFORM
 OF X
SUGGESTIONS FOR USE: AN ALTERNATIVE VERSION
 ''MRFFT1'' REQUIRES LESS STORAGE THAN
 ''MRFFT'' AND HAS A MARGINALLY SLOWER
 EXECUTION TIME: AS A GENERAL RULE THE
 ELEMENTS OF PR (FACTORS OF N) SHOULD BE
 SMALL IN ORDER TO MINIMIZE EXECUTION TIME.
```

CODE FOR MRFFT

```
      ∇  A←PR  MRFFT  X;B;C;CS;I;L;LM;M;M1;N;
         NL;PWR;S;U;V;□IO
[1]   □IO←1
[2]   X←(PR,2)ρ⍉(2,N)ρ(,X),(NL←N←¯1↑ρX)ρ0
[3]   CS←⍉ 2 1 ∘.∘∘(2÷N)×¯1+ιN×PWR←L←×M1←1+
         LM←M←ρPR←,PR
[4]   TOP:X←((C+.×A)-S+.×B),((S←CS[I;2])
         +.×(A←(V,1)↑X))+(C←CS[I←1+N|((N÷PR[L])
         ×U)∘.×U←¯1+ιPR[L];1])+.×B←((V←¯1↓ρX)
         ,¯1)↑X
[5]   I←⍉(1,⌽V)ρ⍉1+N|(PWR×¯1+ιPR[L])∘.×(L↓PR)
         ρ¯1+ιNL←NL÷PR[L]
[6]   X←(LM,(ιLM-1),(LM+ιL-1),M1)⍉((C×A)
         -S×B),((S←CS[1;2])×(A←(V,1)↑X))+
         (C←CS[I;1])×B←(V,¯1)↑X
[7]   →(N>PWR←PWR×PR[¯1+L←M1-LM←LM-1])/TOP
[8]   A←(2,N)ρ,(1⌽ιM1)⍉X
      ∇
```

CODE FOR MRFFT1

```
      ∇  A←PR  MRFFT1  X;CS;I;L;LM;M;M1;N;NL;
         PWR;U;V;□IO
[1]   □IO←1
[2]   X←(PR,2)ρ⍉(2,N)ρ(,X),(NL←N←¯1↑ρX)ρ0
[3]   CS←⍉ 2 1 ∘.∘∘(2÷N)×¯1+ιN×PWR←L←×M1
         ←1+LM←M←ρPR←,PR
[4]   TOP:I←1+N|((N÷PR[L])×U)∘.×U←¯1+ιPR[L]
[5]   X←((CS[I;1]+.×(V,1)↑X)-CS[I;2]+.×
         (V,¯1)↑X),(CS[I;2]+.×(V,1)↑X)+CS[I;1]
         +.×((V←¯1↓ρX),¯1)↑X
[6]   I←⍉(1,⌽V)ρ⍉1+N|(PWR×¯1+ιPR[L])∘.×
         (L↓PR)ρ¯1+ιNL←NL÷PR[L]
[7]   X←(LM,(ιLM-1),(LM+ιL-1),M1)⍉((CS[I;1]
         ×(V,1)↑X)-CS[I;2]×(V,¯1)↑X),(CS[I;2]
         ×(V,1)↑X)+CS[I;1]×(V,¯1)↑X
[8]   →(N>PWR←PWR×PR[¯1+L←M1-LM←LM-1])/TOP
[9]   A←(2,N)ρ,(1⌽ιM1)⍉X
      ∇
```

Function Name: NEWTON

EXTERNAL DOCUMENTATION

PURPOSE

To use Newton's method to find a root of an arbitrary function $f(x)$.

ALGORITHM

Choose a starting value X_0 and compute $f(X_0)/f'(X_0)$. If $f(X)$ were linear, then the root X_1 would be

$$X_1 = X_0 - f(X_0)/f'(X_0)$$

Since $f(X)$ is not linear, X_1 is only an approximation to a zero of $f(X)$, i.e., X_1 is the x intercept of the tangent line at (X), $f(X_0)$. The process is repeated until a root is found. In general,

$$X_{m+1} = X_m - f(X_m)/f'(X_m)$$

When $X_{m+1} = X_m$, $f(X_m) = 0$ and a root has been found.

NOTES

1. The user-supplied function is $f(X)/f'(X)$; the user must also supply the scalar starting value X_0.

2. Newton's method achieves a quadratic rate of convergence in many cases. For a precise statement of the conditions under which convergence is achieved, see Theorem 5.2.1 in the reference cited. It should be noted that $f'(X)$ must not be zero in a neighborhood of the root as one requirement for convergence of the iterates.

REFERENCE

Blum, E. K. (1972). "Numerical Analysis and Computation: Theory and
    Practice," Addison-Wesley, Reading, Massachusetts.

INTERNAL DOCUMENTATION

SYNTAX: ∇RES←FN NEWTON XM
PURPOSE: FINDS A ROOT FOR AN ARBITRARY
 FUNCTION F(X) USING NEWTON'S METHOD
INDEX ORIGIN: 1
AUTHORS: B. BENNETT, P. KRETZMER

```
LAST UPDATE: 11/29/78
RIGHT ARGUMENT: XM←SINGLE SCALAR STARTING
 POINT
LEFT ARGUMENT: FN-CHARACTER STRING
 FN←NAME OF USER DEFINED FUNCTION WHICH
 ACCEPTS A SCALAR INPUT AND RETURNS AS AN
 EXPLICIT RESULT THE VALUE OF THE RATIO
 OF THE FUNCTION TO ITS DERIVATIVE AT THE
 SPECIFIED POINT
RESULT: RES←SCALAR ROOT OF THE FUNCTION F(X)
WARNING: IF POSSIBLE CHOOSE STARTING POINT
 NEAR ROOT. DERIVATIVE CANNOT BE ZERO AT
 STARTING POINT.
```

CODE

```
      ∇ RES←FN NEWTON XM;XN
[1]  ⍝SOLVING FOR ROOT OF FUNCTION USING
     ⍝NEWTON'S METHOD
[2]  ⍝USER DEFINED EXTERNAL FUNCTION
     ⍝IS CALLED
[3]  ⍝LAST UPDATE: 11/29/78
[4]   XN←XM
[5]  LOOP:XM←XN
[6]   XN←XM-(⍎FN,' XM')
[7]   →(XM≠XN)/LOOP
[8]   RES←XN
      ∇
```

Function Name: ROOT

EXTERNAL DOCUMENTATION

PURPOSE

To find a root for an arbitrary function $f(x)$.

ALGORITHM

The technique is a modification of NEWTON, using Newton's method
with an approximation for the derivative. (See external documentation for

NEWTON.) Thus the general iteration becomes

$$x_{m+1} = x_m - f(x_m)(x_m - x_{m-1})/(f(x_m) - f(x_{m-1}))$$

The user-supplied function is simply $f(x)$; the user must also supply a vector of two starting values. Note that convergence will likely be less rapid than for NEWTON.

INTERNAL DOCUMENTATION

```
SYNTAX: ∇RES←FN ROOT XM
PURPOSE: FINDS A ROOT FOR AN ARBITRARY
 FUNCTION F(X) USING A MODIFICATION OF
 NEWTON'S METHOD.
INDEX ORIGNN: 1
AUTHORS: B. BENNETT, P. KRETZMER
LAST UPDATE: 11/29/78
RIGHT ARGUMENT: XM←VECTOR CONTAINING TWO
 STARTING POINTS
LEFT ARGUMENT: FN-CHARACTER STRING
 FN←NAME OF USER DEFINED FUNCTION WHICH
 ACCEPTS A SCALAR INPUT AND RETURNS AS AN
 EXPLICIT RESULT THE VALUE OF THE
 FUNCTION AT THE SPECIFIED POINT.
RESULT: RES←SCALAR ROOT OF THE FUNCTION F(X)
WARNING: IF POSSIBLE CHOOSE THE TWO STARTING
 POINTS IN VICINITY OF ROOT.
```

CODE

```
     ∇ RES←FN ROOT XM;XN
[1] ⍝SOLVING FOR ROOT OF FUNCTION USING A
    ⍝MODIFICATION OF NEWTON'S METHOD
[2] ⍝USER DEFINED EXTERNAL FUNCTION IS CALLED
[3] ⍝LAST UPDATE: 11/29/78
[4]  XN←XM[1]
[5] LOOP:XM[1]←XM[2]
[6]  XM[2]←XN
[7]  XN←XM[2]-(⍎FN,' XM[2]')×(XM[2]-XM[1])
    ÷(⍎FN,' XM[2]')-(⍎FN,' XM[1]')
[8]  →(XM[2]≠XN)/LOOP
[9]  RES←XN
     ∇
```

Function Name: SIMPSON

EXTERNAL DOCUMENTATION

PURPOSE

To compute the numerical approximation of the definite integral

$$\int_a^b f(x)\ dx$$

by Simpson's rule.

The user must supply the function $f(x)$ that accepts input in vector form $(x_0, x_1, \ldots, x_n)$ and that explicitly returns the values ($f(x_0), f(x_1), \ldots, f(x_n)$), also in vector form. In addition, the user must specify in L the lower and upper bounds, a and b, respectively, as well as the number of subintervals n as

$$a = L(1), \qquad b = L(2), \qquad n = L(3)$$

WARNING

n must be an even integer!

ALGORITHM

Divide the interval from a to b into n equal subintervals of length $*x = (b - a)/n$. Since n is even, the integral can be written as

$$\int_a^b f(x)\ dx = \int_{x_0}^{x_2} f(x)\ dx + \int_{x_2}^{x_4} f(x)\ dx + \cdots$$
$$+ \int_{x_{(n-4)}}^{x_{(n-2)}} f(x)\ dx + \int_{x_{(n-2)}}^{x_n} f(x)\ dx$$

Approximate each integral from x_i to $x_{(i+2)}$ by the area under a parabola through the three points $(x_i, f(x_i))$, $(x_{(i+1)}, f(x_{(i+1)}))$, and $(x_{(i+2)}, f(x_{(i+2)}))$, where

$$\int_{x_i}^{x_{(i+2)}} f(x)\ dx \doteq (*x/3)(\ f(xi) + 4f(x(i + 1)) + f(x(i + 2))$$

Summing over all subintervals, we arrive at the formula of Simpson's rule:

$$\int_a^b f(x)\ dx \doteq (b - a)(\ f(x_0) + 4f(x_1) + 2f(x_2) + 4f(x_3) + \cdots$$
$$+ 2f(x_{n-2}) + 4f(x_{n-1}) + f(x_n))$$

NOTE

This program provides the Simpson approximation of the integral, of course, not the exact evaluation. This approximation is often more accurate than that of the trapezoidal rule, however, and becomes more and more precise as the number of subintervals (*n*) increases.

INTERNAL DOCUMENTATION

```
SYNTAX: ∇Z←NAME SIMPSON L∇
PURPOSE: NUMERICAL APPROXIMATION OF A
 DEFINITE INTEGRAL BY SIMPSONS RULE
INDEX ORIGIN: 1
AUTHORS: B. BENNETT, P. KRETZMER
LAST UPDATE: 6/19/79
LEFT ARGUMENT: NAME--A CHARACTER STRING
 NAME←NAME OF USER DEFINED FUNCTION WHICH
 ACCEPTS INPUT IN VECTOR FORM (X[0]=L[1],
 X[2],X[3],...,X[L[3]]=L[2]) AND RETURNS
 AS AN EXPLICIT RESULT THE VALUES F(L[1]),
 F(X[2]),F(X[3]),...,F(L[2]) IN VECTOR FORM.
RIGHT ARGUMENT: L--A VECTOR
 L[1]=LOWER LIMIT OF INTEGRATION
 L[2]=UPPER LIMIT OF INTEGRATION
 L[3]=NUMBER OF SUBINTERVALS; THIS MUST
 BE AN EVEN INTEGER!
RESULT: Z--A SCALAR
```

CODE

```
      ∇ Z←NAME SIMPSON L;N
[1]  ⍝ NUMERICAL APPROXIMATION OF A DEFINITE
     ⍝ INTEGRAL BY SIMPSONS RULE
[2]  ⍝ USER DEFINED EXTERNAL FUNCTION IS
       CALLED
[3]  ⍝ LAST UPDATE: 6/19/79
[4]    Z←(N÷3)×+/(1,((L[3]-1)ρ 4 2),×⍎NAME,
     ' L[1]+0,+\L[3]ρN←(L[2]-L[1])÷L[3]'
     ∇
```

Function Name: SPLINE

EXTERNAL DOCUMENTATION

PURPOSE

To fit a set of cubic polynomials to the data points $t, x(t)$. The result is a continuous curve passing through the points, with each interval fit by a different cubic.

DEVELOPMENT

In order to form a "smooth" set of cubics through the given points, it is necessary (in the development of the cubic spline) that the slope and curvature be the same for the pair of cubics that join at each point.

Over the interval (x_i, y_i) to (x_{i+1}, y_{i+1}) the corresponding equations are

$$y = a_i(x - x_i)^3 + b_i(x - x_i)^2 + c_i(x - x_i) + d_i \qquad (1)$$
$$y' = 3a_i(x - x_i)^2 + 2b_i(x - x_i) + c_i$$
$$y'' = 6a_i(x - x_i) + 2b_i$$

After some algebraic simplification, the second derivative terms reduce to

$$S_i = 2b_i$$
$$S_{i+1} = 6a_i(x_{i+1} - x_i)2b_i$$

Letting $(x_{i+1} - x_i) = h_i$ (not necessarily the same for each interval), the coefficients of Eq. (1) can be shown to be

$$b_i = S_i/2$$
$$a_i = (S_{i+1} - S_i)/6h_i$$
$$c_i = [(y_{i+1} - y_i)/h_i] - [(2h_iS_i + h_iS_{i+1})/6]$$
$$d_i = y_i$$

Avoiding the intermediate steps, the final relationship between the second derivatives S and the y values is

$$h_{i-1}S_{i-1} = (2h_{i-1} + 2h_i)S_i + h_iS_{i+1} = \frac{6(Y_{i+1} - Y_i)}{h_i} - \frac{(Y_i - Y_{i-1})}{h_{i-1}}$$

The above constitutes $n - 2$ equations relating the n values S_i. Two additional equations involving S_1 and S_n can be obtained by specifying conditions pertaining to the end intervals of the spline. The three conditions allowed for in this program are

1. $S_1 = S_n = 0$;

2. $S_1 = S_2$ and $S_n = S_{n-1}$;

3. $(S_2 - S_1)/h_1 = (S_3 - S_2)/h_2$ and analogously for the other end of the spline.

We now have n equations and n unknowns, which can be written in the form of a tridiagonal matrix, thus always enabling us to solve for the vector of Ss. Once these are known, the coefficients can be obtained, and Eq. (1) will be known for each interval. Thus a different cubic has been specified for each interval, with the requirement that the first and second derivatives at the end of one interval are the same as those at the beginning of the next interval.

REFERENCES

Gerald, C. F. (1970). "Applied Numerical Analysis," 2nd ed. pp. 474–482. Addison-Wesley, Reading, Massachusetts.

INTERNAL DOCUMENTATION

SYNTAX: ∇RES←INT SPLINE T×∇
PURPOSE: TO FIT A SET OF CUBIC POLYNOMIALS
 TO THE DATA POINTS T,X(T). THE RESULT IS
 A CONTINUOUS CURVE PASSING THROUGH THE
 POINTS DESIGNATED BY TX, WITH EACH INTERVAL
 FIT BY A DIFFERENT CUBIC.
INDEX ORIGIN: 1
AUTHOR: P. FORCHIONE
LAST UPDATE: 7/30/80
RIGHT ARGUMENT: TX: A 2 BY N ARRAY OF POINTS.
 ROW 1= T (INDEPENDENT VARIABLE)
 ROW 2= X (DEPENDENT VARIABLE)
LEFT ARGUMENT: INT: VECTOR CONTAINING VALUES
 OF T FOR WHICH AN INTERPOLATED X(T) IS
DESIRED. THESE VALUES MUST BE IN RANGE OF T.
 IF NO INTERPOLATIONS ARE DESIRED TYPE
 'NONE' AS THE LEFT ARGUMENT.
RESULT: 1) IF NO INTERPOLATIONS ARE DESIRED,
 RESULT IS AN N ROW CHARACTER MATRIX WHOSE
 ITH ROW READS: INTERVAL [T(I),T(I+1)],
 COEFFICIENTS A(I) B(I) C(I) D(I).
 2) IF INTERPOLATIONS ARE DESIRED, RESULT
 IS AN N BY 2 ARRAY; THE FIRST COL CONSISTING
 OF THE VECTOR INT, AND COL 2 BEING THE

INTERPOLATED X(T)'S FOR EACH CORRESPONDING T.
IF THE USER DESIRES, OUTPUT WILL INCLUDE THE
CHARACTER MATRIX DESCRIBED IN 1) ABOVE (THE
COMPUTER WILL ASK YOU ABOUT THIS).
GLOBAL VARIABLES: COEF: THE MATRIX OF COEF-
FICIENTS OF EACH CUBIC FOR THE PARTICULAR
INTERVAL. EACH ROW OF THE MATRIX IS OF FORM
A(I) B(I) C(I) D(I). FOR EXAMPLE, THE CURVE
WHICH APPLIES TO THE 3RD INTERVAL IS:
Y(3)=A(3)(T-T(3))*3+B(3)(T-T(3))*2+C(3)(T-T
(3))+D(3),
WHERE THE COEFFICIENTS ARE FOUND IN ROW 3 OF
COEF. VALUES: THE MATRIX DESCRIBED AS TX
ABOVE.
SUGGESTED USE: TO FIT A SMOOTH, CONTINUOUSLY
DIFFERENTIABLE CURVE TO A SET OF POINTS, THE
RELATIONSHIP OF WHICH IS NOT WELL KNOWN
(I.E. NO SPECIFIC FUNCTION FITS THE POINTS).
NOTE: THE USER WILL BE INTERROGATED ABOUT
END POINT CONDITIONS AS DESCRIBED IN
EXTERNAL DOCUMENTATION. THE FUNCTION NEEDS
A FUNCTION CALLED PRINT FOR OUTPUT.

CODE

```
        ∇ RES←INT SPLINE TX;H;I;K;S;T;X;M;
        R;TDIF;Z;J;C3;EP
[1]     ⍝ INT REPRESENTS VECTOR OF VALUES TO
        ⍝ BE INTERPOLATED, IN ASCENDING ORDER
[2]     TX[1;]←TX[1;(PERM←⍋TX[1;])]
[3]     TX[2;]←TX[2;PERM]
[4]     VALUES←TX
[5]     'CHOOSE ONE OF THE FOLLOWING ENDPOINT
        CONDITIONS'
[6]     'FOR SECOND DERIVATIVES.'
[7]     '1: END CUBICS APPROACH LINEARITY
        AT EXTREMITIES'
[8]     '2: END CUBICS APPROACH PARABOLAS
        AT EXTREMITIES'
[9]     '3: END CUBICS APPROACH A THIRD DEGREE
        POLYNOMIAL AT EXTREMITIES'
[10]    'NOTE: INPUT 3 IF IT IS BELIEVED THAT
        THE POINTS WERE'
```

```
[11]    'GENERATED BY A THIRD DEGREE POLYNOMIAL.
        IF YOU ARE'
[12]    'UNFAMILIAR WITH THESE CONDITIONS, YOUR
        BEST CHOICE IS 1.'
[13]    INT←⎕,(~INT∊'NONE')/INT
[14]    INT←,INT
[15]    ⍝ TX IS ARRAY OF POINTS TO BE FITTED
        ⍝ BY SPLINE, ROW 1 = T, ROW 2 = X(T).
[16]    T←TX[1;]
[17]    X←TX[2;]
[18]    H←(1↓T)-¯1↓T
[19]    I←¯2+⍴T
[20]    R←6×(((2↓X)-K)÷1↓H)-((K←↓¯1↓X)-¯2↓X)÷¯1↓H
[21]    C3←(2↑H),[1.2]⌽¯2↑H
[22]    →(3 3 ⍴ 1 0 0 0)[INT[1];]/BR1,BR2,BR3
[23]    BR1:M←(¯1↓H),(2×(¯1↓H)+1↓H),[1.5] 1↓H
[24]    →OUT
[25]    BR2:M←(¯1↓H),((3 2 +.×2↑H),(2×(1↓¯2↓H)+
        2↓¯1↓H),2 3+.×¯2↑H),[1.4] ↓H
[26]    →OUT
[27]    BR3:M←1⌽(((C3[;1]+2×C3[;2])×+/C3)÷
        C3[;2]),2×(1↓¯2↓H)+2↓¯1↓H
[28]    M←((¯2↓H),(-/(¯2↑H)*2)÷H[I]),[1.5] M
[29]    M←M,((-/(2↑H)*2)÷-H[2]),2↑H
[30]    OUT:M[I;3]←M[1;1]←0
[31]    ⍝ S IS VECTOR OF SECOND DERIVATIVES
[32]    S←R⌹(2-⍳I)⌽(I,I)↑M
[33]    →(INT[1]≠3)/DN
[34]    EP←⌽(((S[1],S[I])×+/C3)-(C3[1;1],
        C3[2;2])×S[2],S[I-1])÷C3[1;2],C3[2;1]
[35]    →BELOW
[36]    DN:EP←((0 0),[1.2] S[I],S[1])[;INT[1]]
[37]    BELOW:S←1⌽EP,S
[38]    COEF←((((1↓S)-¯1↓S)÷6×H),((¯1↓S)÷2),
        (((((1↓X)-¯1↓X)÷H)-(H×(2×¯1↓S)+1↓S)
        ÷6),[1.4] ¯1↓X
[39]    →(0≠⍴INT←1↓,INT)/PRNT
[40]    ⎕←' '
[41]    RES←TX[1;] PRINT COEF
[42]    →0
[43]    ⍝ TDIF IS COL. VECTOR OF VALUES
```

```
      ⍝ INT(I)-T(I)
[44]  PRNT:⎕←' '
[45]  'ENTER YES OR NO DEPENDING ON WHETHER
      OR NOT YOU'
[46]  'WANT OUTPUT TO INCLUDE THE MATRIX
      OF SPLINE COEFFICIENTS '
[47]  'FOR EACH INTERVAL.'
[48]  →(0=+/⎕∊'YES')/GO
[49]  ⎕←' '
[50]  ⎕←TX[1;] PRINT COEF
[51]  GO:⎕←' '
[52]  'INTERPOLATIONS DESIRED'
[53]  ⎕←' '
[54]  TDIF←((J←⍴INT),1)⍴INT-T[Z←+/INT∘.≥T]
[55]  RES←INT,[1.4]⍉+/((TDIF*3),(TDIF*2),
      TDIF,J⍴1)×COEF[Z;]
      ∇
```

Function Name: SPLINT

EXTERNAL DOCUMENTATION

PURPOSE

To integrate the cubic spline formed in spline over any closed interval (a,b) in the domain of the spline.

ALGORITHM

Remember that, given the knot sequence $T(1)$, $T(2)$, . . . , $T(n)$, the spline is formed by splicing together a cubic defined on $(T(1), T(2))$ with one defined on $((T(2), T(3))$, etc. Since the cubics above are not usually equal (except at common end points), integration over (a,b) must consist of the sum of a series of integrations over all or part of each knot interval wholly or partially covered by (a,b).

EXAMPLE

Given the knot sequence 1,2,3,4, to integrate over (.5,3) integrate over (.5,1), (1,2) and (2,3) and sum.

For the Ith knot interval $(T(I), T(I + 1))$ and any a_1 and b_1 such that $T(I) < a_1 < b_1 < T(I + 1)$:

$$\int_{a_1}^{b_1}(\text{spline in that interval}) = [A(I)/4]((b_1 - T(I))^4 - (a_1 - T(I))^4)$$
$$+ [B(I)/3]((b_1 - T(I))^3 - (a_1 - T(I))^3)$$
$$+ [C(I)/2]((b_1 - T(I))^2 - (a_1 - T(I))^2)$$
$$+ D(I)(b_1 - a_1)$$

Thus once the intervals of integration are determined, the integral over (a,b) is calculated by using the above equation in each interval (with the appropriate values of a_1 and b_1). In the first interval of the example, $a_1 = 0.5$, $b_1 = 1$, and then summing the result over all the intervals used.

INTERNAL DOCUMENTATION

```
SYNTAX: ∇AREA←AB SPLINT VALUES ∇
PURPOSE: INTEGRATION OF CUBIC SPLINE
 OBTAINED BY RUNNING SPLINE
INDEX ORIGIN: 1
AUTHOR: P. FORCHIONE
LAST UPDATE: 4/8/80
LEFT ARGUMENT: INT: BOUNDARIES OVER WHICH
 INTEGRAL WILL BE CALCULATED. CONTAINS
 TWO NUMBERS, T(A) AND T(B), SUCH THAT
 T(1)≤T(A)<T(B)≤T(N), WHERE T(1) AND T(N)
 ARE THE ENDPOINTS OF THE INTERVAL
 DEFINING THE SPLINE.
RIGHT ARGUMENT: VALUES: GLOBAL VARIABLE
 CREATED BY SPLINE CONTAINING THE
 ORIGINAL POINTS T,X(T) USED IN DETERMINING
 THE CUBIC SPLINE.
RESULT: AREA: THE DEFINITE INTEGRAL OVER
 SPECIFIED REGION
NOTE: THE FUNCTION SPLINE MUST BE EXECUTED
 BEFORE USING SPLINT.
```

CODE

```
      ∇ AREA←AB SPLINT VALUES;B1;B;VEC;SET;A;
        INTS;A1;DT;IND;MAT;JOIN;AREA;DIFFS
[1]   B1←+/(B←AB[2])>VEC←VALUES[1;]
[2]   SET←A,VEC[¯1↓INTS←A1+ι1+B1-A1←+/
      VEC≤A←AB[1]],B
```

```
[3]     DT←(1↓SET)-¯1↓SET
[4]     IND←(4ρ(~VEC[1↑¯1+INTS]<1↑SET)),[1]
        ((¯1+ρDT),4)ρ1
[5]     JOIN←IND×((DT*4),(DT*3),(DT*2),[1.3]
        DT)×MAT←((ρDT),4)ρ÷⌽ι4
[6]     AREA←+/+/JOIN×COEF[¯1+INTS;]
[7]     →(IND[1;1]=0)/ADJUST
[8]     →0
[9]     ADJUST:DIFFS←(SET[2]-VEC[1↑¯1+INTS]),
        SET[1]-VEC[1↑¯1+INTS]
[10]    AREA←AREA++/(((DIFFS[1]*⌽1+ι3)-(DIFFS
        [2]*⌽1+ι3)),DT[1])×MAT[1;]×COEF
        [(¯1+INTS[1]);]
        ∇
```

Function Name: SPLDER

EXTERNAL DOCUMENTATION

PURPOSE

To differentiate the cubic spline created in spline. The cubic spline formed in spline is continuously differentiable on its domain. In the Ith interval, denoted $(T(I), T(I + 1))$:

$$y'(T) = 3A(I)(T - T(I))*2 + 2B(I)(T - T(I)) + C(I)$$

where $T(I) \leq T < T(I + 1)$ and $A(I)$, $B(I)$, $C(I)$ are the coefficients of the cubic polynomial in that interval.

To differentiate the spline at any point x in the set $(T(1), T(n))$, find the half-open interval (open on the right) $(T(I), T(I + 1))$, which contains x, and set $T = x$ in the above equation and compute with the appropriate coefficients for that interval.

INTERNAL DOCUMENTATION

```
SYNTAX: ∇VEC←INT SPLDER VALUES∇
PURPOSE: DIFFERENTIATES THE CUBIC SPLINE
 AT POINTS VEC.
INDEX ORIGIN: 1
AUTHOR: P. FORCHIONE
LAST UPDATE: 5/2/80
```

LEFT ARGUMENT: INT: THE POINTS AT WHICH
THE DERIVATIVE IS DESIRED. EACH VALUE OF
INT MUST SATISFY T(1)≤INT(I)<T(N), WHERE
T(1) AND T(N) LOCATE THE ENDPOINTS OF
THE SPLINE.
RIGHT ARGUMENT: VALUES: GLOBAL VARIABLE
CREATED BY SPLINE CONTAINING THE ORIGINAL
POINTS T,X(T) USED IN DETERMINING THE
CUBIC SPLINE.
RESULT: THE DERIVATIVE OF THE SPLINE AT
EACH OF THE POINTS INT.
NOTE: THE FUNCTION SPLINE MUST BE EXECUTED
BEFORE USING SPLDER.

CODE

```
      ∇ VEC←INT SPLDER TX;T;TDIF;J;Z
[1]  ⍝ DIFFERENTIATES THE SPLINE CREATED
     ⍝ IN THE SPLINE PROGRAM
[2]   T←TX[1;]
[3]   TDIF←((J←ρINT),1)ρINT-T[Z←+/INT∘.≥T]
[4]   VEC←⌽+/((3×TDIF*2),(2×TDIF),Jρ1)×
      COEF[Z;⍳¯1+¯1↑ρCOEF]
      ∇
```

Function Name: TRAPEZOIDAL

EXTERNAL DOCUMENTATION

PURPOSE

To compute the numerical approximation of the definite integral

$$\int_a^b f(x)\ dx$$

according to the trapezoidal rule.

The user must supply the function $f(x)$, which accepts input in vector form $(X_0, X_1, \ldots, X_n)$ and which explicitly returns the values $(f(X_0), f(X_1), \ldots, f(X_n))$, also in vector form. In addition, the user must specify in L the lower and upper bounds, a and b, respectively, as well as the number of subintervals desired n as

$$a = L(1), \qquad b = L(2), \qquad n = L(3)$$

ALGORITHM

Approximate the area under the curve $y = f(x)$ from $X_0 = a$ to $X_n = b$ by dividing the interval $b-a$ into n equal subintervals of length $(b-a)/n$ and summing the areas of the trapezoids which are thus formed.

FORMULA

$$\int_a^b f(x)\ dx = ((b-a)/2n)(y_0 + 2y_1 + \cdots + 2y_{n-1} + y_n)$$

NOTE

This program provides the trapezoidal approximation of the integral, not the exact evaluation. As the number of subintervals (n) increases, however, this approximation becomes more and more accurate.

INTERNAL DOCUMENTATION

```
SYNTAX: ∇Z←NAME TRAPEZOIDAL L∇
PURPOSE: NUMERICAL APPROXIMATION OF A
 DEFINITE INTEGRAL BY TRAPEZOIDAL RULE
INDEX ORIGIN: 1
AUTHORS: B. BENNETT, P. KRETZMER
LAST UPDATE: 6/19/79
LEFT ARGUMENT: NAME--CHARACTER STRING
 NAME=NAME OF USER DEFINED FUNCTION WHICH
 ACCEPTS INPUT IN VECTOR FORM (X[0]=L[1],
 X[2],X[3],...,X[L[3]]=L[2]) AND RETURNS
 AS AN EXPLICIT RESULT THE VALUES F(L[1]),
 F(X[2]),F(X[3]),...,F(L[2]) IN VECTOR FORM
RIGHT ARGUMENT: L--A VECTOR
 L[1]=LOWER LIMIT OF INTEGRATION
 L[2]=UPPER LIMIT OF INTEGRATION
 L[3]=NUMBER OF SUBINTERVALS
RESULT: Z--A SCALAR
```

CODE

```
      ∇ Z←NAME TRAPEZOIDAL L;N
[1]  ⍝ NUMERICAL EVALUATION OF A DEFINITE
     ⍝ INTEGRAL BY TRAPEZOIDAL RULE
[2]  ⍝ USER DEFINED EXTERNAL FUNCTION IS
     ⍝ CALLED
[3]  ⍝ LAST UPDATE: 6/19/79
[4]    Z←N×+/(0.5,((L[3]-1)ρ1),0.5)×⍎NAME,
     ' L[1]+0,+\L[3]ρN←(L[2]-L[1])÷L[3]'
      ∇
```

CHAPTER 18

Arithmetic

We give only a few programs for arithmetic purposes: generate primes, factor integers, calculate greatest common divisor and smallest common multiple, continued fractions, and recognize rational numbers. The latter is useful in experiments when some real sequence is believed to consist of rational numbers whose numerators and denominators have some simple rule of generation.

Function Names: CONFRAC and RATIONAL

EXTERNAL DOCUMENTATION FOR CONFRAC

For X in $(0, 1)$ the program finds the expansion in continued fraction

$$\left(B[1] + \cfrac{1}{B[2] + \cdots + \cfrac{1}{B[N]}} \right)^{-1}$$

of order N or lower. The order is lower than the specified value N if the following denominators are very large. This is decided in line [4]. The constant $1E^{-9}$ can be changed by the user when circumstances so demand.

The vector B with entries $B[1], B[2], \ldots$ is a global variable.

Other global variables are the convergents $P[M]$ and $Q[M]$ as well as the approximation $R[M] = P[M] \div Q[M]$. These values are entries in the vectors P, Q, and R.

WARNING

The threshold constant in line [4] may have to be modified in certain cases.

INTERNAL DOCUMENTATION FOR CONFRAC

SYNTAX: N CONFRAC X
PURPOSE: TO EXPAND X IN CONTINUED FRACTION
INDEX ORIGIN: 1
AUTHOR: U. GRENANDER
LAST UPDATE: MARCH 1981
LEFT ARGUMENT: MAXIMUM ORDER OF CONTINUED
 FRACTION
RIGHT ARGUMENT: REAL NUMBER IN (0,1) TO BE
 EXPANDED

CODE FOR CONFRAC

```
       ∇ N CONFRAC X;XNEW;T
[1]    XNEW←÷X
[2]    B←Nρ⌊XNEW
[3]    T←2
[4]    LOOP1:→(1E⁻9>XNEW-B[T-1])/STOP
[5]    XNEW←÷XNEW-B[T-1]
[6]    B[T]←⌊XNEW
[7]    T←T+1
[8]    →(T≤N)/LOOP1
[9]    STOP:N←T-1
[10]   B←N↑B
[11]   R←P←Q←Nρ0
[12]   P[1]←1
[13]   Q[1]←B[1]
[14]   R[1]←P[1]÷Q[1]
[15]   →(N=1)/0
[16]   P[2]←B[2]
[17]   Q[2]←1+B[1]×B[2]
[18]   R[2]←P[2]÷Q[2]
[19]   T←3
[20]   →(T>N)/0
[21]   LOOP2:P[T]←(B[T]×P[T-1])+P[T-2]
[22]   Q[T]←(B[T]×Q[T-1])+Q[T-2]
[23]   R[T]←P[T]÷Q[T]
[24]   T←T+1
[25]   →(T≤N)/LOOP2
       ∇
```

EXTERNAL DOCUMENTATION FOR RATIONAL

This heuristic function tries to represent a given real number X in (0, 1) as a rational number. It tests for large denominators in statement [2]. It calls the auxiliary function CONFRAC.

WARNING

The threshold constant in line [2] should be made smaller when needed.

INTERNAL DOCUMENTATION FOR RATIONAL

```
SYNTAX: RATIONAL X
PURPOSE: TO RECOGNIZE RATIONAL NUMBERS
INDEX ORIGIN: 1
AUTHOR: U. GRENANDER
LAST UPDATE: MARCH 1981
RIGHT ARGUMENT: REAL NUMBER IN (0,1)
```

CODE FOR RATIONAL

```
      ∇ RATIONAL X
[1]   10 CONFRAC X
[2]   I←¯1+(1000000000<B)ι1
[3]   'NUMERATOR IS = ',⍕P[I]
[4]   'DENOMINATOR IS = ',⍕Q[I]
      ∇
```

Function Name: FACTOR

EXTERNAL DOCUMENTATION

PURPOSE

To factor into primes any integer less than or equal to $9.9 \times 10 * 15$.

EXAMPLE

For the integer 18 the output is

$$
\begin{array}{cc}
1 & 2 \\
2 & 3
\end{array}
$$

meaning that the prime factors are 2 and 3 and to obtain 18 we multiply 2 to the first power times 3 to the second power.

ALGORITHM

The program uses the global variable primvec to facilitate calculation. Primvec contains all primes $\leq$ (integer to be factored)$^{1/2}$. When finding prime factors, two all-inclusive cases arise (the integer to be factored is referred to as int):

Case (1)   All factors are less than or equal to the square root of int.

Case (2)   One factor is greater than the square root of int and the others are less than the square root of int.

The algorithm used attempts to minimize storage space by first finding all factors $\leq$ int$^{1/2}$. A check is then run to see if the product of all factors raised to their respective exponents equals int. If they do not, then this is a case (2) situation, so find the last factor.

Find prime factors [with exception of case (2) types]:

(1)   Select all primes from primvec $\leq$ (int)$^{1/2}$.

(2)   Divide int by each and select those that leave no remainder.

Find the upper bound of the set of exponents of factors:

If (int)$^{1/2}$ > max(primvec), then call PRIMGEN to generate necessary primes.

Else.

Do for each factor:

(1)   Multiply all other factors together.

(2)   Divide int by this product.

(3)   Calculate x; factor = result of step 2.

(4)   Max exponent = ceiling x.

End.

Take max of the set of max's (call it sup).

Calculate the exponent of each factor:

(1)   Create a matrix whose ith row is the ith factor (factors are arranged in ascending order) raised to the 1,2,3, . . . , sup.

(2)   Create a matrix of size (1) but with all elements = int

(3)   Divide matrices elementwise and note those elements in each row that leave no remainder.

(4)   In each row find the corresponding exponents of the noted elements and take the max of these (the number found for the ith row is proper exponent of the ith factor).

Check for a case (2) type and calculate last factor if necessary.

WARNING

Because of the number of elements in primvec, it is necessary to log on with 512k.

NOTES

If int is a prime, the message, ' 'int' is prime' will be outputted. If int has no prime factors < largest element in primvec, an appropriate message is outputted.

INTERNAL DOCUMENTATION

```
SYNTAX: RES←FACTOR INT
PURPOSE: TO FACTOR INTO PRIMES ANY INTEGER
 LESS THAN 9.9 × (10*15)
INDEX ORIGIN: 1
AUTHOR: D. MCARTHUR
LAST UPDATE: 7/80
RIGHT ARGUMENT: INTEGER TO BE FACTORED
RESULT: A MATRIX WHOSE SECOND, FOURTH,...
 ROWS ARE THE PRIME FACTORS AND WHOSE
 FIRST, THIRD,... ROWS ARE THE EXPONENTS
 TO WHICH THESE FACTORS ARE RAISED.
```

CODE

```
      ∇ RES←FACTOR INT;FACMAT;SS;POW;SHAP;
        FAC;NTF;PRIM;R
[1]    →(INT≤(¯1↑PRIMVEC)*2)/START
[2]    R←0/R←(1+¯1↑PRIMVEC) PRIMGEN(⌈INT*0.5)
[3]    START:□IO←1
[4]    PRIM←(PRIMVEC≤⌈INT*0.5)/PRIMVEC
[5]    SS←ρFAC←(0=PRIM|INT)/PRIM
```

```
[6]     →(SS=0)/P
[7]     FACMAT←(¯1+ιρFAC)φSHAP←(0 1 +2ρρFAC)
        ρFAC,1
[8]     POW←⌈⌈/FACMAT[;1]⊛INT÷×/ 0 1 ↓FACMAT
[9]     NTF←⌈/((SS,POW)ριPOW)×0=1|((SS,POW)
        ρINT)÷FACMAT[;1]∘.*ιPOW
[10] BACK:RES←NTF,[0.5] FAC
[11]    →(INT≠×/FAC*NTF)/SPEC
[12]    →0
[13] SPEC:FAC←FAC,INT÷×/FAC*NTF
[14]    NTF←NTF,1
[15]    →BACK
[16] P:→(INT<(¯1↑PRIMVEC)*2)/PMM
[17]    RES←(⍕INT),' HAS NO PRIME FACTORS
        LESS THAN ',(⍕¯1↑PRMVEC),' SQUARED'
[18]    →0
[19] PMM:RES←(⍕INT),' IS PRIME'
        ∇
```

Function Names: GCD and SCM

EXTERNAL DOCUMENTATION

PURPOSE

To compute the greatest common divisor and smallest common multiple of two given natural numbers. To be able to handle large values, nonrecursive functions are being used. The greatest common divisor is obtained in the familiar way through a looping program carrying out Euclid's algorithm. To get the smallest common multiple one just observes that it is the same as the product of the two numbers divided by the greatest common divisor.

INTERNAL DOCUMENTATION FOR GCD

```
SYNTAX: Z←N1 GCD N2
PURPOSE: TO COMPUTE THE GREATEST COMMON
 DIVISOR OF TWO NATURAL NUMBERS
INDEX ORIGIN: 1
```

```
AUTHOR: U. GRENANDER
LAST UPDATE: 10/80
RIGHT ARGUMENT: ONE OF THE TWO NATURAL
  NUMBERS
LEFT ARGUMENT: THE OTHER NATURAL NUMBER
RESULT: GREATEST COMMON DIVISOR OF N1 AND N2
```

CODE FOR GCD

```
      ∇ Z←N1 GCD N2
[1]  ⍝COMPUTES GREATEST COMMON DIVISOR OF
     ⍝N1 AND N2
[2]  LOOP:Z←N1
[3]    N1←N1|N2
[4]    N2←Z
[5]    →(N1≠0)/LOOP
     ∇
```

INTERNAL DOCUMENTATION FOR SCM

```
SYNTAX: Z←N1 SCM N2
PURPOSE: TO COMPUTE THE SMALLEST COMMON
  MULTIPLE OF TWO NATURAL NUMBERS
INDEX ORIGIN: 1
AUTHOR: U. GRENANDER
LAST UPDATE: 10/80
RIGHT ARGUMENT: ONE OF THE TWO NATURAL
  NUMBERS
LEFT ARGUMENT: THE OTHER NATURAL NUMBER
RESULT: SMALLEST COMMON MULTIPLE OF N1 AND N2
```

CODE FOR SCM

```
     ∇SCM[□]∇
      ∇ Z←N1 SCM N2
[1]  ⍝COMPUTES SMALLEST COMMON MULTIPLE
     ⍝OF N1 AND N2
[2]  ⍝CALLS FUNCTION GCD
[3]    Z←(N1×N2)÷N1 GCD N2
     ∇
```

Function Names: PRIMGEN and PRIMGN

EXTERNAL DOCUMENTATION

PURPOSE

To find the primes between two specified integers m and n ($n > m$) inclusive.

NOTE

This program uses the global variable primvec, which contains primes up to the largest n specified by any user. Primvec will be extended to include more primes each time a user requests primes past the largest one in primvec.

ALGORITHM

If $m \leq n \leq$ max(primvec), then call PRIMGN to select appropriate primes from primvec.

If $m \leq$ max(primvec) $< n$, then call PRIMGN to find primes that are $\geq m$ and $\leq$ max(primvec). Store these in result. Go to mainbody.

If max(primvec) $< m \leq n$, then Go to mainbody.

MAINBODY

Let max(primvec) $= a$. Create a vector called vec of all the integers $>$ max(a,m) and $\leq n$ that end in 1,3,7, or 9. Partition the interval (a,n) into as many subintervals as possible of the form

$$(a, a^2), \ldots , (a^{2^{k-1}}, n), \qquad k = 1, 2, 3 \ldots$$

Denote intervals of this type by (a_i, a_{i+1}).
Do for each interval above:

Create a vector, vec1, of the elements in vec that lie in this sub-interval.
Do for all elements of primvec $\leq \sqrt{(a_{i+1})}$:

      vec1 = those elements of vec1 not divisible
          by the current prime

End.

Update result and primvec to include those primes just calculated.

End.

WARNING

The input value of *n* must be less than or equal to $9.9 \times 10 * 15$.

INTERNAL DOCUMENTATION

```
SYNTAX: R←M PRIMGEN N
PURPOSE: TO FIND THE PRIMES BETWEEN TWO
 SPECIFIED INTEGERS M AND N (N≥M) INCLUSIVE.
 N MUST BE ≤ 9.9 × (10*15).
INDEX ORIGIN: 0
AUTHOR:
LAST UPDATE: 8/20/80
RIGHT ARGUMENT: THE LARGER OF THE TWO
 SPECIFIED INTEGERS
LEFT ARGUMENT: THE SMALLER OF THE TWO
 SPECIFIED INTEGERS
RESULT: VECTOR: THE PRIMES BETWEEN M AND N
 INCLUDING M OR N IF EITHER IS PRIME.
NOTE: THIS PROGRAM USES THE GLOBAL VARIABLE
 PRIMVEC WHICH CONTAINS PRIMES UP TO THE
 LARGEST N SPECIFIED BY ANY USER. PRIMVEC
 WILL BE EXTENDED TO INCLUDE MORE PRIMES
 EACH TIME A USER REQUESTS PRIMES PAST THE
 LARGEST ONE IN PRIMVEC. CALLS FUNCTION
 PRIMGN.
```

CODE FOR PRIMGEN

```
        ∇ R←M PRIMGEN N;VEC;N1;I;J;VEC1
[1]     R←ι0
[2]     →(M>¯1↑PRIMVEC)/NEWP
[3]     →(N>¯1↑PRIMVEC)/MIX
[4]     →(R=R←M PRIMGN N)/0
[5]     MIX:R←M PRIMGN (¯1↑PRIMVEC)
[6]     M←1+¯1↑PRIMVEC
[7]     NEWP:□IO←0
[8]     VEC←M+ι1+N-M
[9]     N←¯1↑VEC←(~(10⊤VEC)∊ 0 2 4 5 6 8)
        /VEC
```

```
[10]    →(N≤N1←M,(¯1↑PRIMVEC)*2)/SET
[11]  BIGN:N1←N1,(¯1↑N1)*2
[12]    →(N≤¯1↑N1)/SET
[13]    →BIGN
[14]  SET:N1[¯1+ρN1]←N
[15]    I←0
[16]  CONT:J←-1
[17]    VEC1←(~VEC1<1+N1[I])/VEC1←(N1[I+1]
       ≥VEC)/VEC
[18]  CALC:VEC1←(0≠PRIMVEC[J]|VEC1)/VEC1,
       0ρJ←J+1
[19]    →(N1[I+1]>PRIMVEC[J]*2)/CALC
[20]    R←R,VEC1
[21]    PRIMVEC←PRIMVEC,VEC1
[22]    I←I+1
[23]    →((ρN1)>1+I)/CONT
[24]    □IO←1
        ∇
```

CODE FOR PRIMGN

```
      ∇ R←M PRIMGN N
[1] ⍝THIS PROGRAM IS CALLED BY PRIMGEN TO
    ⍝FIND THE PRIMES BETWEEN TWO INTEGERS
[2] ⍝M AND N (N≥M) WHERE N ≤ MAX(PRIMVEC)
[3]   R←(~M>PRIMVEC1)/PRIMVEC1←(N≥PRIMVEC)
      /PRIMVEC
      ∇
```

CHAPTER 19

Asymptotics

The programs given in this chapter correspond to the discussion in Chapter 14 of heuristic asymptotics. The original programs used in Chapter 14, written by the author, have been modified by T. Ferguson in order to correct some of the deficiencies mentioned in the earlier discussion.

In the original version all the results were printed immediately, requiring much more terminal time than is really needed, since only few of the attempts to fit images will succeed. To save terminal time the printout starts only after execution of MON or OSC and the experimenter is interrogated about how many fitted images he wishes to see printed.

In the new version DOMAIN and VALUE ERROR can still occur but much more rarely than earlier. They can be reduced even further by inserting tests and go-to statements in the code similar to lines [12], [13] in CGAUTOR or by testing for large values of variables. It is doubtful whether this is worthwhile. For example, the cost of computing the determinants in CGAUTOR[13] is the price we pay for avoiding DOMAIN ERROR in statement [14] and it may be too high, especially if we add many similar test/go-to statements.

Instead we can bypass any particular image by updating the subscripts I,J,K (in MON) or $I1, J1$ (in OSC) and then reentering the function at the appropriate statement. This was the procedure used for the earlier version, but there it occurred frequently, leading to time-consuming manipulations at the terminal. Now it happens only occasionally.

Also, it should be mentioned that we can avoid most such program interrupts if we choose the generator space carefully. When we know X and F and perhaps have done some preliminary attempts at analysis, we can probably discard some of the generators listed as $G10, G11, \ldots$. Actually that list should not be used uncritically: It is only intended as an example, and experimenters are encouraged to add to or subtract generators from it according to their intuition.

The master program ASYMPTOTICS calls a plotting function with the syntax

$$(N1, N2) \quad PLOT \quad Y \quad VS \quad X$$

Here $N1$ and $N2$ are the appropriate number of lines (height of plot) and number of characters (width of plot) that are desired; Y and X are vectors of the same length and we want the ordinates in Y plotted against the abscissas in X; VS is the name of a function. The standard PLOT function that has been in use in many APL systems has exactly this syntax. If the experimenter does not have access to it and wishes to use his own plot function or none at all, he need only modify statement [6] in ASYMP-TOTICS or delete it altogether.

A function DETERMINANT is called repeatedly to evaluate the determinant of a square matrix. The experimenter could, for example, use the program, also called DETERMINANT, described in Chapter 2.

Function Name: ASYMPTOTICS

EXTERNAL DOCUMENTATION

PURPOSE

To carry out heuristic asymptotic analysis of given function (vector) F for X values in vector X. See Chapter 14 in Part III for detailed description.

INTERNAL DOCUMENTATION

```
SYNTAX:  X ASYMPTOTICS F
PURPOSE:  PERFORMS A HEURISTIC SEARCH FOR
  ASYMPTOTIC PATTERNS IN DATA FROM A
  COMPUTABLE FUNCTION WHOSE ASYMPTOTIC
  BEHAVIOUR IS NOT KNOWN.
INDEX ORIGIN:  1
LAST UPDATE:  4/80
AUTHORS:  U. GRENANDER AND T. FERGUSON
LEFT ARGUMENT:  A VECTOR OF REAL, POSITIVE
  X-VALUES IN INCREASING ORDER.
RIGHT ARGUMENT:  A VECTOR OF RESULTING FUNC-
  TION VALUES, F(X). THERE ARE TWO CASES--
  WHEN THE FUNCTION APPROACHES ZERO MONO-
  TONICALLY AND WHEN IT APPROACHES ZERO
  THROUGH A SERIES OF OSCILLATIONS. IN THE
```

FIRST CASE, F(X) IS MADE NON-NEGATIVE
WITHOUT LOSS OF GENERALITY.
RESULT: A PRINTOUT OF PATTERNS GENERATED
BY THE SEARCH, ORDERED ACCORDING TO
GOODNESS OF FIT. WHEN F(X) IS MONOTONIC,
THE ALGORITHM GENERATES PATTERNS BY PSEUDO-
LINEARIZATION, ASSUMING LINEAR RELATIONSHIPS
PSI[F(X)]=C1[FI1(X)]+C2[FI2(X)] WHERE THE
FUNCTIONS PSI, FI1, AND FI2 HAVE BEEN
GENERATED FROM PRIMITIVES.
WHEN F(X) IS OSCILLATING, THE ALGORITHM
GENERATES PATTERNS OF THE FORM LF[FI(X)]=
CONSTANT × G WHERE L IS A SIMPLE DIFFERENCE
OPERATOR WITH CONSTANT COEFFICIENTS, AND FI
AND G ARE GENERATED FROM PRIMITIVES.
FOR MORE COMPLETE DOCUMENTATION: CONSULT
CHAPTER 14 IN PART III.

WARNING

The set of functions FI and G should be modified to suit the circumstances of the experiment. This can be done by changing the definitions of G10, G11, . . . as well as the matrices FIL, PSIL, GL and the corresponding numbers IMAX, JMAX, KMAX, IMAXO, JMAXO.

CODE

```
       ∇ X ASYMPTOTICS F
[1]    N←ρF
[2]    'WHAT LP-NORM?P='
[3]    P←□
[4]    'WHAT Q-STRESS OF LARGE ARGUMENTS ?Q='
[5]    Q←□
[6]    (30,(30⌊N)) PLOT F VS X
[7]    WEIGHT←FWEIGHT F
[8]    WEIGHT1←WEIGHT
[9]    GENERATORS←''
[10]   ERRORS←ι0
[11]   CSTARS←ι0
[12]   'IS FUNCTION MONOTONIC OR OSCILLATING?'
[13]   ⍕(3↑□),' ','1 1 2'
       ∇
```

Function Name: CDIVDIFF

EXTERNAL DOCUMENTATION

PURPOSE

This function is called by CGAUTOR for the calculation of divided differences.

INTERNAL DOCUMENTATION

```
SYNTAX: RESULT←ORDER CDIVDIFF W
PURPOSE: CALCULATES DIVIDED DIFFERENCES OF
 DATA IN MATRIX W
INDEX ORIGIN: 1
LAST UPDATE: 4/80
AUTHORS: U. GRENANDER AND T. FERGUSON
LEFT ARGUMENT: ORDER OF DESIRED DIFFERENCES
 EITHER 1 OR 2
RIGHT ARGUMENT: 2×N MATRIX, FIRST ROW
 CONTAINS ORDINATES OF DATA, SECOND ROW
 CONTAINS ABSCISSAS
RESULT: MATRIX WITH ORDER+1 ROWS.
 FIRST ROW CONTAINS ORDINATES.
 SECOND ROW CONTAINS FIRST DIFFERENCES.
 IF ORDER IS 2 THIRD ROW WILL CONTAIN
 SECOND DIFFERENCES.
```

CODE

```
        ∇ RESULT←ORDER CDIVDIFF W;N1;U;V
[1]   ⍝COMPUTES CENTRAL DIVIDED DIFFERENCES OF
      ⍝W[1;]
[2]   ⍝W.R.T. W[2;].RESULTS STORED IN MATRIX
      ⍝WITH
[3]   ⍝1+ ORDER ROWS:FUNCTION ITSELF IN ROW
      ONE,
[4]   ⍝FIRST DIFFERENCES IN SECOND ROW,...
[5]   ⍝MATRIX WILL HAVE TWO COLUMNS LESS
      ⍝THAN W
[6]   ⍝ORDER SHALL BE 1 OR 2
[7]    U←W[1;]
```

```
[8]    V←W[2;]
[9]    N1←(ρW)[2]
[10]   X1M←(2↓V)-¯2↓V
[11]   RESULT←((ORDER+1),N1-2)ρ0
[12]   RESULT[1;]←¯1↓1↓U
[13]   RESULT[2;]←((2↓U)-¯2↓U)÷X1M
[14]   →(ORDER=1)/0
[15]   X10←(2↓V)-1↓¯1↓V
[16]   X0M←(1↓¯1↓V)-¯2↓V
[17]   RESULT[3;]←((2↓U)÷X10×X1M)+((-1↓¯1↓U)÷
       X10×X0M)+(¯2↓U)÷X0M×X1M
[18]   RESULT[3;]←2×RESULT[3;]
       ∇
```

Function Name: CGAUTOR

EXTERNAL DOCUMENTATION

PURPOSE

Solves for best (least squares) difference equation that has given data as solution.

INTERNAL DOCUMENTATION

```
SYNTAX: ORDER CGAUTOR W
PURPOSE: TO FIND BEST AUTOREGRESSION SAT-
  ISFIED BY DATA GIVEN IN RIGHT ARGUMENT W.
INDEX ORIGIN: 1
LAST UPDATE: 4/80
AUTHORS: U. GRENANDER AND T. FERGUSON
LEFT ARGUMENT: ORDER OF DIFFERENCE EQUATION,
  EITHER 1 OR 2
RIGHT ARGUMENT: 2×N MATRIX, WHOSE FIRST ROW
  CONTAINS ORDINATES, SECOND CONTAINS
  ABSCISSAS.
RESULT: CSTAR IS COEFFICIENT VECTOR, FSTAR
  IS APPROXIMATION TO FUNCTION REPRESENTED
  BY ORDINATES IN W[1;]
```

CODE

```
        ∇ ORDER CGAUTOR W;DIAGONAL;DIFF;INV;
          INNERV;DETGDIFFEQ
[1]     ⍝COMPUTES BEST GENERAL AUTOREGRESSION FOR
[2]     ⍝GENERAL SPACINGS OF W[2;] OF ORDER
[3]     ⍝GIVEN BY LEFT ARGUMENT.ORDINATES ARE IN
        W[1;]
[4]     ⍝ORDER=1 OR 2
[5]     ⍝USES CENTRAL DIFFERENCES
[6]     N1←¯2+(ρW)[2]
[7]     DIAGONAL←((⍳N1)∘.=⍳N1)×WEIGHT1[⍳N1]
        ∘.×N1ρ1
[8]     DIFF←ORDER CDIVDIFF W
[9]     DIFF←DIFF[ϕ⍳ORDER+1;]
[10]    →(~∧/0=G)/LOOP3
[11]    INV←DIFF[1+⍳ORDER;]+.×DIAGONAL+.×
        ⍉DIFF[1+⍳ORDER;]
[12]    ⍝BYPASSES SINGULAR MATRICES
[13]    →(1E¯13≥DETERMINANT INV)/0
[14]    INV←⌹INV
[15]    INNERV←DIFF[1+⍳ORDER;]+.×DIAGONAL
        +.×DIFF[1;]
[16]    CSTAR←INV+.×INNERV
[17]    CSTAR←CSTAR,0
[18]    →⍙'LOOP',⍕ORDER
[19]    LOOP1:CSTAR[1] CORDER1 W[2;]
[20]    FSTAR←W[1;⍳N1] GDIFFEQ N1ρ0
[21]    →(1E¯13≥DETGDIFFEQ)/0
[22]    →CATENATE
[23]    LOOP2:CSTAR[1 2] CORDER2 W[2;]
[24]    FSTAR←W[1;⍳N1] GDIFFEQ N1ρ0
[25]    →(1E¯13≥DETGDIFFEQ)/0
[26]    →CATENATE
[27]    LOOP3:DIFF←DIFF,[1] G[1+⍳N1]
[28]    INV←DIFF[1+⍳ORDER+1;]+.×DIAGONAL
        +.×⍉DIFF[1+⍳ORDER+1;]
[29]    →(1E¯13≥DETERMINANT INV)/0
[30]    INV←⌹INV
[31]    INNERV←DIFF[1+⍳ORDER+1;]+.×DIAGONAL
        +.×DIFF[1;]
```

```
[32]    CSTAR←INV+.×INNERV
[33]    →⍎'LOOP',⍕ORDER+3
[34]  LOOP4:CSTAR[1] CORDER1 W[2;]
[35]    FSTAR←W[1;⍳N1] GDIFFEQ CSTAR[2]×G
[36]    →(1E¯13≥DETGDIFFEQ)/0
[37]    →CATENATE
[38]  LOOP5:CSTAR[1 2] CORDER2 W[2;]
[39]    FSTAR←W[1;⍳N1] GDIFFEQ CSTAR[3]×G
[40]    →(1E¯13≥DETGDIFFEQ)/0
[41]  CATENATE:GENERATORS←GENERATORS,FI,'   ',
         GFCN
[42]    ERRORS←ERRORS,F RERROR FSTAR
[43]    CSTARS←CSTARS,CSTAR
         ▽
```

Function Names: CONCAVEMAJ and CONVEXMIN

EXTERNAL DOCUMENTATION

PURPOSE

Used to preprocess data in order to find envelope. May be called before oscillating pattern is processed.

INTERNAL DOCUMENTATION

```
SYNTAX: RESULT←U CONCAVEMAJ V AND RESULT←U
 CONVEXMIN V RESPECTIVELY
PURPOSE: COMPUTE SMALLEST CONCAVE MAJORANT
 AND LARGEST CONVEX MINORANT RESPECTIVELY
INDEX ORIGIN: 1
AUTHORS: U. GRENANDER AND T. FERGUSON
LAST UPDATE: 4/80
LEFT ARGUMENT: VECTOR CONTAINING ORDINATES
 OF GIVEN FUNCTION
RIGHT ARGUMENT: VECTOR CONTAINING ABSCISSAS
 OF GIVEN FUNCTION
RESULT: SMALLEST CONCAVE MAJORANT OR LARGEST
 CONVEX MINORANT RESPECTIVELY
```

CODE

```
      ∇ RESULT←U CONCAVEMAJ V
[1]   ACOMPUTES SMALLEST CONCAVE MAJORANT
      AOF V W.R.T. U
[2]   AU SHOULD BE INCREASING
[3]    RESULT←-U CONVEXMIN-V
      ∇


      ∇ RESULT←U CONVEXMIN V;V1;I;I1;J
[1]   ACOMPUTES LARGEST CONVEX MINORANT OF
      AV W.R.T. U
[2]    N1←ρV
[3]    RESULT←N1ρV[1]
[4]    I←1
[5]   LOOP:V1←(V[I+ιN1-I]-V[I])÷U[I+ιN1-I]
      -U[I]
[6]    MIN←⌊/V1
[7]    I1←I+V1ιMIN
[8]    J←I1-I
[9]    RESULT[I+ιJ]←V[I]+MIN×U[I+ιJ]-U[I]
[10]   I←I1
[11]   →(I1<N1)/LOOP
      ∇
```

Function Names: CORDER1 and CORDER2

EXTERNAL DOCUMENTATION

PURPOSE

Used to compute first or second order difference. Operator in terms of central divided differences.

WARNING

Should not be used uncritically for numerical purpose other than the one described in Chapter 14.

INTERNAL DOCUMENTATION

SYNTAX: C CORDER1 X AND C CORDER2 X RE-
 SPECTIVELY
PURPOSE: TO CALCULATE DIFFERENCE OPERATOR
 OF ORDER 1 OR 2 WHEN CENTRAL DIVIDED
 DIFFERENCES ARE AVAILABLE
INDEX ORIGIN: 1
LAST UPDATE: 4/80
AUTHORS: U. GRENANDER AND T. FERGUSON
LEFT ARGUMENT: COEFFICIENTS OF LINEAR
 OPERATOR
RIGHT ARGUMENT: ABSCISSAS
RESULT: GLOBAL VARIABLES L0 AND L, WHERE
 L0 IS VECTOR OF WEIGHTS OF CONSTANT TERM.
 L IS MATRIX WITH TWO ROWS THAT HOLD THE
 WEIGHTS CORRESPONDING TO THE FIRST AND
 SECOND ORDER DIFFERENCES.
NOTE: FUNCTION USE GLOBAL VARIABLES X10,
 X1M.X0M COMPUTED BY CDIVDIFF

CODE

```
      ∇ C CORDER1 X
[1]   ⍝COMPUTES FIRST ORDER DIFFERENCE OPERATOR
[2]   ⍝IN TERMS OF VECTOR L0, LENGTH TWO LESS
      ⍝THAN X
[3]   ⍝AND MATRIX L WITH TWO ROWS
[4]   ⍝USES CENTRAL DIFFERENCES
[5]   ⍝USES GLOBAL VAR. X1M
[6]   N1←ρX
[7]   L0←÷X1M
[8]   L←(2,N1-2)ρ0
[9]   L[1;]←-(N1-2)ρC[1]
[10]  L[2;]←-L0
      ∇


      ∇ C CORDER2 X
[1]   ⍝COMPUTES SECOND ORDER DIFFERENCE
      ⍝OPERATOR
```

```
[2]   AIN TERMS OF VECTOR L0, LENGTH TWO LESS
[3]   ATHAN X AAND MATRIX WITH TWO ROWS.
[4]   AUSES CENTRAL DIFFERENCES AND GLOBAL
      AVARIABLES X10,X1M,X0M
[5]   N1←ρX
[6]   L0←(2÷X10×X1M)-C[1]÷X1M
[7]   L←(2,N1-2)ρ0
[8]   L[1;]←(-2÷X10×X0M)-C[2]
[9]   L[2;]←(2÷X0M×X1M)+C[1]÷X1M
      ∇
```

Function Name: FIT

EXTERNAL DOCUMENTATION

Fits monotonic pattern to ordinates in F, abscissas in X; both X and F
should be global variables. Collects results, coefficients, and error into
global variables GENERATORS, CSTARS, and ERRORS.

INTERNAL DOCUMENTATION

```
SYNTAX: FIT
PURPOSE: TO FIT F-VALUES BY GIVEN FUNCTIONS
  PSI,FI1 AND FI2
INDEX ORIGIN: 1
AUTHORS: U. GRENANDER, T. FERGUSON
LAST UPDATE: 4/80
RESULT: THE FUNCTION DOES NOT PRINT ANY
  RESULT BUT STORES RESULTS FOR LATE USE;
  SEE EXTERNAL DOCUMENTATION.
```

CODE

```
      ∇ FIT;U;V
[1]   AFITS GIVEN MONOTONIC PATTERN
[2]   U←⍕PSI,' ',⍕F
[3]   V←(N,2)ρ0
[4]   V[;1]←⍕FI1,' ',⍕X
[5]   V[;2]←⍕FI2,' ',⍕X
[6]   U LSQ V
```

```
[7]   ⍝BYPASSES IRREGULAR CONFIGURATIONS
[8]   →(((⍋1↓PSIINV)=21,23)∨.∧(N≠+/|×USTAR),
      (N≠+/×USTAR))/0
[9]   FSTAR←⍋PSIINV,' ',⍖USTAR
[10]  GENERATORS←GENERATORS,PSI,FI1,FI2
[11]  ERRORS←ERRORS,FSTAR RERROR F
[12]  CSTARS←CSTARS,CSTAR
      ∇
```

Function Name: FITO

EXTERNAL DOCUMENTATION

PURPOSE

To calculate fit to oscillating pattern. Calls CGAUTOR, which stores results in global variables GENERATORS, CSTARS, and ERRORS. These arrays contain the resulting values of fitted pattern, coefficients, and corresponding errors of approximations.

INTERNAL DOCUMENTATION

```
SYNTAX: FITO ORDER
PURPOSE: TO FIT OSCILLATING PATTERN
INDEX ORIGIN: 1
LAST UPDATE: 4/80
AUTHORS: U. GRENANDER AND T. FERGUSON
RIGHT ARGUMENT: ORDER OF DIFFERENCE EQUATION
RESULT: NO RESULT PRINTED; THEY ARE STORED
 IN GLOBAL VARIABLES GENERATORS, CSTARS, AND
 ERRORS
```

CODE

```
    ∇ FITO ORDER
[1]  ⍝FITS OSCILLATING PATTERN TO GIVEN F
    ⍝AND X
[2]  W←(2,ρF)ρF
[3]  W[2;]←⍋FI,' ',⍖X
[4]  G←⍋GFCN,' ',⍖X
[5]  ORDER CGAUTOR W
    ∇
```

Function Name: FWEIGHT

EXTERNAL DOCUMENTATION

PURPOSE

To compute weight to be used in calculating distance measures for main function asymptotics; see Chapter 14.

INTERNAL DOCUMENTATION

```
PURPOSE: COMPUTES WEIGHT VECTOR THAT IS
  NEEDED FOR DETERMINING THE DISTANCE BETWEEN
  TWO PATTERNS
SYNTAX: RESULT←FWEIGHT F
INDEX ORIGIN: 1
LAST UPDATE: 4/80
AUTHORS: U. GRENANDER AND T. FERGUSON
RIGHT ARGUMENT: ORDINATES OF FUNCTION STORED
  AS A VECTOR
RESULT: WEIGHT VECTOR
```

CODE

```
      ∇ RESULT←FWEIGHT F
[1]   N1←ρF
[2]   RESULT←MAJORANT←MONMAJ|F
[3]   J←MAJORANTι0
[4]   →(J>N1)/DONE
[5]   RESULT[J+0,ιN1-J]←(1+N1-J)ρMAJORANT[J-1]
[6]   DONE:RESULT←÷RESULT
      ∇
```

Function Name: GDIFFEQ

EXTERNAL DOCUMENTATION

PURPOSE

Finds the solution to given inhomogeneous linear difference equation with constant coefficients that is as close as possible to given *F*-vector.

Uses fundamental solution set for homogeneous equation.

INTERNAL DOCUMENTATION

SYNTAX: FSTAR←F GDIFFEQ G
PURPOSE: TO FIND BEST (LEAST SQUARES WITH
WEIGHTS DETERMINED BY FUNCTION FWEIGHT)
SOLUTION OF INHOMOGENEOUS LINEAR DIFFERENTIAL
EQUATION WHOSE RIGHT SIDE IS G.
INDEX ORIGIN: 1
LAST UPDATE: 4/80
AUTHORS: U. GRENANDER AND T. FERGUSON
LEFT ARGUMENT: GIVEN F-VECTOR REPRESENTING
PATTERN STUDIED
RIGHT ARGUMENT: G-VECTOR REPRESENTING RIGHT
SIDE OF INHOMOGENEOUS EQUATION
RESULT: VECTOR FSTAR IS THE DESIRED SOLUTION

CODE

```
        ∇ FSTAR←F GDIFFEQ G;BASIS;FUNDAMENTAL;
          R;I;J;DIAGONAL;INV;INNERV;COEFF
[1]   ⍝COMPUTES BEST LEAST SQUARES APPROXIMA-
[2]   ⍝TION TO F THAT SOLVES LSTAR=G
[3]   ⍝DIFFERENCE OPERATOR L HAS LEADING
      ⍝COEFF.S STORED IN VECTOR LO AND
[4]   ⍝REMAINING COEF.S STORED IN MATRIX WITH
[5]   ⍝R ROWS AND N-R COLUMNS
[6]   R←(ρL)[1]
[7]   N←ρF
[8]   BASIS←(⍳R)∘.=⍳R
[9]   FUNDAMENTAL←(R,N)ρ0
[10]  FUNDAMENTAL[⍳R;⍳R]←BASIS
[11]  INH←Nρ0
[12]  I←1
[13] LOOP1:J←R+1
[14] LOOP2:FUNDAMENTAL[I;J]←-(÷ L0[J-R])×
          L[;J-R]+.×FUNDAMENTAL[I;J-⍳R]
[15]  J←J+1
[16]  →(J≤N)/LOOP2
[17]  I←I+1
[18]  →(I≤R)/LOOP1
```

```
[19]    J←R+1
[20]    LOOP3:INH[J]←(÷LO[J-R])×G[J]-L[;J-R]
        +.×INH[J-ιR]
[21]    J←J+1
[22]    →(J≤N)/LOOP3
[23]    DIAGONAL←((ιN)∘.=ιN)×WEIGHT[ιN]∘.×Nρ1
[24]    INV←FUNDAMENTAL+.×DIAGONAL+.×⍉
        FUNDAMENTAL
[25]    →(1E‾13≥DETGDIFFEQ←DETERMINANT INV)/0
[26]    INV←⊞INV
[27]    INNERV←FUNDAMENTAL+.×DIAGONAL+.×F-INH
[28]    COEFF←INV+.×INNERV
[29]    FSTAR←INH++/[1](COEFF∘.×Nρ1)×
        FUNDAMENTAL
        ∇
```

Function Names: G10,G11,G12,G13,G21,G22,G23,G24,G25,G26,G27

EXTERNAL DOCUMENTATION

PURPOSE

Set of generators (primitive patterns) from which image algebra is derived as described in Chapter 14. This is done by letting the functions *PSI*, *FI*1, *FI*2, *FI*, and *G* be defined as one of the above list *G*10, *G*11, . . . and by pseudolinearization.

WARNING

The list is only intended as an example and should be modified by the experimenter in order to get a sufficiently flexible and suitable image algebra. If this is done, one must remember to redefine the numbers IMAX, JMAX, KMAX, IMAX0, JMAX0 as well as the matrices PSIL, FIL, GL, FI1L, FI2, PSIINVL. The choice should also take into account possible domain and value errors.

INTERNAL DOCUMENTATION

```
SYNTAX: Z←G10 X, Z←G11 X, ETC.
PURPOSE: TO DEFINE GENERATORS FROM WHICH
 IMAGES ARE FORMED
```

INDEX ORIGIN: 1
AUTHORS: *U. GRENANDER AND T. FERGUSON*
RIGHT ARGUMENT: *ABSCISSA VALUE OF FUNCTION*
RESULT: *ORDINATE OF FUNCTION*

CODE

```
      ∇ Z←G10X              ∇ Z←G23 X
[1]   Z←(ρX)ρ0        [1]   Z←⊛X
      ∇                     ∇


      ∇ Z←G11 X             ∇ Z←G24 X
[1]   Z←(ρX)ρ1        [1]   Z←X*0.5
      ∇                     ∇


      ∇ Z←G12 X             ∇ Z←G25 X
[1]   Z←X             [1]   Z←X*⁻0.5
      ∇                     ∇


      ∇ Z←G13 X             ∇ Z←G26 X
[1]   Z←X*2           [1]   Z←÷X*2
      ∇                     ∇


      ∇ Z←G21 X             ∇ Z←G27 X
[1]   Z←÷X            [1]   Z←÷1+⊛X
      ∇                     ∇


      ∇ Z←G22 X
[1]   Z←*X
      ∇
```

Function Name: LSQ

EXTERNAL DOCUMENTATION

PURPOSE

To calculate least squares fit of vector U in terms of column vectors of matrix V. This is done in the metric corresponding to vector WEIGHT1

INTERNAL DOCUMENTATION

SYNTAX: U LSQ V
PURPOSE: COMPUTES LINEAR COMBINATION OF
  COLUMN VECTORS OF MATRIX V CLOSEST IN
  THE SENSE OF THE METRIC DEFINED BY VECTOR
  WEIGHT1.
INDEX ORIGIN: 1
LAST UPDATE: 4/80
AUTHORS: U. GRENANDER AND T. FERGUSON
LEFT ARGUMENT: PATTERN TO BE FITTED
RIGHT ARGUMENT: MATRIX WHOSE COLUMNS ARE THE
  VECTORS SPANNING GIVEN SUBSPACE
RESULT: THE RESULTING COEFFICIENT VECTOR IS
  GLOBAL VARIABLE CSTAR AND BEST APPROXIMATION
  TO U IS GLOBAL VARIABLE USTAR.

CODE

```
      ∇ U LSQ V;DIAGONAL;INV;INNERV
[1]   N←ρU
[2]   ⍝COMPUTES LEAST SQUARES FIT OF U-VECTOR
[3]   ⍝AS LINEAR COMBINATION OF COLUMNS OF
      ⍝V-MATRIX
[4]   ⍝RESULTING COEFF.S STORED IN CSTAR-VECTOR
[5]   DIAGONAL←((ιN)∘.=ιN)×WEIGHT1∘.×Nρ1
[6]   INV←⌹(⍉V)+.×DIAGONAL+.×V
[7]   INNERV←(⍉V)+.×DIAGONAL+.×U
[8]   CSTAR←INV+.×INNERV
[9]   USTAR←V+.×CSTAR
      ∇
```

Function Name: MON

EXTERNAL DOCUMENTATION

PURPOSE

Master program executing search for monotonic pattern. It computes
image algebra from given generators, determines distance to closest image
by called function FIT.

WARNING

Value and domain errors can occur unless set of generators is chosen carefully.

INTERNAL DOCUMENTATION

```
SYNTAX: MON START
PURPOSE: TO SEARCH FOR MONOTONIC PATTERNS
INDEX ORIGIN: 1
LAST UPDATE: 4/80
AUTHORS: U. GRENANDER AND T. FERGUSON
RIGHT ARGUMENT: DECIDES WHERE IN GENERATOR
 SPACE SEARCH SHALL BE STARTED AS DESCRIBED
 BY 3-VECTOR MEANING THE SUBSCRIPT I,J,K.
RESULT: FUNCTION FIT CALLED BY MON STORES
 RESULTS WITHOUT IMMEDIATE PRINTOUT; SEE
 DESCRIPTION OF FIT.
```

CODE

```
        ∇ MON START;I;J;K
[1]     ⍝GENERATES AND FITS MONOTONIC PATTERNS
[2]     I←START[1]
[3]     J←START[2]
[4]     K←START[3]
[5]     LOOPM:PSI←'G',(⍕PSIL[1;I]),⍕PSIL[2;I]
[6]     FI1←'G',(⍕FI1L[1;J]),⍕FI1L[2;J]
[7]     FI2←'G',(⍕FI2L[1;K]),⍕FI2L[2;K]
[8]     PSIINV←'G',(⍕PSIINVL[1;I]),⍕PSIINVL[2;I]
[9]     FIT
[10]    K←K+1
[11]    →(K≤KMAX)/LOOPM
[12]    K←1+J←J+1
[13]    →(J≤JMAX)/LOOPM
[14]    I←I+1
[15]    K←1+J←1
[16]    →(I≤IMAX)/LOOPM
[17]    PRINTOUT
        ∇
```

Function Names: MONMAJ and MONMIN

EXTERNAL DOCUMENTATION

PURPOSE

To compute smallest nonincreasing majorant and largest nondecreasing minorant.

INTERNAL DOCUMENTATION

```
SYNTAX: Z←MONMAJ U OR Z←MONMIN U
PURPOSE: CALCULATES MONOTONIC MAJORANT
 OR MINORANT OF VECTOR U
INDEX ORIGIN: 1
LAST UPDATE: 4/80
AUTHORS: U. GRENANDER AND T. FERGUSON
RIGHT ARGUMENT: VECTOR OF ORDINATES
RESULT: CORRESPONDING MONOTONIC MAJORANT
 OR MINORANT
```

CODE

```
      ∇ Z←MONMAJ U
[1]  ⍝COMPUTES SMALLEST NONINCREASING MAJORANT
     ⍝OF U
[2]   Z←⌽⌈\⌽U
      ∇


      ∇ Z←MONMIN U
[1]  ⍝COMPUTES LARGEST NONDECREASING MINORANT
     ⍝OF U
[2]   Z←-MONMAJ-U
      ∇
```

Function Name: NORM

EXTERNAL DOCUMENTATION

PURPOSE

Calculates LP-norm of vector with Q-stress on arguments. Metric defined by global vector-weight.

INTERNAL DOCUMENTATION

```
SYNTAX: RESULT←U NORM V
PURPOSE: TO COMPARE LP-NORM OF VECTOR U.
 P IS GLOBAL VARIABLE WITH 0 MEANING SUPNORM.
 Q ALSO GLOBAL EXPRESSES STRESS ON LARGE
 ABSCISSAS
INDEX ORIGIN: 1
LAST UPDATE: 4/80
AUTHORS: U. GRENANDER AND T. FERGUSON
LEFT ARGUMENT: VECTOR OF ABSCISSAS
RIGHT ARGUMENT: VECTOR OF ORDINATES
RESULT: NORM OF VECTOR
```

CODE

```
     ∇ RESULT←U NORM V
[1]  ⍝COMPUTES LP-NORM OF V WITH Q-STRESS ON
[2]  ⍝LARGE VALUES OF ARGUMENTVALUES IN U
[3]  →(P>0)/FINITEP
[4]  RESULT←⌈/((⌊U)*Q)×WEIGHT×(⌊V)
[5]  →0
[6]  FINITEP:RESULT←(+/((⌊U)*Q)×WEIGHT×
     (⌊V)*P)*÷P
     ∇
```

Function Name: OSC

EXTERNAL DOCUMENTATION

PURPOSE

Master program executing search for oscillating pattern. It computes image algebra from given generators, determines distance to closest image by calling function FITO.

NOTE

Can sometimes successfully find also nonoscillating pattern.

WARNING

Value and domain errors can occur unless set of generators is chosen carefully.

INTERNAL DOCUMENTATION

SYNTAX: OSC START
PURPOSE: TO SEARCH FOR OSCILLATING PATTERNS
INDEX ORIGIN: 1
LAST UPDATE: 4/80
AUTHORS: U. GRENANDER AND T. FERGUSON
RIGHT ARGUMENT: DECIDES WHERE IN GENERATOR
 SPACE SEARCH SHALL BE STARTED AS DESCRIBED
 BY 2-VECTOR MEANING THE SUBSCRIPT I1 AND J1.
RESULT: FUNCTION FITO CALLED BY OSC STORES
 RESULT WITHOUT IMMEDIATE PRINTOUT; SEE
 DESCRIPTION OF FITO.

CODE

```
      ∇ OSC START;I1;J1
[1]   ⍝SEARCHES FOR OSCILLATING PATTERN OF
      ⍝ORDER 2
[2]    I1←START[1]
[3]    J1←START[2]
[4]   LOOPO:FI←'G',(⍕FIL[1;I1]),⍕FIL[2;I1]
[5]    GFCN←'G',(⍕GL[1;J1]),⍕GL[2;J1]
[6]    FITO 2
[7]    J1←J1+1
[8]    →(J1≤JMAXO)/LOOPO
[9]    J1←1
[10]   I1←I1+1
[11]   →(I1-IMAXO)/LOOPO
[12]   PRINTOUT
      ∇
```

Function Name: PRINTOUT

EXTERNAL DOCUMENTATION

PURPOSE

To print out results of functions MON and OSC.

INTERNAL DOCUMENTATION

SYNTAX: PRINTOUT
PURPOSE: PRINTS RESULTS OF MON OR OSC USES
GLOBAL VARIABLES GENERATORS, CSTARS, ERRORS
INDEX ORIGIN: 1
LAST UPDATE: 4/80
AUTHOR: T. FERGUSON
RESULT: PRINTS RESULTS AFTER INTERROGATING
EXPERIMENTER ABOUT DESIRED NUMBER OF BEST
FITTING IMAGES.

CODE

```
        ∇ PRINTOUT;Q;NE;FITS;TYPE;MONPRINT;
          OSCPRINT
[1]     NE←ρERRORS
[2]     ORDER←⍋ERRORS
[3]     ERRORS←ERRORS[ORDER]
[4]     GENERATORS←(⍉(NE,9)ρGENERATORS)[;ORDER]
[5]     CSTARS←((NE,(ρCSTARS)÷NE)ρCSTARS)
          [ORDER;]
[6]     'PATTERNS ARE ORDERED ACCORDING TO
          GOODNESS OF FIT'
[7]     'HOW MANY SHOULD BE PRINTED OUT?
          (MAX=',(⍕NE),')'
[8]     FITS←⎕
[9]     MONPRINT← 3 6 ρ'PSI = ','FI1 = ',
          'FI2 = '
[10]    OSCPRINT← 3 5 ρ'FI = ','AND ','G = '
[11]    TYPE←(3+⁻6×(GENERATORS[4;1]=' '))↑
          'MONOSC'
[12]    Q←1
[13]  LOOPRINT:'FOR PATTERN WITH'
[14]    (⍕TYPE,'PRINT'), 3 3 ρGENERATORS[;Q]
[15]    'THE RELATIVE ERROR IS'
[16]    ERRORS[Q]
[17]    '(CSTAR = ',(⍕CSTARS[Q;]),')'
[18]    Q←Q+1
[19]    ι0
```

```
[20]   →(Q≤FITS)/LOOPRINT
[21]   GENERATORS←''
[22]   ERRORS←ι0
[23]   CSTARS←ι0
       ∇
```

Function Name: RERROR

EXTERNAL DOCUMENTATION

PURPOSE

To calculate the relative distance from $V1$ to $V2$ w.r.t. $V1$.

INTERNAL DOCUMENTATION

```
SYNTAX: Z←V1 RERROR V2
PURPOSE: COMPUTE DISTANCE FROM V1 TO V2
 DIVIDED BY NORM OF V1.
INDEX ORIGIN: 1
LAST UPDATE: 4/80
AUTHORS: U. GRENANDER AND T. FERGUSON
LEFT ARGUMENT: VECTOR V1
RIGHT ARGUMENT: VECTOR V2
RESULT: RELATIVE ERROR
```

CODE

```
    ∇ Z←V1 RERROR V2
[1] ⍝COMPUTES RELATIVE ERROR OV V2 W.R.T. V1
[2] ⍝USES WEIGHT-VECTOR
[3] ⍝ASSUMES X-VECTOR GIVEN
[4]  Z←(X NORM V1-V2)÷X NORM V1
    ∇
```

CHAPTER 20

Geometry

Most of the programs in this chapter are designed for operations on convex sets in the plane. They were originally written in APL/360 by A. Tarr and R. A. Vitale and were recently modified by G. Podorowsky. The modification was necessary to convert the code to VS APL. More importantly, some of the algorithms were replaced by fast ones, improving the speed of the programs considerably.

They deal with convex, bounded closed polygons. The polygons are represented in three ways:

(1)   Raw form: If the polygon has n faces (n vertices), the raw form is a $2 \times n$ matrix of vertices in counterclockwise order, with X values in the first row and Y values in the second row (the choice of initial vertex is immaterial). For example, a triangle with vertices at $(-1,0)$, $(0.5,0)$, and $(-1,1)$ would be represented as

$$\text{RAWFORM} \quad \begin{array}{rrr} -1 & 0 & 0.5 \\ 0 & -1 & 1 \end{array}$$

(2)   Standard form: a $3 \times n$ matrix. The first two rows list the vertices in counterclockwise order as in the raw form. However, the initial vertex must be chosen as follows: Choose the vertex with the smallest Y value; if there is more than one, choose the one that has the smallest X value. The third row contains a list of angles A_j, where A_j is the angle to the horizontal of the vector $(X_{j+1}, Y_{j+1}) - (X_j, Y_j)$ and, for $j = n$, of the vector $(X_1, Y_1) - (X_n, Y_n)$.

NOTE

$0 \le A1 < A2 < \cdots < 2\pi$. For the same triangle as above, the representation would be

$$\text{STANDARDFORM} \quad \begin{array}{rrr} 0 & 0.5 & -1 \\ -1 & 1 & 0 \\ 1.3258 & 3.7296 & 5.4978 \end{array}$$

(3)   Support form: The support function of a polygon is given by

$$s(A) = \max X \cos(A) + Y \sin(A), \qquad (X, Y) \text{ a vertex}$$

In support form we shall use as the first row the (signed) distance from the origin to the line that contains a face. The second row will be angles $0 <$ $B1 < B2 < \cdots < 2\pi$ that correspond to the normals.

SUPPORTFORM
$$\begin{array}{ccc} 0.5547 & 0.70711 & 0.24254 \\ 2.1588 & 3.927 & 6.0382 \end{array}$$

An experimenter who plans to carry out large experiments is advised to consult Shamos (1976) for faster algorithms. The following programs are adequate for small and moderate sizes of the experiment.

We also include a few programs with an entirely different purpose: to compute objects in a Riemannian geometry. In particular, one of these programs calculates geodesic curves.

Function Names: ADDC and ADDC1

EXTERNAL DOCUMENTATION

Both of these functions are designed to add convex sets in the sense of Minkowsky addition

$$C_1 + C_2 = \{z = z_1 + z_2 | z_1 \in C_1, z_2 \in C_2\}$$

ADDC adds two convex sets while ADDC1 adds an arbitrary number of sets with the same number of vertices.

WARNING

The algorithm is fairly slow; faster ones can be designed.

INTERNAL DOCUMENTATION FOR ADDC

```
SYNTAX:K←K1 ADDC K2
PURPOSE: TO ADD TWO CONVEX POLYGONS
INDEX ORIGIN: 1
AUTHOR: A. TARR, R.A. VITALE
LAST UPDATE: 3/8/80
LEFT ARGUMENT: K1: A CONVEX POLYGON, IN
 STANDARD FORM
```

RIGHT ARGUMENT: *K2*: *A CONVEX POLYGON, IN*
STANDARD FORM
RESULT: *K*: *A CONVEX POLYGON, ALSO IN*
STANDARD FORM

CODE FOR ADDC

```
       ∇ K←K1 ADDC K2;F;G;D;ORD;MAX;I;KK
[1]    F←K1[1 2 ;]
[2]    G←K2[1 2 ;]
[3]    D← 0 ¯1 ↓(((1ΦF)-F),((1ΦG)-G))[;⍋
       ORD←(K1[3;],K2[3;])]
[4]    KK← 2 1 ρ(F[;1]+G[;1])
[5]    MAX←(ρD)[2]
[6]    I←0
[7]    LP:→10×⍳MAX<I←I+1
[8]    KK←KK, 2 1 ρD[;I]+,KK[;¯1↑ρKK]
[9]    →LP
[10]   K←REMFVT(3,ρKK[1;])ρ(,KK),ORD[⍋ORD]
       ∇
```

INTERNAL DOCUMENTATION FOR ADDC1

SYNTAX: *K←ADDC*1 *FIG*
PURPOSE: *TO ADD ANY NUMBER OF POLYGONS*
 WHICH HAVE THE SAME NUMBER OF VERTICES
INDEX ORIGIN: 1
AUTHOR: *A. TARR, R.A. VITALE*
LAST UPDATE: *3/8/80*
RIGHT ARGUMENT: *FIG*: *A STACKED STANDARD*
 FORM ARRAY
RESULT: *K*:*STANDARD FORM SUM OF POLYGONS*
 IN FIG

CODE FOR ADDC1

```
       ∇ K←ADDC1 FIG;D;DX;DY;F;A;B;MAX;I;
       KK;N
[1]    D←(1ΦF)-F←FIG[B[⍋B←(¯2+3×⍳N),(¯1+3
       ×⍳N←(ρFIG[;1])÷3)];]
```

```
[2]    DX←,D[(¯1+2×ιN);]
[3]    DY←,D[(2×ιN);]
[4]    A←,FIG[(3×ιN);]
[5]    KK← 2 1 ρ(+/F[(¯1+2×ιN);1]),(+/F
       [(2×ιN);])
[6]    D← 0 ¯1 ↓(((2,ρDX)ρDX,DY)[;⍋A])
[7]    MAX←(ρD)[2]
[8]    I←0
[9]  LOOP:→12×ιMAX<I←I+1
[10]   KK←KK, 2 1 ρD[;I]+,KK[;¯1↑ρKK]
[11]   →LOOP
[12]   K←REMFVT(3,ρKK[1;])ρ(,KK),A[⍋A]
       ∇
```

Function Name: AREA

EXTERNAL DOCUMENTATION

PURPOSE

To compute the area of a convex polygon given in raw or standard form.

INTERNAL DOCUMENTATION

```
SYNTAX: A←AREA P
PURPOSE: TO FIND THE AREA OF P
INDEX ORIGIN: 1
AUTHOR: A. TARR, R.A. VITALE
LAST UPDATE: 3/8/80
RIGHT ARGUMENT: P: A CONVEX POLYGON, IN
  EITHER RAW OR STANDARD FORM
RESULT: A: AREA OF P
```

CODE

```
      ∇ A←AREA P;X;Y
[1]    A←0.5×VADD(X×1ϕY),-(Y←P[2;])×1ϕX←P[1;]
       ∇
```

Function Name: ATAN

EXTERNAL DOCUMENTATION

PURPOSE

To compute the angle between two straight lines. One line given by two points on, the other line horizontal through given point.

INTERNAL DOCUMENTATION

```
SYNTAX: A←X0Y0 ATAN XY
PURPOSE: TO COMPUTE THE ANGLE BETWEEN THE
 LINE THROUGH X0Y0 AND XY AND THE POSITIVE
 HORIZONTAL THROUGH X0Y0
INDEX ORIGIN: 1
AUTHOR: A. TARR, R.A. VITALE
LAST UPDATE: 3/8/80
LEFT ARGUMENT: X0Y0: THE COORDINATES OF THE
 ORIGINAL REFERENCE POINT (A VECTOR OF LENGTH
 2)
RIGHT ARGUMENT: XY: 2×N MATRIX OF POINTS
 NOT CONTAINING X0Y0
RESULT: A: VECTOR OF LENGTH N, CONTAINING FOR
 EACH POINT OF XY, THE ANGLE BETWEEN THE LINE
 THROUGH X0Y0 AND THAT POINT, AND THE
 POSITIVE HORIZONTAL THROUGH X0Y0
```

CODE

```
      ∇ A←X0Y0 ATAN XY;X;Y;DX;DY
[1]   X←XY[1;]
[2]   Y←XY[2;]
[3]   DX←X-X0Y0[1]
[4]   DY←Y-X0Y0[2]
[5]   A←((○(DY=0)×DX<0)+(DY≠0)×(○1-0.5×DY)
      -‾3○DX÷DY+DY=0)
      ∇
```

Function Names: AXF and CIRCUM

EXTERNAL DOCUMENTATION

PURPOSE

To compute a rectangle circumscribing a given polygon, and to circumscribe a polygon by one with given angles.

INTERNAL DOCUMENTATION FOR AXF

```
SYNTAX: FENCE←AXF F
PURPOSE: TO CIRCUMSCRIBE A CONVEX POLYGON
 BY A RECTANGLE
INDEX ORIGIN: 1
AUTHOR: A. TARR, R.A. VITALE
LAST UPDATE: 3/8/80
RIGHT ARGUMENT: F: A CONVEX POLYGON, IN
 RAW OR STANDARD FORM
RESULT: FENCE: STANDARD FORM RECTANGLE
 CIRCUMSCRIBED ABOUT F; THE SIDES OF FENCE
 ARE ORIENTED WITH THE AXES
```

CODE FOR AXF

```
      ∇ FENCE←AXF F;MX;MY;XM;YM
[1]   →(((ρF[2;])=2)∧(F[1;1]=F[1;2])∨F[2;1]=
      F[2;2])/LN
[2]   FENCE← 3 4 ρMX,XM,(XM←⌈/F[1;]),(MX←⌊/F
      [1;]),MY,(MY←⌊/F[2;]),YM,(YM←⌈/F[2;]),
      ○0,0.5,1,1.5
[3]   →0
[4]   LN:FENCE←STFORM F
      ∇
```

INTERNAL DOCUMENTATION FOR CIRCUM

```
SYNTAX: FENCE← PHI CIRCUM FIG
PURPOSE: TO CIRCUMSCRIBE A POLYGON WITH
 GIVEN FACE ANGLES ABOUT ANOTHER CONVEX
 POLYGON
```

```
INDEX ORIGIN: 1
AUTHOR: A. TARR, R.A. VITALE
LAST UPDATE: 3/8/80
LEFT ARGUMENT: PHI: THE VECTOR OF FACE
  ANGLES
RIGHT ARGUMENT: FIG: THE POLYGON TO BE
  CIRCUMSCRIBED, IN STANDARD FORM
RESULT: FENCE: POLYGON CIRCUMSCRIBED ABOUT
  FIG WITH FACE ANGLES PHI; IN STANDARD FORM
```

CODE FOR CIRCUM

```
     ∇ FENCE←PHI CIRCUM FIG;S
[1]    FENCE←POLY PHI SUPPORT SFACE FIG
     ∇
```

Function Names: DCNP, DECOMP, DEQUAD

EXTERNAL DOCUMENTATION

PURPOSE

These functions perform triangulation of convex figures.

NOTE

For special cases the user should rotate figure first (to avoid domain error) and then rotate back.

If a convex polygon has two parallel sides, it is the sum of a degenerate polygon (a line segment) and another polygon. Those line segments should be removed before using programs.

INTERNAL DOCUMENTATION FOR DCNP

```
SYNTAX: TR←DCNP FIG
PURPOSE: TO FIND A SET OF TRIANGLES WHOSE
  SUM IS SOME TRANSLATION OF A GIVEN POLYGON
  (WHICH HAS NO PARALLEL SIDES)
INDEX ORIGIN: 1
AUTHOR: A. TARR, R.A. VITALE
```

LAST UPDATE: 5/20/80
RIGHT ARGUMENT: FIG: THE ORIGINAL POLYGON,
 IN STANDARD FORM; THE POLYGON MUST HAVE
 NO PARALLEL SIDES
RESULT: TR: STACKED STANDARD FORM ARRAY
 OF TRIANGLES SUCH THAT THEIR SUM IS SOME
 TRANSLATION OF THE ORIGINAL FIGURE
NOTE: THERE ARE ((NUMBER OF SIDES OF FIG)-2)
 TRIANGLES

CODE FOR DCNP

```
       ∇ TR←DCNP F;T;I;J;L;RF;TH;X;Y;C;D;
         A;MIN;S
[1]    TR← 0 3 ρ0
[2]    LP:→((ρF)[2]=3)/END
[3]    I←(L=MIN←⌊/L)/ιρL←(((((1⌽F[1;])-F[1;])
       *2)+((1⌽F[2;])-F[2;])*2)*0.5
[4]    RF←(-A←F[3;I[1]]) ROTATE F
[5]    X←RF[1;]
[6]    Y←RF[2;]
[7]    TH←RF[3;]
[8]    J←(Y=⌈/Y)/ιρY
[9]    X[J]←(C←X[J])+MIN×S÷(((3○TH[J]×S
       ←X[J]≠X[(1⌽ιρX)[J]])÷(3○TH[J-1]))-1
[10]   Y[J]←(D←Y[J])+(S×(3○TH[J]×S)×(X[J]-C))
       -|(~S)×MIN÷3○(TH[J-1]-○○.5)
[11]   TR←TR,[1] T←A ROTATE 3 3 ρX[J],(MIN+X
       [J]),C,Y[J],Y[J],Y[J],D,○,TH[J-1],TH[J]
[12]   X[1+ιJ-2]←X[1+ιJ-2]-MIN
[13]   F←A ROTATE 0 1 ↓(3,ρX)ρX,Y,TH
[14]   →LP
[15] END:TR←TR,[1] F
       ∇
```

INTERNAL DOCUMENTATION FOR DECOMP

SYNTAX: TR←DECOMP FIG
PURPOSE: TO DECOMPOSE A CONVEX POLYGON INTO
 THE SUM OF TRIANGLES (AND/OR LINE SEGMENTS)
INDEX ORIGIN: 1

AUTHOR: A. TARR, R.A. VITALE
LAST UPDATE: 3/8/80
RIGHT ARGUMENT: FIG: THE ORIGINAL POLYGON,
 IN STANDARD FORM WITH NO PARALLEL SIDES.
RESULT: TR: STACKED STANDARD FORM ARRAY OF
 TRIANGLES (AND/OR LINE SEGMENTS) SUCH
 THAT THEIR SUM IS THE ORIGINAL FIGURE

CODE FOR DECOMP

```
        ∇ TR←DECOMP FIG;A;C;D;F;H;I;II;J;L;
          LI;LIJ;LJ;N;Q;TAN;U;V;W;S;X;Y;T
[1]     F←FIG
[2]     TR← 0 3 ρ0
[3]     H←TAN[Q←⍋TAN←3○FIG[3;]]
[4]     →(0=+/V←1E⁻10>|(1⌽H)-H)/CALL
[5]     C←V/⍳ρV
[6]     I←Q[C]⌊Q[C+1]
[7]     J←Q[C]⌈Q[C+1]
[8]     L←((((1⌽X)-X←FIG[1;])*2)+((1⌽Y)-
          Y‹FIG[2;])*2)*0.5
[9]     LIJ←U/⍳ρU←L[I]=L[J]
[10]    L←V/⍳ρV←L[I]>L[J]
[11]    LJ←W/⍳ρW←L[I]<L[J]
[12]    S←(I×(U∨W))+J×V
[13]    II←0
[14]  LOOP:II←II+1
[15]    N←I[II]+⍳J[II]-I[II]
[16]    TR←TR,[1] T[;1],T←(3 2 ρX[S[II]],
          X[L],Y[S[II]],Y[L←1↑S[II]⌽⍳ρX],A)
          [;⍋A←F[3;S[II]],(○2)|F[3;S[II]]+○1]
[17]    X[N]←X[N]+(×F[3;I[II]]-○0.5)×|X[L]-
          X[S[II]]
[18]    Y[N]←Y[N]-|Y[L]-Y[S[II]]
[19]    →(II<ρI)/LOOP
[20]    V←(ρX)ρ1
[21]    V[I[LIJ],J[LIJ],J[LI],I[LJ]]←0
[22]    F←V/(3,ρX)ρX,Y,FIG[3;]
[23]    →(3>(ρF)[2])/FUDGE
[24]  CALL:TR←TR,[1] DCNP F
[25]  FUDGE:D←(((+/TR[(1+N);1]),(+/TR[(2+N←3
          ×⁻1+⍳N);1])))-FIG[1 2 ;1])÷N←(ρTR)[1]÷3
```

```
[26]    TR[(1+N);]←TR[(1+N);]-D[1]
[27]    TR[(2+N);]←TR[(2+N);]-D[2]
        ∇
```

INTERNAL DOCUMENTATION FOR DEQUAD

SYNTAX: TRI←DEQUAD QUAD
PURPOSE: TO DECOMPOSE A QUADRILATERAL INTO
THE SUM OF TRIANGLES (AND/OR LINE SEGMENTS)
INDEX ORIGIN: 1
AUTHOR: A. TARR, R.A. VITALE
LAST UPDATE: 3/8/80
RIGHT ARGUMENT: QUAD: THE QUADRILATERAL, IN
STANDARD FORM
RESULT: TRI: STACKED STANDARD FORM ARRAY OF
TRIANGLES (AND/OR LINE SEGMENTS) SUCH THAT
THEIR SYM IS QUAD

CODE FOR DEQUAD

```
        ∇ TRI←DEQUAD QUAD;N;V;TAN;TH;X;Y;W;
          C;MIN;T1;T2;L;Q;B;R;S;T;D;I;J;H
[1]     TAN←3○TH←QUAD[3;]
[2]     X←QUAD[1;]
[3]     Y←QUAD[2;]
[4]     →((0=N),1=N←+/W←1E¯10≥|TAN[1 2]-TAN[3
        4])/ZERO,ONE
[5]     'PARALLELOGRAM'
[6]     T1← 3 2 ρX[1],X[2],Y[1],Y[2],TH[1],TH[3]
[7]     T2← 3 2 ρX[1],X[4],Y[1],Y[4],TH[2],TH[4]
[8]     →FUDGE
[9]   ONE:'TRAPEZOID'
[10]    I←W/ιρW
[11]    C←(L=MIN←⌊/L)/ιρL←(((((1⌽X)-X)*2)+
        (((1⌽Y)-Y)*2))*0.5)[I,I+2]
[12]    Q←(¯1+J←,(2 2 ρι4)[C;I])⌽ι4
[13]    T1←T1[;1],T1←(3 2 ρX[J],X[Q[2]],
        Y[J],Y[Q[2]],H)[;⍋H←(TH[J],TH[Q[3]])]
[14]    H←TH[Q[2]],TH[Q[3]],TH[Q[4]]
[15]    N←MIN÷(1+TAN[I]*2)*0.5
[16]    T2←(3 3ρX[J],(X[Q[3]]+N××X[Q[4]]-
        X[Q[3]]),X[Q[4]],Y[J],(Y[Q[3]]+
```

```
        (×(Y[Q[4]]-Y[Q[3]])))×N×|TAN[I]),
        Y[Q[4]],H)[;⍙H]
[17]    →FUDGE
[18]    ZERO:'QUADRILATERAL'
[19]    C←B/⍳⍴B←D=⌈/D←L+1⌽L←((((1⌽X)-X)*2)+
        ((1⌽Y)-Y)*2)*0.5
[20]    R←X[I←(Q←(C-2)⌽⍳4)[1]]
[21]    S←Y[I]
[22]    V←(Y[J]+(R×D)-(S+T×X[J]))÷((D←⌽T)-
        T←TAN[J←Q[2 3]])
[23]    W←S+D×(V-R)
[24]    T1←(3 3 ⍴R,X[Q[2]],V[1],S,Y[Q[2]],
        W[1],H)[;⍙H←TH[I],TH[Q[2]],TH[Q[3]]]
[25]    T2←(3 3 ⍴R,V[2],X[Q[4]],S,W[2],Y[Q[4]],
        H)[;⍙H←TH[Q[2]],TH[Q[3]],TH[Q[4]]]
[26]    FUDGE:D←0.5×T1[1 2 ;1]+T2[1 2 ;1]-
        QUAD[1 2 ;1]
[27]    TRI←T1,[1] T2
[28]    TRI[1 4 ;]←TRI[1 4 ;]-D[1]
[29]    TRI[2 5 ;]←TRI[2 5 ;]-D[2]
        ∇
```

Function Names: DIST, DISTANCE, DISTPT

EXTERNAL DOCUMENTATION

PURPOSE

The functions DIST and DISTPT compute Hausdorff distances between convex sets and between a point and a convex set. The auxiliary function distance is called by DIST.

WARNING

The program is fairly slow for a large number of vertices; faster algorithms exist.

INTERNAL DOCUMENTATION FOR DIST

SYNTAX: D←K1 DIST K2
PURPOSE: TO COMPUTE THE HAUSDORFF DISTANCE
BETWEEN K1 AND K2

INDEX ORIGIN: 1
AUTHOR: A. TARR, R.A. VITALE
LAST UPDATE: 5/20/80
LEFT ARGUMENT: K1: A POLYGON, IN STANDARD
FORM
RIGHT ARGUMENT: K2: A POLYGON, IN
STANDARD FORM
RESULT: D: THE HAUSDORFF DISTANCE BETWEEN
K1 AND K2

CODE FOR DIST

```
      ∇ D←K1 DIST K2
[1]   D←⌈/(K1 DISTANCE K2),K2 DISTANCE K1
      ∇
```

CODE FOR DISTANCE

```
      ∇ D←K1 DISTANCE K2;M;S1;S2;A1;A2;
        A;I;PHI;B;B1;B2
[1]   S1←SFACE K1
[2]   S2←SFACE K2
[3]   K1←(¯1+1↑I←(K1[3;]≥○0.5)/⍳ρK1[3;])
      ⌽K1
[4]   K2←(¯1+1↑I←(K2[3;]≥○0.5)/⍳ρK2[3;])
      ⌽K2
[5]   PHI←(0,0) ATAN(2,×/ρM)ρ(,K2[1;]∘.-K1
      [1;]),(,M←K2[2;]∘.-K1[2;])
[6]   A1←((ρA[1;])ρS1[2;(ρS1[2;])]-○2),
      [1] A←⍉(⌽ρM)ρS1[2;]
[7]   A2←((ρA[;1])ρS2[2;(ρS2[2;])]-○2),
      B←(ρM)ρS2[2;]
[8]   B1←¯1⌽((ρA[1;])ρ(S1[2;1]+○2)),[1] A
[9]   B2←¯1⊖((ρA[;1])ρ(S2[2;]+○2)),'B
[10]  I←((PHI≥, ¯1 0 ↓A1)∨(PHI≥, ¯1 0 ↓B1))
      ∧((PHI≤, 1 0 ↓A1)∨(PHI≤, 1 0 ↓B1))
[11]  I←(I∧((PHI≥, 0 ¯1 ↓A2)∨(PHI≥, 0 ¯1
      ↓B2))∧((PHI≤, 0 1 ↓A2)∨(PHI≤, 0 1 ↓B2)))
      /⍳ρPHI
[12]  D←⌈/|(A SUPPORT S1)[1;]-((A←PHI[I],
      S1[2;],S2[2;]) SUPPORT S2)[1;]
      ∇
```

INTERNAL DOCUMENTATION FOR DISTPT

```
SYNTAX: D←K DISTPT PT
PURPOSE: TO FIND THE HAUSDORFF DISTANCE
 BETWEEN A POINT AND A POLYGON
INDEX ORIGIN: 1
AUTHOR: A. TARR, R.A. VITALE
LAST UPDATE: 5/20/80
LEFT ARGUMENT: K: A POLYGON, IN RAW OR
 STANDARD FORM
RIGHT ARGUMENT: PT: A VECTOR OF LENGTH TWO;
 THE POINT (X,Y)
RESULT: D: THE MAXIMUM DISTANCE FORM PT
 TO A VERTEX OF K
```

CODE FOR DISTPT

```
     ∇ D←K DISTPT PT
[1]   D←(⌈/(((K[1;]-PT[1])*2)+((K[2;]-PT[2])
      *2)))*0.5
     ∇
```

Function Names: GEODESIC, CHRISTOFFEL, GEOAUX

EXTERNAL DOCUMENTATION

PURPOSE

To compute geodesics for Riemannian geometry on an N-dimensional manifold. This is done by the function GEODESIC, which is supported by CHRISTOFFEL, which computes the Christoffel symbol of the second kind, and the auxiliary function GEOAUX, which sets up the differential equation for the geodesics.

For S = arclength the O.D.E. can be written as

$$\frac{d^2x_i}{ds^2} = -\sum_{j,k} \{{}^i_{jk}\} \frac{dx^j}{ds}\frac{dx^k}{ds}$$

where the 3-array in curly brackets is the Christoffel symbol.

The Christoffel symbol

$$\{^i_{jk}\} = \frac{1}{2} \sum_h (g^{-1})_{ih} \left(\frac{\partial g_{kh}}{\partial x_j} + \frac{\partial g_{jh}}{\partial x_k} - \frac{\partial g_{jk}}{\partial x_h} \right)$$

requires the metric tensor $G = \{g_{ij}\}$ and its partial derivatives. The program uses the convention that in this 3-array the first axis stands for the subscript corresponding to the variable w.r.t. which differentiation is performed. Also, in the 3-array representing the Christoffel symbol, the first axis stands for the upper of the three subscripts. To solve the O.D.E., the main problem called KUTTA; see Chapter 3.

REFERENCES

Sokolnikoff, I. S. (1951). "Tensor Analysis," (Section 58). Wiley, New York.

Graustein, W. (1951). "Differential Geometry," (Section 68). Dover, New York.

INTERNAL DOCUMENTATION FOR GEODESIC

```
SYNTAX:GEODESIC
PURPOSE:TO COMPUTE GEODESICS FOR RIEMANNIAN
 GEOMETRY
INDEX ORIGIN:1
AUTHOR:U. GRENANDER
LAST UPDATE:10/80
NOTE:CALLS FUNCTIONS KUTTA AND CHRISTOFFEL
 AND GEOAUX
 USER MUST SUPPLY TWO FUNCTIONS. ONE THAT HAS
 THE METRIC TENSOR AS EXPLICIT RESULT, THE
 OTHER ONE THE DERIVATIVES OF THE METRIC
 TENSOR
```

CODE FOR GEODESIC

```
        ∇ GEODESIC;X0;V0;STEP;S
[1]   ⍝COMPUTES GEODESICS FOR RIEMANNIAN
      ⍝GEOMETRY
[2]   ⍝CALLS FUNCTIONS GEOAUX, CHRISTOFFEL
      ⍝AND KUTTA
[3]    'WHAT IS DIMENSION OF SPACE?'
[4]    N1←⎕
```

```
[5]      'WHAT IS NAME OF FUNCTION THAT
          COMPUTES METRIC TENSOR?'
[6]      G←⎕
[7]      'WHAT IS NAME OF FUNCTION THAT
          COMPUTES DERIVATIVES OF METRIC
          TENSOR?'
[8]      GPRIME←⎕
[9]      'WHAT IS STARTING POINT?'
[10]     X0←⎕
[11]     'WHAT IS VECTOR OF STARTING DIRECTION?'
[12]     V0←⎕
[13]     V0←V0÷(V0+.×(⍀G,' ',⍏X0)+.×V0)*0.5
[14]     'WHAT IS DESIRED STEPSIZE?'
[15]     STEP←⎕
[16]     'WHAT IS DESIRED TOTAL ARC LENGTH?'
[17]     S←⎕
[18]     RES←'GEOAUX' KUTTA 0,X0,V0,STEP,S
          ∇
```

INTERNAL DOCUMENTATION FOR CHRISTOFFEL

```
SYNTAX:Z←M CHRISTOFFEL MPRIME
PURPOSE:COMPUTES CHRISTOFFEL SYMBOL OF
 SECOND KIND FOR RIEMANNIAN GEOMETRY
INDEX ORIGIN:1
AUTHOR:U. GRENANDER
LAST UPDATE:10/80
LEFT ARGUMENT:METRIC TENSOR
RIGHT ARGUMENT:3-ARRAY OF PARTIAL DERIVATIVES
 OF METRIC TENSOR
RESULT:CHRISTOFFEL SYMBOL AS 3-ARRAY
```

CODE FOR CHRISTOFFEL

```
     ∇ Z←M CHRISTOFFEL MPRIME
[1]  ⍝COMPUTES CHRISTOFFEL SYMBOL OF THE
     ⍝SECOND KIND
[2]  ⍝ M IS METRIC TENSOR
[3]  ⍝ MPRIME IS TENSOR OF DERIVATIVES OF
     ⍝ M WITH FIRST
```

```
[4] ⍝SUBSCRIPT FOR SUBSCRIPT OF INDEPENDENT
[5] ⍝VARIABLE W.R.T. WHICH DIFFERENTIATION IS
    ⍝PERFORMED
[6] ⍝IN RESULT (3-ARRAY) FIRST AXIS IS
[7] ⍝SUPERSCRIPT IN CHRISTOFFEL SYMBOL
[8]  Z←0.5×(⌹M)+.×(3 1 2 ⍉MPRIME)+(3 2 1
    ⍉MPRIME)-MPRIME
    ∇
```

INTERNAL DOCUMENTATION FOR GEOAUX

```
SYNTAX:Z←GEOAUX Y
PURPOSE:AUXILIARY FUNCTION THAT COMPUTES
 RIGHT HAND SIDE FUNCTION IN O.D.E. FOR
 FUNCTION GEODESIC
RIGHT ARGUMENT:VALUE OF DEPENDENT VARIABLES
 AND THEIR DERIVATIVES
RESULT:FUNCTION VALUES ON RIGHT HAND SIDE
 OF THE O.D.E
```

CODE FOR GEOAUX

```
    ∇ Z←GEOAUX Y;CHRISTOFF
[1]   Y←1↓Y
[2]   Z←(2×N1)⍴0
[3]   Z[⍳N1]←Y[N1+⍳N1]
[4]   CHRISTOFF←(⍎G,' ',⍕Y[⍳N1]) CHRISTOFFEL
      ⍎GPRIME,' ',⍕Y[⍳N1]
[5]   Z[N1+⍳N1]←-(CHRISTOFF+.×Y[N1+⍳N1])
      +.×Y[N1+⍳N1]
      ∇
```

Function Name: HULL1

EXTERNAL DOCUMENTATION

PURPOSE

Calculates the convex hull of given set of points in the plane.

NOTE

See Shamos (1976) for discussion of algorithms.

INTERNAL DOCUMENTATION

SYNTAX: $Z \leftarrow HULL1\ X$
PURPOSE: TO FIND THE CONVEX HULL OF A SET
 OF POINTS
INDEX ORIGIN: 1
AUTHOR: G. PODOROWSKY
LAST UPDATE: 5/18/80
RIGHT ARGUMENT: X: THE SET OF POINTS, IN
 EITHER RAW OR STANDARD FORM
RESULT: Z: THE POINTS ON THE CONVEX HULL OF
 X, IN RAW FORM

CODE

```
       ∇ Z←HULL1 X;COUNT1;MINY;MINS;RS;THETS;
         X1;X2;COUNT2;ZER;T;TT;I
[1]    X←X[1 2 ;]
[2]    COUNT1←ρX[1;]
[3]    MINY←⌊/X[2;]
[4]    →(1<+/MINY=X[2;])/FIND
[5]    MINS←X[;X[2;]ιMINY]
[6]    REMIN:MINS←⍉((ρX)[2],2)ρMINS
[7]    X←X-MINS
[8]    RS←+/X*2
[9]    →((+/X[1;]∊0)>1)/ZEROS
[10]   THETS←X[2;]÷X[1;]
[11]   LINK:X←(X,[1] THETS),[1] RS
[12]   X1←((THETS≥0)∧0≠+/X[1 2 ;])/X
[13]   X2←(THETS<0)/X
[14]   X1←X1[;⍋X1[3;]]
[15]   X2←X2[;⍋X2[3;]]
[16]   X←(X1,X2),(4 1 ρ0)
[17]   LP:→((X[3;2]=X[3;3])∧(0≠X[3;2])∧(1≠(X[3;
         2]=9999)∧0=+/X[1;]<0))/SAME
[18]   →(((X[1;2]-X[1;1])×(X[2;3]-X[2;1]))
         <(X[2;2]-X[2;1])×(X[1;3]-X[1;1]))/DROP
```

```
[19]    X←1⌽X
[20]    →((,X[1 2 ;1])≠ 0 0)/LP
[21]    COUNT2←ρX[1;]
[22]    →(COUNT2=COUNT1)/END
[23]    COUNT1←COUNT2
[24]    →LP
[25]    END;X←X[1 2 ;]+MINS←(2,TT)ρ((TTρMINS
        [1;]),((TT←(ρX.)[2])ρMINS[2;]))
[26]    Z←X
[27]    →0
[28]    DROP:X←X[;T←(T≠2)/T←ι(ρX)[2]]
[29]    →LP
[30]    FIND:MINS←(⌊/X[1;(X[2;]∈MINY)/ιρX[2;]]),
        MINY
[31]    →REMIN
[32]    ZEROS:ZER←(X[1;]∈0)/ιρX[2;]
[33]    DUM←X[1;]
[34]    DUM[ZER]←1
[35]    THETS←X[2;]÷DUM
[36]    THETS[ZER]←9999
[37]    →LINK
[38]    SAME:OUT←1+X[4; 2 3]ι⌊/X[4; 2 3]
[39]    X←X[;I←(I≠OUT)/I←ι(ρX)[2]]
[40]    →LP
        ∇
```

Function Name: LINSERIES

EXTERNAL DOCUMENTATION

PURPOSE

To calculate a Minkowsky polynomial of a given convex polygon.

INTERNAL DOCUMENTATION

SYNTAX: KSTAR←L LINSERIES K
PURPOSE: TO COMPUTE THE 'POLYNOMIAL'
 DESCRIBED BELOW
INDEX ORIGIN: 1
AUTHOR: A. TARR, R.A. VITALE
LAST UPDATE:3/8/80

```
LEFT ARGUMENT: L: A 2×2 MATRIX
RIGHT ARGUMENT: K: A CONVEX POLYGON, IN
 STANDARD FORM
RESULT: KSTAR: A POLYGON IN STANDARD FORM
 SUCH THAT :
                        2              N
 KSTAR = K + LK + L K + ... + L K
NOTE: PROGRAM WILL ASK USER FOR MAX POWER N
```

CODE

```
      ∇ KSTAR←L LINSERIES K;N;MAT;I
[1]   'HOW FAR SHOULD SUM GO -- MAX POWER
      OF L?'
[2]   N←□
[3]   I←0
[4]   MAT←K
[5]   LP:I←I+1
[6]   MAT←MAT,[1] L MATRANS MAT[(¯2+3×I),
      (¯1+3×I);]
[7]   →(I<N)/LP
[8]   KSTAR←ADDC1 MAT
      ∇
```

Function Name: MATRANS

EXTERNAL DOCUMENTATION

PURPOSE

Calculates linear transformation of given convex polygon in standard form.

INTERNAL DOCUMENTATION

```
SYNTAX: XY←M MATRANS P
PURPOSE: TO COMPUTE AN ARBITRARY LINEAR
 TRANSFORMATION OF A CONVEX POLYGON
INDEX ORIGIN: 1
AUTHOR: A. TARR, R.A. VITALE
LAST UPDATE: 3/8/80
```

*LEFT ARGUMENT: M: A 2×2 MATRIX, THE LINEAR
 OPERATOR
RIGHT ARGUMENT: P: THE CONVEX POLYGON, IN
 STANDARD FORM
RESULT: XY: THE TRANSFORMED FIGURE (MP),
 IN STANDARD FORM*

CODE

```
      ∇ XY←M MATRANS P;T;C;Z;DET
[1]   →(0=DET←(M[1;1]×M[2;2])-M[1;2]×M[2;1])
      /SING
[2]   T←M+.×P[1 2 ;]
[3]   →(DET>0)/OK
[4]   T←⌽T
[5]   OK:XY←STFORM T
[6]   →0
[7]   SING:C←(÷((C[1]*2)+(C[2]*2))*0.5)×C←M[(1+
      ∧/0=M[1;]);]
[8]   T← 2 1 ρC×⌈/Z←(P[1;]×C[1])+P[2;]×C[2]
[9]   XY←STFORM T, 2 1 ρC×⌊/Z
      ∇
```

Function Name: PERIM

EXTERNAL DOCUMENTATION

PURPOSE

Calculates the perimeter of given convex polygon.

INTERNAL DOCUMENTATION

*SYNTAX: Z←PERIM P
PURPOSE: TO COMPUTE THE PERIMETER OF A
 CONVEX POLYGON OR LINE SEGMENT
INDEX ORIGIN: 1
AUTHOR: A. TARR, R.A. VITALE
LAST UPDATE: 3/8/80
RIGHT ARGUMENT: P: CONVEX POLYGON OR LINE*

```
 SEGMENT, IN RAW OR STANDARD FORM
RESULT: Z: THE PERIMETER OF P
NOTE: THE PERIMETER OF A LINE SEGMENT IS
 TWICE THE APPARENT LENGTH
```

CODE

```
      ∇ Z←PERIM P;X;Y
[1]   Z←VADD(((((1ϕX)-X←P[1;])*2)+((1ϕY)
      -Y←P[2;])*2)*0.5
      ∇
```

Function Names: POLY, SFACE, STFORM

EXTERNAL DOCUMENTATION

PURPOSE

Transform support form of given convex polygon to standard form, the inverse, or from raw form to standard form.

INTERNAL DOCUMENTATION FOR POLY

```
SYNTAX: VERT←POLY SUP
PURPOSE: TO CONVERT FROM SUPPORT TO
 STANDARD FORM
INDEX ORIGIN: 1
AUTHOR: A. TARR, R.A. VITALE
LAST UPDATE: 3/8/80
RIGHT ARGUMENT: SUP: SUPPORT FORM OF
 A POLYGON
RESULT: VERT: STANDARD FORM REPRESENTATION
 OF SUP
```

CODE FOR POLY

```
      ∇ VERT←POLY SUP;X;Y;SIN;COS;SIND;I;
        J;A;S;PHI
```

```
[1]    X←((S[J]×SIN[I])-(S[I]×(SIN←1○PHI)
       [J←1ΦI]))÷(SIND←1○(PHI-(1ΦPHI←SUP
       [2;])))[I←ιρS←SUP[1;]]
[2]    Y←((S[I]×COS[J])-(S[J]×(COS←2○PHI)
       [I]))÷SIND[I]
[3]    VERT←((3,ρS)ρ(¯1ΦX),(¯1ΦY),A)[;ΔA←(○2)
       |PHI+○0.5]
       ∇
```

INTERNAL DOCUMENTATION FOR SFACE

```
SYNTAX: S←SFACE XY
PURPOSE: TO CONVERT TO SUPPORT FORM FROM
 RAW OR STANDARD FORM
INDEX ORIGIN: 1
AUTHOR: A. TARR, R.A. VITALE
LAST UPDATE: 3/8/80
RIGHT ARGUMENT: XY: A POLYGON (OR LINE
 SEGMENT) IN RAW OR STANDARD FORM
RESULT: S: SUPPORT FORM REPRESENTATION OF
 XY
```

CODE FOR SFACE

```
       ∇ S←SFACE XY;A
[1]    →((ρXY)[1]=3)/CONT
[2]    XY←STFORM XY
[3]    CONT:A←(○2)|XY[3;]-○0.5
[4]    S←((2,ρA)ρ(((XY[1;]×(2○A))+XY[2;]
       ×1○A),A))[;ΔA]
       ∇
```

INTERNAL DOCUMENTATION FOR STFORM

```
SYNTAX: NEWF←STFORM F
PURPOSE: TO CONVERT TO STANDARD FORM FROM
 RAW FORM
INDEX ORIGIN: 1
AUTHOR: A. TARR, R.A. VITALE
LAST UPDATE: 3/8/80
```

RIGHT ARGUMENT: F: RAW FORM REPRESENTATION
OF POLYGON OR LINE SEGMENT
RESULT: NEWF: STANDARD FORM REPRESENTATION
OF F

CODE FOR STFORM

```
     ∇ NEWF←STFORM F;A
[1]  NEWF←((3,ρF[1;])ρ(,F),A)[;⍋A←(0,0)
     ATAN(1⌽F)-F]
     ∇
```

Function Name: SUPPORT

EXTERNAL DOCUMENTATION

PURPOSE

For a convex polygon given in support form and for given angles,
calculate the values of the support function.

INTERNAL DOCUMENTATION

SYNTAX: SUP←PHI SUPPORT SF
PURPOSE: TO CALCULATE THE SUPPORT FUNCTION AT
 ANY GIVEN ANGLES [0,2PI)
INDEX ORIGIN: 1
AUTHOR: A. TARR, R.A. VITALE
LAST UPDATE: 5/18/80
LEFT ARGUMENT: PHI: THE ANGLES AT WHICH
 SUPPORT FUNCTION IS TO BE EVALUATED (ON
 INTERVAL [0,2PI)
RIGHT ARGUMENT: SF: SUPPORT FORM
RESULT: SUP: MATRIX WITH ANGLES PHI
 INCREASING IN [0,2PI) IN THE SECOND ROW,
 WITH CORRESPONDING VALUES OF THE SUPPORT
 FUNCTION IN THE FIRST ROW

CODE

```
      ∇ SUP←PHI SUPPORT SF;I;J;II;N
[1]   PHI←PHI[⍋PHI←,PHI]
[2]   I←+/PHI∘.≥SF[2;]
[3]   II←(I=0)/⍳PI
[4]   I[II]←N←⍴SF[1;]
[5]   J←(1ϕ⍳N)[I]
[6]   SUP←(2,⍴PHI)⍴(((SF[1;I]×1○(SF[2;J]
      -PHI))+(SF[1;J]×1○(PHI-SF[2;I])))
      ÷1○SF[2;J]-SF[2;I]),PHI
      ∇
```

Function Names: ROTATE, SCALE, TRANS

EXTERNAL DOCUMENTATION

PURPOSE

Applies Euclidean transformations to convex polygon given in standard form.

INTERNAL DOCUMENTATION FOR ROTATE

```
SYNTAX: RF←T ROTATE FIG
PURPOSE: TO COMPUTE THE ROTATION OF A
 POLYGON
INDEX ORIGIN: 1
AUTHOR: A. TARR, R.A. VITALE
LAST UPDATE: 3/8/80
LEFT ARGUMENT: T: ANGLE THROUGH WHICH FIGURE
 IS TO BE ROTATED (T IS IN RADIANS)
RIGHT ARGUMENT: FIG: POLYGON TO BE ROTATED,
 IN STANDARD FORM
RESULT: RF: THE ROTATED POLYGON, IN STANDARD
 FORM
```

CODE FOR ROTATE

```
      ∇ RF←T ROTATE FIG;A
[1]   A←(○2)|FIG[3;]+T
```

```
[2]   RF←((ρFIG)ρ((FIG[1;]×2○T)-(FIG[2;]
      ×1○T)),((FIG[1;]×1○T)
[3]   RF←RF[;⍋A]
      +(FIG[2;]×2○T)),A)
      ∇
```

INTERNAL DOCUMENTATION FOR SCALE

```
SYNTAX: NEW←N SCALE OLD
PURPOSE: TO SCALE POLYGONS BY AN ARBITRARY
  SCALING CONSTANT
INDEX ORIGIN: 1
AUTHOR: A. TARR, R.A. VITALE
LAST UPDATE: 3/8/80
LEFT ARGUMENT: N: A SCALING CONSTANT
RIGHT ARGUMENT: OLD: POLYGON(S) TO BE
  SCALED, IN (STACKED) STANDARD FORM ARRAY
RESULT: NEW: OLD SCALED BY N (EACH POLYGON
  IN OLD IS SCALED BY N
```

CODE FOR SCALE

```
      ∇ NEW←N SCALE OLD;R
[1]   OLD[(1+R),(2+R);]←OLD[(1+R),(2+R←3×¯1
      +ι(ρOLD)[1]÷3);]×N
[2]   NEW←OLD
      ∇
```

INTERNAL DOCUMENTATION FOR TRANS

```
SYNTAX: NEWF←V TRANS FIG
PURPOSE: TO TRANSLATE A FIGURE IN X AND/OR
  Y DIRECTIONS
INDEX ORIGIN: 1
AUTHOR: A. TARR, R.A. VITALE
LAST UPDATE: 3/8/80
LEFT ARGUMENT: V: VECTOR OF ΔX, ΔY THROUGH
  WHICH FIGURE IS TO BE TRANSLATED
RIGHT ARGUMENT: FIG: STANDARD FORM OF FIGURE
  TO BE TRANSLATED
```

*RESULT: NEWF: TRANSLATED FIGURE, IN STANDARD
FORM*

CODE FOR TRANS

```
      ∇ NEWF←V TRANS FIG
[1]   NEWF←(ρFIG)ρ(FIG[1;]+1↑,V),(FIG[2;]
      +¯1↑,V),FIG[3;]
      ∇
```

Function Name: RMVFVT

EXTERNAL DOCUMENTATION

PURPOSE

Cleans given representation of convex polygon in standard form by
removing false vertices.

INTERNAL DOCUMENTATION

*SYNTAX: XYTH←REMFVT UV
PURPOSE: TO REMOVE ANY FALSE VERTICES FROM
 A POLYGON REPRESENTATION
INDEX ORIGIN: 1
AUTHOR: A. TARR, R.A. VITALE
LAST UPDATE: 3/8/80
RIGHT ARGUMENT: UV: A COMPLEX POLYGON, IN
 STANDARD FORM
RESULT: XYTH: ESSENTIALLY THE SAME POLYGON,
 BUT WITH ANY FALSE VERTICES REMOVED
NOTE: FALSE VERTICES ARISE MOST OFTEN IN THE
 ADDITION OF CONVEX FIGURES IN WHICH A SIDE
 OF ONE IS PARALLEL TO A SIDE OF THE OTHER*

CODE

```
      ∇ XYTH←REMFVT UV;W;N
[1]   W←(N←ρUV[3;])ρ1
[2]   W[(UV[3;]=¯1⌽UV[3;])/ιN]←0
```

```
[3]   XYTH←W/UV
      ∇
```

Function Name: VADD

EXTERNAL DOCUMENTATION

PURPOSE

To achieve high accuracy in adding the elements of a numeric vector.

INTERNAL DOCUMENTATION

```
SYNTAX: Z←VADD V
PURPOSE: TO ADD THE ELEMENTS OF THE VECTOR V
INDEX ORIGIN: 1
AUTHOR: A. TARR, R.A. VITALE
LAST UPDATE: 5/18/80
RIGHT ARGUMENT: V: A VECTOR
RESULT: Z: A SCALAR, THE SUM OF THE ELEMENTS
  OF V
NOTE: TO OBTAIN MINIMUM ERROR, V IS SEPARATED
  INTO POSITIVE AND NEGATIVE TERMS AND THEN
  ADDED THE SMALLEST ELEMENTS IN ABSOLUTE
  VALUE FIRST
```

CODE

```
      ∇ Z←VADD V;VP;VM;I
[1]   Z←(+/VM[⍋VM←(~I)/V])+(+/VP[⍒VP←(I←V≥0)
      /V])
      ∇
```

CHAPTER 21

Graphs

This chapter consists only of a few programs for computing connected components and the reachability matrix of graphs. Most important computing tasks for graphs are therefore not presented here. This is particularly unfortunate because fast algorithms have recently been constructed for efficient computational treatment of graph problems.

Function Name: CONNECT

EXTERNAL DOCUMENTATION

PURPOSE

To find all components of an undirected graph.

The output of CONNECT consists of a vector of index numbers (VINDX) such that the same index number is assigned to all vertices that are in the same connected component. The input must consist of a set of ordered pairs indicating which vertices are connected.

EXAMPLE

The input

$$E = \begin{matrix} 1 & 3 \\ 2 & 4 \end{matrix}$$

implies four vertices, where vertices 1 and 3 are connected and 2 and 4 are connected.

Output will be

$$\text{VINDX} = 1 \quad 2 \quad 1 \quad 2$$

indicating that vertices 1 and 3 make up one set of connected components and 2 and 4 make up a second set.

ALGORITHM

Find the lowest numbered vertex that is not part of an existing component.

Search the edge set E for all vertices connected to that vertex, V. Place these in STACK. (If 0 = shape of STACK, find next component.) For each vertex in STACK, mark it and search E for all vertices connected to it. Place item in STACK as well. Thus, once each vertex in STACK has been culled, every vertex directly or indirectly connected to V has been marked.

Then assign the next component index number and find the next component.

NOTES

(1) All vertices between 1 and (max/,E) that do not appear in E are assumed to be separate components of size one and are assigned different component numbers.

(2) The vertex pair (i, j) may appear in a row of E as i j or j i. Both i j and j i can appear in E any number of times. The algorithm actually requires that if i j is in E, then j i is also (done in initialization by concatenating the reverse of all existing rows in E to the original argument E).

REFERENCE

Reingold, E. M., Nievergelt, J., and Deo, N. (1977). "Combinatorial Algorithms: Theory and Practice," pp. 327–331. Prentice-Hall, Englewood Cliffs, New Jersey.

INTERNAL DOCUMENTATION

```
SYNTAX: VINDX←CONNECT E
PURPOSE: TO FIND ALL CONNECTED COMPONENTS
 OF AN UNDIRECTED GRAPH
INDEX ORIGIN: 1
AUTHOR: P. FLANAGAN
LAST UPDATE: 7/79
RIGHT ARGUMENT: E, AN N BY 2 MATRIX WHOSE
 ELEMENTS ARE VERTEX NUMBERS (LABELS). EACH
 ROW OF E REPRESENTS A CONNECTED PAIR OF
 VERTICES.
```

RESULT: VINDX, A VECTOR OF INDEX NUMBERS
SUCH THAT ALL VERTICES WHICH ARE IN THE
SAME CONNECTED SUB-COMPONENT HAVE THE SAME
INDEX NUMBER. VINDX[K] IS THE INDEX NUMBER
OF VERTEX K.
NOTES:
1) ALL VERTICES BETWEEN 1 AND ⌈/,E WHICH
DO NOT APPEAR IN E ARE ASSUMED TO BE
SEPARATE COMPONENTS OF SIZE ONE AND ARE
ASSIGNED DIFFERENT COMPONENT NUMBERS.
2) THE VERTEX PAIR MAY APPEAR IN A ROW OF
E AS EITHER I J OR J I. BOTH I J AND
J I CAN APPEAR IN E ANY NUMBER OF TIMES.

CODE

```
        ∇ VINDX←CONNECT E;STACK;V;C
[1]     STACK←0ρVINDX←(⌈/,E)ρC←0
[2]     E←E,[1]⌽E
[3]     FINDCOMP:→((ρVINDX)<V←VINDXιC-C←C+1)/0
[4]     MARKV:VINDX[V]←C
[5]      STACK←STACK,(E[;1]=V)/E[;2]
[6]     NXTV:→(0=ρSTACK)/FINDCOMP
[7]      V←STACK[ρSTACK]
[8]      STACK←STACK[ι-1-ρSTACK]
[9]      →(VINDX[V]=0)/MARKV
[10]     →NXTV
        ∇
```

Function Names: CONVERT, MATMAK

EXTERNAL DOCUMENTATION

PURPOSE

To go from list of directed lines of digraph to adjacency matrix or inverse. See external documentation for REACH.

INTERNAL DOCUMENTATION FOR CONVERT

SYNTAX: Z←CONVERT BITMAT
PURPOSE: TO OUTPUT THE CODED VECTOR

REPRESENTATION OF ALL DIRECTED LINES X
OF D.
INDEX ORIGIN: 1
AUTHOR: D. MCARTHUR
LAST UPDATE: 8/13/80
RIGHT ARGUMENT: THE GLOBAL VARIABLE BITMAT
CONSISTING OF THE ADJACENCY MATRIX OF D
(CREATED IN MATMAK).
RESULT: AN N×2 ARRAY WHERE N IS THE NUMBER
OF DIRECTED LINES OF D. EACH ROW OF THE
ARRAY IS THE CODED VECTOR REPRESENTATION
OF A LINE X OF D.
NOTE: THE TERM 'CODED VECTOR REPRESENTATION'
IS DEFINED IN DMATMAK.

CODE FOR CONVERT

```
      ∇  Z←CONVERT BITMAT;N;V;T
[1]   N←(ρBITMAT)[1]
[2]   →(0=N)/NONE
[3]   V←(V≠0)/V←(ιN*2)×,BITMAT
[4]   T←N|V
[5]   Z←(⌈V÷N),[1.5] T,0ρT[(0=T)/ιρT]←N
[6]   →0
[7]   NONE;'NOTHING TO CONVERT'
      ∇
```

INTERNAL DOCUMENTATION FOR MATMAK

SYNTAX: MATMAK
PURPOSE: TO CREATE THE ADJACENCY MATRIX OF
A DIGRAPH, D.
INDEX ORIGIN: 1
AUTHOR: D. MCARTHUR
LAST UPDATE: 8/13/80
TO USE: CODE THE SET OF POINTS $V=(V_1,V_2,\ldots,V_P)$
OF D BY THE SET (1,2,...,P). ENTER,
AS SEPARATE INPUTS, EACH DIRECTED LINE X
OF D IN ITS CODED VECTOR FORM (I,J)
REPRESENTING $X=(V_I,V_J)$. TO END INPUT LOOP
TYPE 0

RESULT: THE P×P ADJACENCY MATRIX OF D.
GLOBAL VARIABLES: BITMAT: THE ADJACENCY
MATRIX OF D.
NOTE: CONSULT THE EXTERNAL DOCUMENTATION
OF REACH IF THE TERMINOLOGY USED ABOVE IS
UNFAMILIAR OR UNCLEAR.

CODE FOR MATMAK

```
       ∇  Z←MATMAK;EM;LIN;M;N;J;MAX;PEM
[1]    EM← 0 2 ρ0
[2]    'ENTER, AS SEPARATE INPUT, THE CODED
        VECTOR REPRESENTATION OF EACH
[3]    DIRECTED LINE OF THE DIGRAPH.'
[4]    LBL1;'ENTER'
[5]    LIN←□
[6]    →(0=(,LIN)[1])/MAK
[7]    0ρ(M←1↑ρEM),0ρN←‾1↑ρEM
[8]    N←(ρLIN)⌈N
[9]    EM←((M,N)↑EM),[1] N↑LIN
[10]   →LBL1
[11]  MAK:J←1
[12]   MAX←⌈/⌈/EM
[13]   BITMAT←(2ρMAX)ρ0
[14]  LBL2:PEM←(0≠EM[J;])/EM[J;]
[15]   BITMAT[PEM[1];1↓PEM]←1
[16]   →((1↑ρEM)≥J←J+1)/LBL2
[17]   Z←BITMAT
       ∇
```

Function Names: REACH, COMREACH

EXTERNAL DOCUMENTATION

PURPOSE

REACH computes the reachability matrix of a digraph. COMREACH computes the commutative reachability matrix (defined below). The following definitions should be read by all users of these programs since no universal set of terminology exists in graph theory.

DEFINITIONS

(1)   A digraph D consists of a finite nonempty set $V = V(P)$ of p points together with a prescribed set X of q unordered pairs of distinct points of V. Each pair $x = (u,v)$ of points in X is a directed line in the direction of v. x is said to join u and v. We write $x = uv$ and say that u and v are adjacent points. Point u and line x are said to be incident with each other as are v and x.

(2)   A walk of a digraph D is an alternating sequence of points and lines $v_1, x_1, v_2, \ldots, v_{n-1}, x_{n-1}, v_n$ beginning and ending with points in which each line is incident with the two points immediately preceding and following it. The length of a walk is the number of directed lines used in defining the walk.

(3)   A point v_j is reachable from a point v_i if there is at least one walk from v_i to v_j. If there is a walk of length ≥ 2 from v_i to itself, v_i is reachable from itself.

(4)   A pair (v_i, v_j) is called commutatively reachable if v_j is reachable from v_i and v_i is reachable from v_j.

THEORY

Every digraph can be represented by a $p \times p$ matrix called the adjacency matrix, denoted A and defined as

$$a_{ij} = \begin{cases} 1 & \text{if } v_i, v_j \text{ is a directed line of } D \ (1 \leq i, j \leq p) \\ 0 & \text{otherwise} \end{cases}$$

Associated with the adjacency matrix is the reachability matrix denoted R and defined as

$$r_{ij} = \begin{cases} 1 & \text{if } v_j \text{ is reachable from } v_i \ (1 \leq i, j \leq p) \\ 0 & \text{otherwise} \end{cases}$$

A close relation to the reachability matrix is the commutative reachability matrix, denoted C, and defined as

$$c_{ij} = \begin{cases} 1 & \text{if } v_i \text{ is reachable from } v_j \text{ and } v_j \\ & \text{is reachable from } v_i \ (1 \leq i, j \leq p) \\ 0 & \text{otherwise} \end{cases}$$

The following two theorems display the association between A and R.

THEOREM 1

The i, j entry $a_{ij}^{(n)}$ of A^n is the number of walks of length n from v_i to v_j.

THEOREM 2

$r_{ij} = 1$ if and only if for some n, $a_{ij}^{(n)} > 0$.

REACH:

Define $A_n = \sum_{k=1}^{n} (A^k)$ and denote the elements of A_n by a a_{ijn} for $1 \leq i, j \leq p$. Define B and E to be $p \times p$ matrices and denote their elements by b_{ij} and $e_{ij}, 1 \leq i, j \leq p$. Their elements will be specified in the algorithm.

ALGORITHM

$n = 1$:
Loop:

$$b_{ij} = \begin{cases} 1 & \text{if} \quad a_{ijn} \geq 1 \\ 0 & \text{otherwise} \end{cases}$$

$$e_{ij} = \begin{cases} 1 & \text{if} \quad a_{ij(n+1)} \geq 1 \\ 0 & \text{otherwise} \end{cases}$$

$$n = n + 1$$

If (B is not equal to E), then

Goto Loop

Else $R = E$
End.

COMREACH:

ALGORITHM

Recall that $r_{ij} = 1$ if v_j is reachable from v_i and $r_{ji} = 1$ if v_i is reachable from v_j. Thus $c_{ij} = 1$ if and only if $r_{ij} = 1$ and $r_{ji} = 1$. So $C = R \times R'$ where R' is the transpose of R.

INTERNAL DOCUMENTATION FOR REACH

```
SYNTAX: Z←REACH NUMAT
PURPOSE: TO CONSTRUCT THE REACHABILITY
 MATRIX OF A DIGRAPH
INDEX ORIGIN: 1
AUTHOR: D. MCARTHUR
LAST UPDATE: 8/14/80
```

RIGHT ARGUMENT: THE P×P ADJACENCY MATRIX
OF THE DIGRAPH. IF MATMAK WAS USED TO CREATE
THE ADJACENCY MATRIX THEN USE THE GLOBAL
VARIABLE BITMAT.
RESULT: THE P×P (P=NUMBER OF POINTS IN THE
DIGRAPH) REACHABILITY MATRIX.

CODE FOR REACH

```
      ∇ Z←REACH NUMAT;RIMAT;MAT
[1]   RIMAT←NUMAT+IDENT 1↑ρNUMAT
[2]   AGN:MAT←NUMAT
[3]   NUMAT←MAT∨.∧RIMAT
[4]   →(~∧/∧/NUMAT=MAT)/AGN
[5] DONE:Z←NUMAT
      ∇
```

INTERNAL DOCUMENTATION FOR COMREACH

SYNTAX: Z←COMREACH MAT
PURPOSE: TO COMPUTE THE COMMUTATIVE
REACHABILITY MATRIX C OF A DIGRAPH. C IS
DEFINED AS FOLLOWS: C   *=1 IF V*   *REACHABLE*
                      IJ   *J*
FROM V AND V IS REACHABLE FROM V , OTHERWISE
     I   I   J
C   =0.
 IJ
INDEX ORIGIN: 1
AUTHOR: D. MCARTHUR
LAST UPDATE: 8/14/80
RIGHT ARGUMENT: THE P×P ADJACENCY MATRIX
OF THE DIGRAPH. IF MATMAK WAS USED TO CREATE
THE ADJACENCY MATRIX THEN USE THE GLOBAL
VARIABLE BITMAT.
RESULT: THE P×P ARRAY (P=NUMBER OF POINTS IN
THE DIGRAPH) DEFINED AS C ABOVE.

CODE FOR COMREACH

```
     ∇ Z←COMREACH MAT;T
[1]  T←REACH MAT
[2]  Z←T×⍉T
     ∇
```

AUXILIARY CODE TO COMPUTE N X N IDENTITY MATRIX

```
     ∇ Z←IDENT N
[1]  Z←(2ρN)ρ1,(2ρN)ρ0
     ∇
```

CHAPTER 22

Probability

There are mainly two types of programs in this chapter. One type computes frequency functions, distribution functions, and other characteristics of probability measures. The other type simulates the measure, generating a sample of specified shape from it.

In addition to functions of these types, we include programs that compute convolutions and the equilibrium distribution of a Markov chain.

Function Name: BERN

EXTERNAL DOCUMENTATION

PURPOSE

To generate random numbers from the Bernoulli distribution, whose density function is

$$f(x) = \begin{cases} p, & \text{for } x = 1 \\ 1 - p, & \text{for } x = 0 \\ 0, & \text{otherwise} \end{cases}$$

$$= \begin{cases} xp + (1 - x)(1 - p), & \text{for } x = 0, 1 \\ 0, & \text{otherwise} \end{cases}$$

The user specifies the probability $p = PAR$. If PAR is an array of dimension SS, then, in the resulting array z, the ith component $Z(i)$ will have distribution $F(p(i))$.

ALGORITHM

Generate SS random numbers from $R(0, 1)$. (The procedure for doing so is the same as that used in the program UNIF; see that discussion for further details.) Since $P(u(i) < p) = p$, let

$$Z(i) = \begin{cases} 1 & \text{if} \quad u(i) \leq p \\ 0 & \text{if} \quad u(i) > p \end{cases}$$

Then, clearly, $P(Z(i) = 1) = p$ and $P(Z(i) = 0) = 1 - p$.

INTERNAL DOCUMENTATION

SYNTAX: Z←SS BERN PAR
PURPOSE: SIMULATION OF THE BERNOULLI
 DISTRIBUTION.
INDEX ORIGIN: 1
AUTHOR: E. SILBERG
LAST UPDATE: 1/79
RIGHT ARGUMENT: A SCALAR, 0≤PAR≤1, WHICH
 IS THE PROBABILITY OF THE OUTCOME '1';
 I.E., P(1)=PAR, P(0)=1-PAR.
LEFT ARGUMENT: (VECTOR OR SCALAR) THE
 DESIRED SAMPLE SIZE.
RESULT: RANDOM NUMBERS FROM THE BERNOULLI
 DISTRIBUTION WITH THE SPECIFIED PROBABILITY.

CODE

```
      ∇ Z←SS BERN PAR
[1]  ⍝   BERNOULLI RANDOM NUMBER GENERATOR
[2]    Z←PAR≥(÷1+10000000000)×?SSρ10000000000
      ∇
```

Function Name: BETA

EXTERNAL DOCUMENTATION

PURPOSE

To generate random numbers from the beta distribution, whose density function is

$$f(x) = \begin{cases} \dfrac{1}{B(a, b)} \, x^{a-1}(1 - x)^{b-1}, & 0 < x < 1 \\ 0, & \text{otherwise} \end{cases}$$

where $B(a,b) = \Gamma(a)\Gamma(b)/\Gamma(a + b)$.

ALGORITHM

Let Y and W be independent random variables with Γ distributions as follows: Y distributed $\Gamma(a)$, W distributed $\Gamma(b)$. Then it can be shown that the new random variable defined as $X = Y/(Y + W)$ has a B distribution with parameters a and b. Therefore, to obtain a sample value $Z(i)$ distributed $B(a,b)$, call the program GAMMA twice, to generate $Y(i)$ distributed $\Gamma(a)$ and $W(i)$ distributed $\Gamma(b)$, and set $Z(i) + Y(i)/(Y(i) + W(i))$, for $i = 1, 2, \ldots, SS$.

INTERNAL DOCUMENTATION

```
SYNTAX: Z←SS BETA PAR
PURPOSE: SIMULATION OF THE BETA DISTRI-
 BUTION
INDEX ORIGIN: 1
AUTHOR: E. SILBERG
LAST UPDATE: 1/79
RIGHT ARGUMENT: THE PARAMETERS, A=PAR[1]>0,
 B=PAR[2]>0, OF THE BETA DISTRIBUTION WITH
 THE FOLLOWING DENSITY FUNCTION:(÷C)×(X*
 (A-1))×(1-X)*(B-1),0<X<1, WHERE C IS THE
 VALUE BETA (A,B) OF THE MATHEMATICAL BETA
 FUNCTION.
LEFT ARGUMENT: (SCALAR) THE DESIRED SAMPLE
 SIZE.
RESULT: RANDOM NUMBERS FROM THE BETA
 DISTRIBUTION DESCRIBED ABOVE.
```

CODE

```
     ∇ Z←SS BETA PAR;V
[1]  ⍝  BETA RANDOM NUMBER GENERATOR
[2]  ⍝  CALLS ∇Z←SS GAMMA PAR∇
[3]    Z←V÷(SS GAMMA PAR[2])+V←SS GAMMA PAR[1]
     ∇
```

Function Name: BIN

EXTERNAL DOCUMENTATION

PURPOSE

To generate random numbers from the binomial distribution, whose density function is

$$b(x; n, p) = \begin{cases} (n!x)p^x(1 - p)^{n-x}, & x = 0, 1, 2, \ldots, n \\ 0, & \text{otherwise} \end{cases}$$

This is the distribution of a discrete random variable X representing the total number of ones in a sequence of n independent Bernoulli trials for which $P(1) = p$ and $P(0) = 1 - p$. The user specifies the number of trials $n = PAR(1)$ and the probability $p = PAR(2)$.

ALGORITHM

Call the program BERN to generate an array of dimension $(SS, PAR(1))$, whose entries are Bernoulli numbers with parameter $p = PAR(2)$. Then take the summation of the ones along the last axis, to obtain an array Z of binomial numbers, whose dimension is SS.

INTERNAL DOCUMENTATION

```
SYNTAX: Z←SS BIN PAR
PURPOSE: SIMULATION OF THE BINOMIAL
 DISTRIBUTION.
INDEX ORIGIN: 1
AUTHOR: E. SILBERG
LAST UPDATE: 1/79
RIGHT ARGUMENT: PAR[1]=NO. OF TRIALS;
 PAR[2]= PROBABILITY OF THE OUTCOME '1'
 IN EACH TRIAL.
LEFT ARGUMENT: (VECTOR OR SCALAR) THE
 DESIRED SAMPLE SIZE.
RESULT: RANDOM NUMBERS FROM THE BINOMIAL
 DISTRIBUTION WITH THE SPECIFIED
 PARAMETERS.
```

CODE

```
      ∇ Z←SS BIN PAR
[1] ⍝   BINOMIAL RANDOM NUMBER GENERATOR.
    ⍝   CALLS ∇Z←SS BERN PAR∇.
[2]   Z←+/(SS,PAR[1]) BERN PAR[2]
      ∇
```

Function Names: BINDF and BINFR

EXTERNAL DOCUMENTATION

PURPOSE

To calculate individual and cumulative binomial probabilities.

The program BINFR computes the binomial probability

$$f(x) = (n!x)p^x(1 - p)^{n-x}$$

associated with each positive integer x between the integer limits RANGE, inclusively, for the binomial distribution with the specified parameters $PAR = (n,p)$. For example, the command 5   10   BINFR 10   .2 will yield a (2 × 6) matrix, the first row of which is (5,   6,   . . . , 10) and the second row of which is

$$(10!x)   (0.2)^x   (0.8)^{10-x},   x = 5, 6, . . . , 10$$

It is permissible to request the probability for only one point; in this case, RANGE will be a single integer, and the result will be a 2 × 1 array. For example, 5   BINFR   10   .2 will yield

$$5$$
$$P(5)$$

If the range is large, it is recommended that the APL utility function PRINTAB be used in conjunction with BINFR, in order to produce neat, tabular output. (See discussion of PRINTAB.)

The program BINDF is the cumulative counterpart to BINFR. The syntax is identical, but the cumulative probability

$$\sum_{x=0}^{k} (n!x)p^x(1 - p)^{(n-x)}$$

will replace the point probability in the output array, for the kth value between the specified limits RANGE. For example, 5   10   BINDF   10 .2 will yield a 2 × 6 matrix, the first row of which is (5,   6,   . . . ,   10) and the second row of which is

$$\sum_{x=0}^{k} (10!x)   (0.2)^x   (0.8)^{10-x}   \text{for}   k = 5, 6, . . . , 10$$

BINDF calls BINFR in order to calculate point probabilities in the summation.

WARNING

If N is large, one should use instead a looping program employing recurrence relations.

INTERNAL DOCUMENTATION FOR BINDF

```
SYNTAX: Z←RANGE BINDF PAR
PURPOSE: CALCULATION OF CUMULATIVE BINOMIAL
  PROBABILITIES (D.F.).
INDEX ORIGIN: 1
AUTHOR: E. SILBERG
LAST UPDATE: 1/79
LEFT ARGUMENT: INTEGER (≥0) OR PAIR OF
  INTEGERS (0≤RANGE[1]<RANGE[2]) INDEXING THE
  BEGINNING AND END VALUES FOR THE LIST OF
  PROBABILITIES TO BE COMPUTED. (ONE INTEGER
  MEANS ONE VALUE IS REQUESTED.)
RIGHT ARGUMENT: VECTOR (N,P) OF THE
  PARAMETERS OF THE BINOMIAL DISTRIBUTION.
RESULT: 2×K MATRIX, THE FIRST ROW OF WHICH
  IS THE SET OF INTEGERS RANGE[1], 1+RANGE[1],
  ...,RANGE[2], AND THE SECOND ROW OF WHICH
  CONTAINS THE CORRESPONDING CUMULATIVE
  PROBABILITIES(SUMMED FROM P(0)), WHERE
  K=1+RANGE[2]-RANGE[1] (OR K=1).
```

CODE FOR BINDF

```
      ∇ Z←RANGE BINDF PAR;K
[1] ค   CUMULATIVE BINOMIAL PROBABILITIES
    ค   (D.F.).
[2] ค   CALLS ∇Z←RANGE BINFR PAR∇.
[3]   K←RANGE[1]↓0,ιRANGE[ρRANGE←,RANGE]
[4]   Z←(2,ρK)ρK,(-ρK)↑+\((0,RANGE[ρRANGE])
      BINFR PAR)[2;]
      ∇
```

INTERNAL DOCUMENTATION FOR BINFR

```
SYNTAX: Z←RANGE BINFR PAR
PURPOSE: CALCULATION OF BINOMIAL PROB-
  ABILITIES (FREQUENCY FUNCTION).
```

```
INDEX ORIGIN: 1
AUTHOR: E. SILBERG
LAST UPDATE: 1/79
LEFT ARGUMENT: INTEGER (≥0) OR PAIR OF
 INTEGERS (0≤RANGE[1]<RANGE[2]) INDEXING
 THE BEGINNING AND END VALUES FOR THE LIST
 OF PROBABILITIES TO BE COMPUTED. (ONE
 INTEGER MEANS ONE VALUE IS REQUESTED.)
RIGHT ARGUMENT: VECTOR (N,P) OF THE PARA-
 METERS OF THE BINOMIAL DISTRIBUTION.
RESULT: 2×K MATRIX, THE FIRST ROW OF
 WHICH IS THE SET OF INTEGERS RANGE[1],
 1+RANGE[1],...,RANGE[2], AND THE
 SECOND ROW OF WHICH CONTAINS THE
 CORRESPONDING PROBABILITIES, WHERE K=
 1+RANGE[2]-RANGE[1] (OR K=1).
```

CODE FOR BINFR

```
      ∇ Z←RANGE BINFR PAR;K
[1] ⍝  BINOMIAL PROBABILITIES (FREQUENCY
    ⍝  FUNCTION).
[2]   K←RANGE[1]↓0,⍳RANGE[ρRANGE←,RANGE]
[3]   Z←(2,ρK)ρK,(K!PAR[1])×((1-PAR[2])
      *PAR[1]-K)×PAR[2]*K
      ∇
```

Function Name: CHI2

EXTERNAL DOCUMENTATION

PURPOSE

To generate random numbers from the χ^2 distribution, whose density function is

$$f(x) = \begin{cases} \dfrac{1}{2^{n/2}G(n/2)}\, x^{n/2-1}e^{-x/2}, & x > 0 \\ 0, & x \leq 0 \end{cases}$$

where $G(n/2) = \Gamma(n/2)$. The user specifies n = degrees of freedom = *PAR*.

ALGORITHM

Let each sample value be $Z(i) = $ sum from $j = 1$ to n of $(X_{ij})^2$, where the X_{ij}s are i.i.d., $N(0,1)$. (The normal random numbers are generated by calling the program GAUSS.) It follows that $Z(i)$ has a χ^2 distribution with n degrees of freedom ($i = 1,2, \ldots , SS$).

INTERNAL DOCUMENTATION

```
SYNTAX: Z←SS CHI2 PAR
PURPOSE: SIMULATION OF THE CHI-SQUARE
 DISTRIBUTION
INDEX ORIGIN: 1
AUTHOR: E. SILBERG
LAST UPDATE: 1/79
RIGHT ARGUMENT: (SCALAR, INTEGER) DEGREES
 OF FREEDOM.
LEFT ARGUMENT: DESIRED SAMPLE SIZE.
RESULT: RANDOM NUMBERS FROM THE STANDARD
 CHI-SQUARE DISTRIBUTION WITH THE SPECIFIED
 DEGREES OF FREEDOM.
```

CODE

```
      ∇ Z←SS CHI2 PAR
[1] ⍝   CHI-SQUARE RANDOM NUMBER GENERATOR.
[2] ⍝   CALLS ∇Z←SS GAUSS PAR∇.
[3]   Z←+/((SS,PAR) GAUSS 0 1)*2
      ∇
```

Function Name: CHISQ

EXTERNAL DOCUMENTATION

PURPOSE

To calculate the probability that a $\chi^2(X_f^2)$ random variable with f degrees of freedom is greater than some specified value x. This is mathematically expressed as

$$P(X_f^2 > x) = \frac{1}{2^{f/2}G(f/2)} \int_x^\infty z^{f/2-1}e^{-z/2} \, dz$$

where $(x \geq 0, d \geq 1)$ and G is the gamma function.

ALGORITHM

(1) If $f = 1$ or 2, then calculate probability explicitly.

For $f = 1$, $P(X_1^2 > x) = 2z$, where

$$z = \frac{1}{\sqrt{(2\pi)}} \int_{\sqrt{(x)}}^\infty e^{-z^2/2} \, dz$$

is 1-cumulative probability of the standard normal distribution of $\sqrt{x}$.

For $f = 2$, $P(X_2^2 > x) = \exp(-x/2)$.
(2) When f does not initially equal 1 or 2, the routine uses the recurrence formula:

$$P(X_f^2 > x) = P(X_{f-2}^2 > x) + \frac{(x/2)^{f/2-1}e^{-x/2}}{G(f/2)}$$

REFERENCES

Hill, I. D., and Pike, M. C. (1978). "Collected Algorithms," Algorithm No. 299, P1-R1-P3-0. Association for Computing Machinery, New York.

INTERNAL DOCUMENTATION

```
SYNTAX: F CHISQ X
PURPOSE: CALCULATES PROBABILITY THAT
 CHI-SQUARE WITH F DEGREES OF FREEDOM IS
 GREATER THAN X
INDEX ORIGIN: 1
AUTHOR: B. BUKIET
LAST UPDATE: 5/13/80
RIGHT ARGUMENT: X-- A NUMBER OR VECTOR
LEFT ARGUMENT: F-- A NATURAL NUMBER
RESULT: Z←PROB(CHI-SQUARE>X) GIVEN F DEGREES
 OF FREEDOM
```

CODE

```
      ∇ Z←F CHISQ X
[1]   ⍝THIS PROGRAM CALCULATES PROBABILITIES
      ⍝THAT CHI SQUARE WITH F DEGREES
[2]   ⍝OF FREEDOM IS GREATER THAN X. X
      ⍝MAY BE A VECTOR.
[3]   Z←0
[4]   BEGIN:→(F≤2)/ENDS
[5]   Z←Z+(((0.5×X)*¯1+0.5×F)××*¯0.5×X)
         ÷!¯1+0.5×F
[6]   F←F-2
[7]   →BEGIN
[8]   ENDS→(F=1)/TWO
[9]    →(Z=Z←Z+*¯0.5×X)/0
[10]  TWO:Z←Z+2×1-((X*0.5) GAUSSDF 0 1)[2;]
      ∇
```

Function Name: CONVOL

EXTERNAL DOCUMENTATION

PURPOSE

To compute the convolution vector $Z = P * Q$ of two vectors of probabilities $P = (p_1, p_2, \ldots, p_m)$ and $Q = (q_1, q_2, \ldots, q_n)$, where m and n are positive integers.

NOTE

In this discussion, the symbol $*$ is the conventional mathematical notation for the convolution; it does not represent the APL operation of exponentiation.

By definition, $0 \le p_i \le 1$ for all i, and $0 \le q_j \le 1$ for all j, and sum of p_i from 1 to m = sum of q_j from 1 to n = 1. If X and Y are discrete random variables taking values $\{1, 2, \ldots, m\}$ and $\{1, 2, \ldots, n\}$, respectively, with corresponding probability vectors P and Q, then $Z = P * Q$ is the vector of probabilities for the discrete random variable

$W = X + Y$; i.e., W takes values $\{2, 3, \ldots, m + n\}$, with probabilities $Z(1)$, $Z(2)$, $\ldots$, $Z(m + n - 1)$, respectively. (W cannot $= 1$, since $X \geq 1$ and $Y \geq 1$.) Since $Z(i) = P(W = i) = P(X + Y = i)$, where X is in the set $\{1, 2, \ldots, m\}$ and Y is in the set $\{1, 2, \ldots, n\}$, it follows that

$$Z(i) = \Sigma \, p_r q_s$$

over all r, s such that r is in the set $\{1, 2, \ldots, m\}$, s is in the set $\{1, 2, \ldots, n\}$, and $r + s = i$.

ALGORITHM

Construct the following matrix:

$$
\begin{matrix}
p_1q_1 & p_1q_2 & p_1q_3 & \cdots & p_1q_n & 0 & 0 & \cdots & 0 \\
p_2q_1 & p_2q_2 & p_2q_3 & \cdots & p_2q_n & 0 & 0 & \cdots & 0 \\
& & & \vdots & & & & \vdots & \\
p_mq_1 & p_mq_2 & p_mq_3 & \cdots & p_mq_n & 0 & 0 & \cdots & 0
\end{matrix}
$$

Reshape the above matrix to the following form, and then sum along the columns:

$$
\begin{matrix}
p_1q_1 & p_1q_2 & p_1q_3 & \cdots & p_1q_n & 0 & 0 & \cdots & 0 \\
0 & p_2q_1 & p_2q_2 & \cdots & p_2q_{n-1} & p_2q_n & 0 & \cdots & 0 \\
0 & 0 & p_3q_1 & \cdots & p_3q_{n-2} & p_3q_{n-1} & & & 0 \\
& & & \vdots & & & & \vdots & \\
0 & 0 & 0 & \cdots & 0 & 0 & & \cdots & p_mq_n
\end{matrix}
$$

The result will be

$$
\begin{aligned}
Z &= (p_1q_1, \; p_1q_2 + p_2q_1, \ldots) \\
&= (P(W = 2), \; P(W = 3), \ldots, P(W = m + n))
\end{aligned}
$$

WARNING

If P and Q are long vectors (or if either P or Q is long and the other is of medium length), then the storage required for the above matrix may be excessive, resulting in a full work space during execution. If this occurs, then a looping algorithm may be the only alternative, although looping will be very costly in this case, because of the dimensions of P and Q. In general, it is best to avoid such extensive calculations in APL. (It is difficult to set maximum dimensions for P and Q, but for the purposes of this discussion, a vector of length 50 would be considered long.) For optimal storage efficiency, the user should choose P as the shorter of the two vectors of probabilities.

INTERNAL DOCUMENTATION

SYNTAX: Z←P CONVOL Q
PURPOSE: CONVOLUTION OF TWO VECTORS OF
 PROBABILITIES.
INDEX ORIGIN: 1
AUTHOR: E. SILBERG
LAST UPDATE: 1/79
LEFT ARGUMENT: FIRST VECTOR OF PROBABILITIES.
 (+/P)=1.
RIGHT ARGUMENT: SECOND VECTOR OF
 PROBABILITIES. (+/Q) =1
RESULT: THE VECTOR WHICH IS THE CONVOLUTION
 OF P AND Q.

CODE

```
     ∇ Z←P CONVOL Q
[1]  ⍝  CONVOLUTION OF TWO VECTORS OF
     ⍝  PROBABILITIES.
[2]   Z←+/[1](((ρP),(¯1+(ρP)+ρQ))ρ(P∘.×Q),
     ((ρP)×1,1)ρ0)
     ∇
```

Function Name: DISC

EXTERNAL DOCUMENTATION

PURPOSE

To generate random numbers from the general discrete distribution, whose density function is

$$f(i) = \begin{cases} p_i, & i = 1, 2, 3, \ldots, n \\ 0, & \text{otherwise} \end{cases}$$

where $(p_1, \ldots, p_n) = PAR$ is a vector of probabilities supplied by the user.

(The distribution on any other vector of numbers X can be obtained by using this program to generate the indices: X(SS DISC PAR) yields random numbers whose frequency is

$$f(x_i) = p_i \quad \text{for} \quad i = 1, 2, \ldots, n$$
$$f(x) = 0 \quad \text{for} \quad x \neq x_i$$

ALGORITHM

Use the APL scan operator to create the vector $V = (p_1, p_1 + p_2, \ldots, \Sigma p_i)$ (where sum is from $i = 1$ to n).

Generate SS uniform random numbers $u_1, \ldots, u_{SS}$ from $R(0, 1)$ by calling the program UNIF:

$$P(v_{k-1} < u_j \leq v_k) = P(u_j \leq v_k) - P(u_j \leq v_{k-1})$$

$$= \sum_{i=1}^{k} p_i - \sum_{i=1}^{k-1} p_i = p_k$$

But $(v_{k-1} < u_j < \text{or } v_k)$ implies

$$u_j \begin{cases} > v_i, & i = 1, 2, \ldots, k - 1 \\ \leq v_i, & i = k, k + 1, \ldots, n \end{cases}$$

Therefore, the interval (v_{k-1}, v_k) in which u_j is located can be obtained by adding 1 to the number of end points $v_1, \ldots, v_n$ that are exceeded by u_j.

REPEATED SIMULATIONS

If one expects to do subsequent simulations with the same vector of probabilities, then it is wise to make V a global variable in the first simulation and then omit the creation of V in each additional simulation.

LARGE SAMPLE SIZES

See the program DISCRETE for a method that is much faster for large sample sizes.

INTERNAL DOCUMENTATION

```
SYNTAX: Z←SS DISC PAR
PURPOSE: SIMULATION OF THE GENERAL DISCRETE
 DISTRIBUTION.
INDEX ORIGIN: 1
AUTHOR: E. SILBERG
LAST UPDATE: 1/79
RIGHT ARGUMENT: THE VECTOR OF PROBABILITIES.
LEFT ARGUMENT: (VECTOR OR SCALAR) THE DESIRED
 SAMPLE SIZE.
```

*RESULT: A RANDOM SAMPLE FROM THE DISCRETE
  DISTRIBUTION ON 1,2,3,...,(ρPAR), FOR WHICH
  P(1)=PAR[1],...,P(ρPAR)=PAR[ρPAR]*

CODE

```
      ∇  Z←SS DISC PAR
[1]  ⍝  GENERAL DISCRETE RANDOM NUMBER
     ⍝  GENERATOR
[2]  ⍝  CALLS: ∇Z←SS UNIF PAR∇
[3]  ⍝  LAST UPDATE: 6/27/79
[4]     Z←(⍳ρPAR)[1++/(SS UNIF 0 1)∘.>V←+\PAR]
      ∇
```

Function Name: DISCRETE

EXTERNAL DOCUMENTATION

PURPOSE

To generate a random sample from a general discrete distribution on the integers 1, 2, . . . , N, by using the alias method.

ALGORITHM

If the right argument of input (PR) is a scalar, use previous values for tables (go to B).

(A)   Create two tables FPR and LPR [(shape FPR) = (shape LPR) = N]:

  1.   Set $FPR = N \times PR$.
  2.   Define sets S and G, where

       G denotes the set of i for which $FPR(i) > 1$;
       S denotes the set of i for which $FPR(i) < 1$.

  3.   Do until the set S is empty.

       (a)   Choose elements j of G and k of S.
       (b)   Set $LPR(K) = j$.

(c)   Set $FPR(j) = FPR(j) - (1 - FPR(k))$.
(d)   If $FPR(j) < 1$,

set $G(j) = 0$
set $S(j) = 1$
else set $S(k) = 0$

(B)   Generate the random variables with the desired distribution:

1.   For each random variable desired:

(a)   Generate U with distribution uniform on $(0, 1)$.
(b)   Generate Z with distribution uniform on $(1, 2, . . . , N)$.
(c)   Get the (index of the) random variable Z with the desired distribution.

1.   If $U < FPR(i)$, set $Z = I$.
2.   If $U > FPR(i)$, set $Z = LPR(I)$.

NOTE

This program is particularly efficient for large sample sizes.

REFERENCE

Kronmal, R. A., and Peterson, A. V. (1978). On the alias method for generating random variables from a discrete distribution, Technical Report No. 17. Dept. of Biostatistics, School of Public Health and Community Medicine, Seattle, Washington.

INTERNAL DOCUMENTATION

```
SYNTAX: ∇Z←SS DISCRETE PR∇
PURPOSE: GENERATE A RANDOM SAMPLE FROM A
  GENERAL DISCRETE DISTRIBUTION ON THE
  INTEGERS A,2,...,N, FOR ARBITRARY N.
METHOD: THE ALIAS METHOD OF A.J. WALKER,
  AS MODIFIED BY R.A. KRONMAL AND A.V.
  PETERSON
AUTHOR: D.E. MCCLURE
LAST UPDATE: 11/15/78
INDEX ORIGIN: 0 OR 1
LEFT ARGUMENT: SAMPLE SIZE SS IS A SCALAR
  OR ANY VECTOR OF POSITIVE INTEGER VALUES.
```

RIGHT ARGUMENT: PR MAY BE A VECTOR OF
PROBABILITIES OR A SCALAR.
(1) WHEN PR IS A VECTOR HAVING MORE THAN
ONE COMPONENT, THE PROGRAM GENERATES A
SAMPLE SIZE SS FROM THE DISTRIBUTION
ASSOCIATING PROBABILITY PR[J] WITH OUTCOME
J, FOR J=1,2,...,ρPR. PR MUST DESCRIBE A
PROPER PROBABILITY DISTRIBUTION. THE
PROGRAM CREATES GLOBAL VARIABLES FPR, LPR
AND NPR USED BY THE ALIAS METHOD.
(2) WHEN PR IS A SCALAR, THE PROGRAM
SKIPS CREATION OF THE GLOBAL VARIABLES
FPR, LPR AND NPR AND WILL IMPLEMENT THE
ALIAS METHOD WITH THE PREVIOUSLY CREATED
VALUES OF THESE GLOBAL VARIABLES. THE
PROGRAM WILL GENERATE A RANDOM SAMPLE OF
SIZE SS FROM A DISTRIBUTION ON 1,...,NPR.
THE RESPECTIVE PROBABILITIES OF THESE
OUTCOMES ARE THE ONES SPECIFIED ON THE
PREVIOUS CALL OF THE PROGRAM DISCRETE
WHEN THE CURRENT VALUES OF FPR, LPR, AND
NPR WHERE CREATED.
RESULT: Z IS THE GENERATED SAMPLE. IF SS
IS A SCALAR, THEN Z IS A VECTOR WHOSE
COMPONENTS ARE THE SS SAMPLE VALUES. IF
SS IS A VECTOR, THEN Z IS AN ARRAY WHOSE
RANK IS SS AND WHOSE ENTRIES ARE THE SAMPLE
VALUES.

CODE

```
        ∇ Z←SS DISCRETE PR;S;F;K;J
[1]   ⍝ GENERATE SAMPLE FROM GENERAL DISCRETE
      ⍝ DISTRIBUTION
[2]   ⍝ ALIAS METHOD
[3]   ⍝ GLOBAL VARIABLES FPR, LPR AND NPR
      ⍝ ARE CREATED WHEN THE ARGUMENT PR IS
[4]   ⍝ A VECTOR AND 2≤ρPR
[5]   ⍝ SKIP GENERATION OF FPR, LPR AND NPR
      ⍝ WHEN PR IS SCALAR
[6]    →(1≥ρ,PR)/RVGEN
[7]   ⍝ GENERATE TABLES FPR AND LPR
```

```
[8]    FPR←(NPR←ρPR)×PR
[9]    LPR←NPRρ0
[10]   S←~G←FPR≥1
[11]   TOP:→((NPR+□IO)=K←Sι1)/ENDTABLE
[12]   LPR[K]←J←Gι1
[13]   →(1≤FPR[J]←FPR[J]-1-FPR[K])/DELETE
[14]   S[J]←1+G[J]←0
[15]   DELETE:S[K]←0
[16]   →TOP
[17]   ENDTABLE:FPR←□IO+1000000×FPR
[18]   LPR←LPR-ιNPR
[19]   ⍴ GENERATE SAMPLE
[20]   RVGEN:Z←Z+LPR[Z]×(?SSρ1000001)>FPR
       [Z←?SSρNPR]
       ∇
```

Function Name: EQDIST

EXTERNAL DOCUMENTATION

PURPOSE

Given an $n \times n$ stochastic matrix P, which is irreducible and aperiodic, determine a unique u such that $uP = u$, $u_i > 0$ and $\Sigma u_i = 1$ (sum from $i = 1$ to n). The vector u is the equilibrium distribution of the Markov chain with stochastic matrix P.

ALGORITHM

The following algorithm determines the row vector u:

(1) Let v satisfy $vP = v$ and represent the n-vector v as $v = (1, w)$.
(2) Solve for the $(n - 1)$-vector w: Let P be represented as

$$
\begin{array}{c|ccc}
P_{11} & P_{12} & \cdots & P_{1n} \\
\hline
P_{21} & & & \\
\vdots & & P^* & \\
P_{n1} & & &
\end{array}
$$

Then $vP = v$ implies $\$ + wP^* = w$, where $\$$ is the $(n - 1)$-vector $(P_{12}, P_{13},$

. . . , P_{1n}). Solving for w we get

$$w = \$(I - P^*)^{-1} \qquad \text{and} \qquad (I - P^*)^{-1} \text{ exists}$$

since P is irreducible and aperiodic.

(3)   Normalize v and set $u = v/\Sigma\, v_i$ (sum from $i = 1$ to n).

WARNING

If conditions (see PURPOSE) are not satisfied, DOMAIN ERROR may occur.

INTERNAL DOCUMENTATION

```
SYNTAX: Z←EQDIST P
PURPOSE: TO FIND THE EQUILIBRIUM DISTRIBUTION
 OF AN APERIODIC, IRREDUCIBLE MARKOV CHAIN
INDEX ORIGIN: 1
AUTHOR: J. ROSS
LAST UPDATE: 6/27/79
RIGHT ARGUMENT: P IS AN N×N APERIODIC,
 IRREDUCIBLE STOCHASTIC MATRIX.
RESULT: Z IS AN N-VECTOR DESCRIBING THE
 EQUILIBRIUM DISTRIBUTION OF P.
```

CODE

```
      ∇ Z←EQDIST P;V;N
[1] ℝ COMPUTES THE EQUILIBRIUM OF A MARKOV
    ℝ CHAIN
[2] ℝ LAST UPDATE: 6/27/79
[3]   V←1,(1↓P[1;])+.×⊞((N,N)ρ1,(N←((1↑ρP)
    -1))ρ0)- 1 1 ↓P
[4]   Z←V÷(+/V)
      ∇
```

Function Name: EXPON

EXTERNAL DOCUMENTATION

PURPOSE

To generate random numbers from the exponential distribution, whose density function is

$$f(x) = \begin{cases} pe^{-px}, & x > 0 \\ 0, & x \le 0 \end{cases}$$

where $p > 0$ is real. The user specifies the parameter $PAR = p$.

ALGORITHM

The method used to generate the random sample is a straightforward application of the probability integral transformation:

Invert the distribution function:

$$F(x) = 1 - e * (-px) = u \text{ implies } e * (-px) = 1 - u$$
$$\text{implies } -px = \ln(1 - u)$$
$$\text{implies } x = -(1/p) \ln(1 - u).$$

One could generate u in $R(0, 1)$ and substitute u into the above formula for x; then x would be exponential. However, note that u is $R(0, 1)$ implies $(1 - u)$ is $R(0, 1)$, so it is more efficient to generate $x(i) = -(1/p) \ln u(i)$, where u is in $R(0, 1)$. The independent uniform random numbers $u(i)$, $i = 1, 2, \ldots, SS$, are generated by calling the program UNIF.

INTERNAL DOCUMENTATION

```
SYNTAX: Z←SS EXPON PAR
PURPOSE: SIMULATION OF THE EXPONENTIAL
  DISTRIBUTION.
INDEX ORIGIN: 1
AUTHOR: E. SILBERG
LAST UPDATE: 1/79
RIGHT ARGUMENT: THE PARAMETER P OF THE
  EXPONENTIAL DISTRIBUTION WITH THE FOLLOWING
  DENSITY FUNCTION: P × *‾PX
LEFT ARGUMENT: THE DESIRED SAMPLE SIZE
RESULT: RANDOM NUMBERS FROM THE EXPONENTIAL
  DISTRIBUTION DESCRIBED ABOVE.
```

CODE

```
    ∇ Z←SS EXPON PAR
[1] ⍝EXPONENTIAL RANDOM NUMBER GENERATOR.
[2] ⍝CALLS ∇Z←SS UNIF PAR∇.
```

[3] $Z \leftarrow (\div PAR) \times - \circledast SS$ $UNIF$ 0 1
       ∇

Function Name: FISHER

EXTERNAL DOCUMENTATION

PURPOSE

To generate random numbers from the Fisher (F) distribution, whose density function is

$$f(x) = \begin{cases} \dfrac{G(.5(m + n))m^{m/2}n^{n/2}x^{m/2-1}}{G(m/2)G(n/2)(mx + n)^{(m+n)/2}} & x > 0 \\ 0 & x \leqq 0 \end{cases}$$

This is the distribution of the random variable defined as $X = (Y/m)/(W/n)$ where Y and W are independent chi-square random variables with m and n degrees of freedom, respectively.

ALGORITHM

Let each sample value $Z(i)$ be the ratio $nY(i)/mW(i)$, where $Y(i)$ distributed chi-square with m degrees of freedom and $W(i)$ distributed chi-square with n degrees of freedom are independent random numbers generated by calling the program CHI2 twice, with parameters m and n, respectively ($i = 1, 2, \ldots, SS$).

INTERNAL DOCUMENTATION

```
SYNTAX: Z←SS FISHER PAR
PURPOSE: SIMULATION OF THE FISHER (F)
 DISTRIBUTION.
INDEX ORIGIN: 1
AUTHOR: E. SILBERG
LAST UPDATE: 1/79
RIGHT ARGUMENT: A PAIR OF INTEGERS (PAR[1],
 PAR[2]), INDICATING DEGREES OF FREEDOM.
LEFT ARGUMENT: THE DESIRED SAMPLE SIZE.
RESULT: RANDOM NUMBERS FROM THE FISHER
 DISTRIBUTION WITH THE SPECIFIED DEGREES
 OF FREEDOM.
```

CODE

```
     ∇  Z←SS FISHER PAR
[1]  ⍝FISHER (F) RANDOM NUMBER GENERATOR.
[2]  ⍝CALLS ∇Z←SS CHI2 PAR∇.
[3]   Z←(PAR[2]÷PAR[1])×(SS CHI2 PAR[1])
      ÷(SS CHI2 PAR[2])
     ∇
```

Function Name: GAMMA

EXTERNAL DOCUMENTATION

PURPOSE

To generate random numbers from the gamma distribution, whose density function is

$$f(x) = \begin{cases} \dfrac{1}{G(r)}\, x^{r-1} e^{-x}, & x \geqq 0 \\[2mm] 0, & x < 0 \end{cases}$$

where $(r > 0)$ is real and $G(r) = \Gamma(r)$.

BASIC THEOREM

If $r = \Sigma\, r(i)$ from $i = 1$ to k, then the sum of k independent Γ random variables with parameters $r(1), r(2), \ldots , r(k)$, respectively, is a random variable with the above gamma distribution with parameter $f = \Sigma\, r(i)$ from $i = 1$ to k (see Feller, p. 47). In the program GAMMA, the variable *PAR* represents r.

ALGORITHM

Write *PAR* as $(N/2) + P$, where N is a nonnegative integer, and $0 \leq P < 0.5$. The possible cases and corresponding procedures for obtaining the random sample Z of size SS are outlined in the accompanying table. Further descriptions of the procedures follow.

Generation of random numbers from the distribution with density h:

$$h(x) = \begin{cases} px^{p-1}, & 0 < x < 1 \\ 0, & \text{otherwise} \end{cases}$$

Case	Procedure for generating Z
$P = 0$	X is chi-square with n degrees of freedom. Call CHI2, and exit from GAMMA.
$P > 0$	
$N = 0$	
if $0 < P < 0.001$	Let Z be a sample from $$h(x) = \begin{cases} px^{p-1}, & 0 \leq x < 1 \\ 0, & \text{otherwise} \end{cases}$$
if $0.001 \leq P < 0.5$	Obtain Z by the acceptance–rejection scheme (see discussion).
$N > 0$	First generate numbers from chi-square with n degrees of freedom, then:
if $0 < P < 0.001$	Add a sample from h, as defined above, or
if $0.001 \leq P < 0.5$	add a sample obtained by the acceptance–rejection scheme.

Use the inversion technique provided by the probability integral transformation:

$$H(x) = \int_0^x pt^{p-1}\, dt = x^p \qquad \text{implies} \quad H^{-1}(y) = y^{(1/p)}$$

Hence, if $u(i)$ is distributed $R(0, 1)$, then

$$H^{-1}(u_i) = u_i^{(1/p)}$$

has distribution H.

Generation of random numbers from the gamma distribution by the acceptance–rejection scheme:

1. SS can be written as a sum of two positive integers: $SS = \text{LEFT} + \text{RIGHT}$, where LEFT = (no. of sample points less than or equal to 1), and RIGHT = (no. of sample points greater than 1). LEFT and RIGHT are not known in advance, since the sample values are random.

2. The probability that a value $x(i)$ will be less than or equal to 1 is AREA = area under the gamma density curve from 0 to 1:

$$\frac{1}{G(p)} \int_0^1 x^{p-1} e^{-x}\, dx$$

If P were greater than or equal to 1, then this integral could be approximated by using the series expansion for the incomplete gamma function (see Olver, p. 45):

$$\int_0^a t^{p-1}e^{-t}\,dt = \sum_{k=0}^{\infty} \frac{(-1)^k}{k!(p+k)}\, a^{p+k} \text{ at } a = 1 = \sum_{k=0}^{\infty} \frac{(-1)^k}{k!(p+k)}$$

However, we know that $0.001 < P < 0.5$. Therefore, use integration by parts to express the desired integral in terms of the incomplete g function for $(p+1)$:

$$\int_0^1 t^{p-1}e^{-t}\,dt = ((1/p)t^p e^{-t})\big|_{t=0}^1 + (1/p)\int_0^1 t^p e^{-t}\,dt$$

$$= (1/p)\left(e^{-1} + \int_0^1 t^p e^{-t}\,dt\right)$$

Now divide by $G(p)$ to obtain

$$\text{AREA} = (1/pG(p))\left(e^{-1} + \int_0^1 t^p e^{-t}\,dt\right)$$

$$= (1/G(p+1))\left(e^{-1} + \int_0^1 t^p e^{-t}\,dt\right)$$

$$= (1/G(p+1))\left(e^{-1} + \sum_{k=0}^{\infty} ((-1)^k/k!(p+1+k))\right)$$

(In APL, $!P = G(P+1)$, where $G(x) = \Gamma(x)$.)

3. Determine LEFT by generating a single binomial number, from EIN(SS,AREA). Then RIGIIT is also known: RIGHT $-$ SS $-$ LEFT.

4. To generate the samples from the left and right portions of the distribution:

(a) LEFT:

SAMP1 = a set of values less than or equal to 1: Generate random numbers from

$$h(x) = \begin{cases} px^{p-1}, & 0 \leq x < 1 \\ 0, & \text{otherwise} \end{cases}$$

(See discussion in earlier part of the section.) Suppose $x(i)$ is a generated value. Accept $x(i)$ with probability $e^*(-x(i))$; reject $x(i)$ with probability $1 - e^*(-x(i))$. This is achieved by a Bernoulli simulation with $P(\text{success}) = e^*(-x(i))$. The total number of values generated is NGEN(1) > LEFT. (See description of NGEN.) But since some values will be rejected, it may be necessary to loop in order to achieve (shape of SAMP1) greater than or equal to LEFT.

(b) RIGHT:

SAMP2 = a set of values > 1: Generate random numbers from the distribution with density:

$$b(x) = \begin{cases} e^{-(x-1)}, & x > 1 \\ 0, & x < 1 \end{cases}$$

(The d.f. can be inverted:

$$B(x) = 1 - e^{-(x-1)}$$
$$B^{-1}(y) = 1 - \ln(1 - y)$$

Let $x_i = 1 - \ln u_i$, u_i is in $R(0, 1)$.)

Accept x_i with probability $x_i * (p - 1)$; reject x_i with probability $1 - x_i * (p - 1)$. This is achieved by a Bernoulli simulation with $P(\text{success}) = x * (p - 1)$. The total number of values generated is NGEN(2) > RIGHT. (See description of NGEN.) But since some values will be rejected, it may be necessary to loop in order to achieve (shape of SAMP2) greater than or equal to RIGHT.

    5.   Now take LEFT values from SAMP1, and RIGHT values from SAMP2, and randomize them with respect to one another. This completes the acceptance–rejection simulation procedure.

DESCRIPTION OF NGEN AND PACCEPT

    NGEN is a vector whose two components are the number of values to generate in each looping for the left and right in the acceptance–rejection procedure. NGEN is based on a rough approximation using PACCEPT, the vector whose two components are the probabilities of acceptance for the left and right:

For values less than or equal to 1:

$$P(\text{Acceptance}) = \int_0^1 px^{p-1}e^{-x} \, dx = pG(p), \qquad \text{AREA} = \text{PACCEPT}(1)$$

For values greater than 1,

$$P(\text{Acceptance}) = \int_1^\infty x^{p-1}e^{-(x-1)} \, dx$$
$$= e\left(\int_0^\infty x^{p-1}e^{-x} \, dx - \int_0^1 x^{p-1}e^{-x} \, dx\right)$$
$$= eG(p)(1 - \text{AREA}) = \text{PACCEPT}(2)$$

(For numerical reasons, the above calculation of PACCEPT(2) will yield a negative result for $p < $ roughly $10*(-5)$; however, note that the acceptance–rejection scheme is only used for $P >$ or $= 0.001$.)

    To calculate NGEN:

(In the following discussion, ' = = ' is used to indicate "is approximately equal to".) LEFT = (no. of successes on left) = = (NGEN(1)) (PACCEPT (1)) implies NGEN(1) = = LEFT/P (ACCEPT(1)). Now add an extra quantity: Var(LEFT) = (NGEN(1)) (PACCEPT(1)) (1-PACCEPT(1)) $\le$ (NGEN(1)) (PACCEPT(1)) = = LEFT. Therefore, take NGEN(1) = (LEFT/PACCEPT(1)) + 5$\sqrt{\text{(LEFT)}}$. Similarly, take NGEN(2) = (RIGHT/PACCEPT(2)) + 5$\sqrt{\text{(RIGHT)}}$.

WARNING

For each value of *PAR*, computing time is a random variable whose mean is a linear function of sample size. Therefore, for very large sample sizes, the method used in this program is impractical; discretizition is a suggested alternative. Although it is difficult to set a definite cutoff point, it is clear that this program would be appropriate for a sample size of 100, but not for a sample size of 1000.

REFERENCES

Feller, W. (1971). "An Introduction to Probability Theory and Its Applications," Vol. 2, p. 47. Wiley, New York.

Olver, F. W. J. (1974). "Introduction to Asymptotics and Special Functions," p. 45. Academic Press, New York.

Tadikamalla, P. (1978a). Computer generation of gamma random variables, *Comm. ACM,* **21,** 419–422.

Tadikamalla, P. (1978b). Computer generation of gamma random variables II, *Comm. ACM,* **21,** 925–928.

INTERNAL DOCUMENTATION

```
SYNTAX: Z←SS GAMMA PAR
PURPOSE: SIMULATION OF THE GAMMA DISTRI-
  BUTION.
INDEX ORIGIN: 1
AUTHOR: E. SILBERG
LAST UPDATE: 12/78
RIGHT ARGUMENT: THE SINGLE PARAMETER OF A
  GAMMA DISTRIBUTION WITH STANDARDIZED
  EXPONENTIAL FACTOR; I.E., WITH DENSITY
  (÷!PAR-1)×(X*PAR-1)×(*-X),FOR X>0
LEFT ARGUMENT: (SCALAR) THE DESIRED SAMPLE
  SIZE.
RESULT: RANDOM NUMBERS FROM THE GAMMA
  DISTRIBUTION.
```

CODE

```
      ∇ Z←SS GAMMA PAR;N;P;AREA;RIGHT;LEFT
        PACCEPT;SAMP1;SAMP2;NGEN;TRIALS
[1]   ⍝ GAMMA RANDOM NUMBER GENERATOR.
[2]   ⍝ CALLS:  ∇Z←SS CHI2 PAR∇,∇Z←SS BIN
      ⍝ PAR∇,∇Z←SS UNIF PAR∇,∇Z←SS BERN PAR∇
[3]   →(0=Z←PAR-P←0.5×(2×PAR)-N←⌊2×PAR)/TESTP
[4]   Z←0.5×SS CHI2 N
[5]   TESTP:→((P=0)∨P≥0.001)/PART2×P≠0
[6]   →0×⍴Z←Z+(SS UNIF 0 1)*÷P
[7]   PART2:AREA←(÷!P)×(*⁻1)+-/÷(P+1+0,⍳8)
      ×!0,⍳8
[8]   RIGHT←SS-LEFT←1 BIN(SS,AREA)
[9]   PACCEPT←(!P)×AREA,(÷P)×(*1)×1-AREA
[10]  SAMP2←SAMP1←0/NGEN←⌈((LEFT,RIGHT)
      ÷PACCEPT)+5×(LEFT,RIGHT)*0.5
[11]  LOOP1:TRIALS←(NGEN[1] UNIF 0 1)*÷P
[12]  →(LEFT>⍴SAMP1←SAMP1,(NGEN[1] BERN*
      -TRIALS)/TRIALS)/LOOP1
[13]  LOOP2:TRIALS←1+-⍟NGEN[2] UNIF 0 1
[14]  →(RIGHT>⍴SAMP2←SAMP2,(NGEN[2] BERN
      TRIALS*P-1)/TRIALS)/LOOP2
[15]  Z←Z+((LEFT↑SAMP1),RIGHT↑SAMP2)[SS?SS]
      ∇
```

Function Name: GAUSSDF

EXTERNAL DOCUMENTATION

PURPOSE

To calculate cumulative Gaussian probabilities for each value x_i, where $X = (x_1, x_2, \ldots, x_n)$ is a vector such that $-\inf < x_1 < x_2 < \cdots < x_n < \inf$. The user specifies X and $PAR = (m = \text{mean}, V*2 = \text{variance})$. The result is the vector $Z = (P(Y < \text{or} = x_1), \ldots, P(Y < \text{or} = x_n))$, where Y is distributed $N(m, V*2)$; i.e.,

$$Z(i) = \int_{-\infty}^{x_i} \frac{1}{V\sqrt{2\pi}} \exp(-1/2)((k - m)/v) * 2 \, dk = \phi((x_i - m)/V),$$

where ϕ is the distribution function of the standard normal distribution $(N(0, 1))$.

ALGORITHM

The APL function ERF (the error function) calculates the quantity

$$\text{Error}(k) = (2/\sqrt{\pi}) \int_0^k \exp(-m * 2) \, dm$$

For $a \geqq 0$,

$$\phi(a) = P(X < 0) + P(0 \leqq X \leqq a)$$
$$= \tfrac{1}{2} + (1/\sqrt{2\pi}) \int_0^a \exp[-(m * 2/2)] \, dm$$

Let $j = m/\sqrt{(2)}$, so that $dj = (1/\sqrt{(2)}) \, dm$. Then

$$\phi(a) = \tfrac{1}{2}(1 + (2/\sqrt{2\pi})\int(\sqrt{(2)}) \exp(-j * 2) \, dj \text{ (integral from 0 to } a/\sqrt{(2)})$$
$$= \tfrac{1}{2}(1 + \text{ERF}(a/\sqrt{(2)}))$$

This is the formula used in the program, for $a = (x - m)/v >$ or $=$ to 0. For $a < 0$, use the fact that ϕ is symmetric:

$$\phi(-|a|) = 1 - \phi(|a|) = 1 - \tfrac{1}{2} (1 + \text{ERF}(|a|/\sqrt{(2)})$$

NOTE

Function ERF can be found in special functions, Chapter 23.

INTERNAL DOCUMENTATION

```
SYNTAX: Z←X GAUSSDF PAR
PURPOSE: CALCULATION OF CUMULATIVE GAUSSIAN
 PROBABILITIES PHI(X) (I.E., DISTRIB. FN.)
INDEX ORIGIN: 1
AUTHOR: E. SILBERG
LAST UPDATE: 1/79
LEFT ARGUMENT: VECTOR OF X VALUES (ORDERED)
 FOR WHICH PHI(X) IS TO BE COMPUTED
RIGHT ARGUMENT: (MEAN, VARIANCE) OF THE
 GAUSSIAN DISTRIBUTION
RESULT: A 2×(ρX) MATRIX, THE FIRST ROW IS
 X AND THE SECOND ROW CONTAINS THE
 CORRESPONDING VALUES PHI(X).
```

CODE

```
     ∇ Z←X GAUSSDF PAR;X1;N
[1]  ⍴  CUMULATIVE GAUSSIAN PROBABILITIES
     ⍴  (D.F.);CALLS ERF.
[2]  X1←(X-PAR[1])÷PAR[2]*0.5
[3]  Z←(2,0.5×⍴Z)⍴Z←X,N+((~N)-N←X1<0)×0.5
     ×1+ERF|X1÷2*0.5
     ∇
```

Function Names: GAUSS and SGAUSS

EXTERNAL DOCUMENTATION

PURPOSE

To produce an iid sample from a univariate Gaussian distribution.

Let x_i be an element from the sample. Then

$$f(x_i) = \frac{1}{(v)\sqrt{2\pi}} \exp(-(x_i - u)^2/2v^2)$$

where u is the mean of the Gaussian distribution and v^2 is the variance. If X is the complete sample of size N, then

$$f(X) = \frac{1}{(2\pi)^{(N/2)}v^N} \exp\left(- \sum_{i=1}^{N} (x_i - u)^2/2v^2\right)$$

ALGORITHM

To generate a sample element:

(1) 1,000,001 is randomized 12 times.

(2) The randomizations are summed.

(3) -12 is added to adjust the randomization from 1,000,001, index origin 1 to 1,000,000 index origin 0.

(4) Divide by 1,000,000 to change interval to (0, 12). Appeal to central limit theorem; the sum of 12 uniform distributions on (0, 1) is approximately normally distributed with mean 6 and variance 1.

(5) Subtract 6 to approximate standard normal distribution.

(6) Multiply by standard deviation and add mean to achieve desired normal distribution.

$$x_i = v\left(\left(\sum_{j=1}^{12} y_j\right) - 6\right) + u$$

where y_j is uniform on $(0, 1)$.

EXAMPLE

2 2 GAUSS 0 2

1.654	1.717
−1.017	0.833

SGAUSS produces an iid sample from a standard normal distribution by setting mean to 0, variance to 1, and calling GAUSS.

INTERNAL DOCUMENTATION FOR GAUSS

```
SYNTAX: Z←SS GAUSS PAR
PURPOSE: GENERATE AN IID SAMPLE FROM THE
  UNIVARIATE GAUSSIAN (NORMAL) DISTRIBUTION
INDEX ORIGIN: 1
AUTHORS: B. BENNETT, P. KRETZMER
LAST UPDATE: 10/7/78
LEFT ARGUMENT: SS CAN BE EITHER A SCALAR
  OR A VECTOR.
IF A SCALAR, THEN Z WILL BE A VECTOR OF SIZE
SS;
IF A VECTOR, THEN Z WILL BE AN ARRAY WITH SS
AS ITS DIMENSIONS.
RIGHT ARGUMENT: PAR, A TWO-ELEMENT VECTOR:
PAR[1]=MEAN; PAR[2]=VARIANCE (OF DESIRED
UNDERLYING DIST.)
RESULT: Z, A VECTOR OR ARRAY OF SHAPE SS,
WHICH IS AN IID SAMPLE FROM THE SPECIFIED
GAUSSIAN DISTRIBUTION.
RELATED ROUTINE: Z←SGAUSS SS
SAMPLE IS DRAWN FROM A STANDARD DIST. (I.E.,
 SS GAUSS 0 1)
```

CODE FOR GAUSS

```
    ∇ Z←SS GAUSS PAR
[1] ⍝SAMPLES FROM A UNIVARIATE NORMAL
    ⍝(GAUSSIAN) DIST
```

```
[2]   ⍝SS←SAMPLE SIZE (VECTOR HERE WOULD
      ⍝SPECIFY ⍴Z)
[3]   ⍝PAR←MEAN,VARIANCE OF UNDERLYING
      ⍝DISTRIBUTION
[4]   ⍝NO EXTERNAL FUNCTIONS
[5]   Z←PAR[1]+(PAR[2]*÷2)×⁻6+1E⁻6×⁻12++/
      ?(SS,12)⍴1000001
      ∇
```

INTERNAL DOCUMENTATION FOR SGAUSS

```
SYNTAX: Z←SGAUSS SS
PURPOSE: GENERATES AN IID SAMPLE FROM THE
 UNIVARIATE STANDARD NORMAL DISTRIBUTION
INDEX ORIGIN: 1
AUTHORS: B. BENNETT, P. KRETZMER
LAST UPDATE: 10/7/78
RIGHT ARGUMENT: SS CAN BE EITHER A SCALAR
 OR A VECTOR.
 IF A SCALAR, THEN Z WILL BE A VECTOR OF
 SIZE SS;
 IF A VECTOR, THEN Z WILL BE AN ARRAY WITH
 SS AS ITS DIMENSIONS.
RESULT: Z, A VECTOR OR ARRAY OF SHAPE SS
```

CODE FOR SGAUSS

```
      ∇ Z←SGAUSS SS
[1]   ⍝SAMPLES FROM A STANDARD NORMAL
      ⍝(GAUSSION) DIST.
[2]   ⍝SS←SAMPLE SIZE (VECTOR HERE WOULD
      ⍝SPECIFY ⍴Z)
[3]   ⍝CALLS ∇Z←SS GAUSS PAR∇
[4]   Z←SS GAUSS 0 1
      ∇
```

Function Name: GEOM

EXTERNAL DOCUMENTATION

PURPOSE

To generate random numbers from the geometric distribution, whose density function is as follows:

$$f(x) = \begin{cases} p(1 - p)^x, & x = 0, 1, 2, \ldots \\ 0, & \text{otherwise} \end{cases}$$

This is the distribution of the number of failures before the first success, in a sequence of independent Bernoulli trials for which P (success) $= p$ in each trial. The user specifies this probability $p = PAR$.

ALGORITHM

The object is to generate SS sequences of Bernoulli trials (with parameter p), each ending in the first success. Let $N(i)$ be the number of trials needed in the ith sequence, and let $N = N(1) + N(2) + \cdots + N(SS)$ be the total number of trials needed in the entire simulation. Then $X(i) = $ no. of failures in the ith sequence $= N(i) - 1$. Since $X(i)$ is geometric, $E(X(i)) = (1 - p)/p$. Therefore,

$$E(N(i)) = E(1 + X(i)) = 1 + E(X(i)) = 1 + (1 - p)/p = 1/p$$

Therefore, $E(N) = E(\Sigma N(i)) = \Sigma F(N(i)) = SS/p$ (Σ runs from $i = 1$ to SS). Also, $\text{Var}(X(i)) = (1 - p)/p * 2$, so $\text{Var}(N(i)) = \text{Var}(1 + X(i)) = \text{Var}(X(i)) = (1 - p)/p * 2$. Since the $X(i)$s are independent, $\text{Var}(N) = \text{Var}(\Sigma N(i)) = SS \text{Var}(N(i)) = SS(1 - p/p * 2$. It is highly probable that $N < \text{or} = E(N) + 5\sqrt{(\text{Var}(N))} = SS/p + (5/p)\sqrt{(SS(1 - p))}$. Therefore, generate $N = \max(1/p) \ SS + 5\sqrt{(SS(1 - p))}$ Bernoulli numbers. To count the number of 0s before each 1, use the vector of 0s and 1s to compress the vector (index of N) of trial numbers to a list of the positions of the 1s. Then find the differences between these position numbers, and subtract 1 to obtain the number of 0s in each sequence. (This subtraction of 1 is not done until the last step, so that the user can easily switch to the other version of the geometric distribution, in which the random variable is $Y(i) = 1 + X(i) = $ no. of trials needed $= 1 + $ (no. of failures).)

In theory, looping may be necessary to obtain SS sample values; in this case, N is reset to (SS (shape of Z))$/p = $ (expected number of trials now needed), and the program loops back for additional sample values.

INTERNAL DOCUMENTATION

```
SYNTAX: Z←SS GEOM PAR
PURPOSE: SIMULATION OF THE GEOMETRIC
 DISTRIBUTION.
INDEX ORIGIN: 1
AUTHOR: E. SILBERG
LAST UPDATE: 1/79
```

RIGHT ARGUMENT: A SCALAR, 0≤PAR≤1,
REPRESENTING P(SUCCESS).
LEFT ARGUMENT: THE DESIRED SAMPLE SIZE.
RESULT: RANDOM NUMBERS FROM THE GEOMETRIC
DISTRIBUTION; EACH VALUE IS THE NUMBER OF
FAILURES IN A SEQUENCE ENDING IN THE FIRST
SUCCESS.

CODE

```
      ∇ Z←SS GEOM PAR;V
[1] ⍝   GEOMETRIC RANDOM NUMBER GENERATOR.
[2] ⍝   CALLS ∇Z←SS BERN PAR∇.
[3]   Z←0/N←⌈(÷PAR)×SS+5×(SS×1-PAR)*0.5
[4]   LOOP:Z←Z,-/(2,ρV)ρ(V←(N BERN PAR)/⍳N),0
[5]   →(0<N←(SS-ρZ)÷PAR)/LOOP
[6]   Z←SS↑Z-1
      ∇
```

Function Name: MARKOV

EXTERNAL DOCUMENTATION

PURPOSE

To generate a Markov chain.

MARKOV generates a Markov chain Z of length SS, beginning with a state determined by the given initial vector of probabilities $P(0) = PAR$ (1;), and having subsequent states determined by the probability transition matrix $P*$, where $P*$ is the $(n \times n)$ matrix (1 0 drop of PAR), specified by the user:

ALGORITHM

1. Scan the rows of the matrix $p*$, to create the following matrix:

$$
\begin{matrix}
p_{11} & p_{11} + p_{12} & \cdots & \Sigma\, p_{1i} \\
p_{21} & p_{21} + p_{22} & \cdots & \Sigma\, p_{2i} \\
& & \vdots & \\
p_{n1} & p_{n1} + p_{n2} & \cdots & \Sigma\, p_{ni}
\end{matrix}
$$

where the sum is from $i = 1$ to n.

2.   Scan the initial probability vector $P(0)$ and use it to generate an initial state $Z(1)$, by the method for simulation of a general discrete distribution (UNIF is called; see the description of the program DISC for the details of this method):

$$(+ \text{ scan of } P_0) = (p_{01}, p_{01} + p_{02}, \ldots, \Sigma \, p_{0j})$$

3.   Set $I = 2$.

4.   Loop: Use the $Z(I - 1)$st row of P to generate the Ith state $Z(I)$ by the method of simulating a general discrete distribution.

5.   Increase I by 1 and return to step 4. (Repeat until $I = 1 + SS$.)

NOTE

For a program which computes the equilibrium distribution of a particular transition matrix, see the program EQDIST.

INTERNAL DOCUMENTATION

```
SYNTAX: Z←SS MARKOV PAR
PURPOSE: SIMULATION OF A MARKOV PROCESS.
INDEX ORIGIN: 1
AUTHOR: E. SILBERG
LAST UPDATE: 1/79
LEFT ARGUMENT: (INTEGER>0) DESIRED SAMPLE
 SIZE.
RIGHT ARGUMENT: (N+1)×N MATRIX. THE 1ST ROW
 IS THE VECTOR OF INITIAL PROBABILITIES.
 THE REST OF THE MATRIX IS THE N×N
 TRANSITION MATRIX.
RESULT: A VECTOR OF POSITIVE INTEGERS
 SPECIFYING THE CLASSES, FOR A MARKOV
 CHAIN OF LENGTH SS.
```

CODE

```
      ∇ Z←SS MARKOV PAR
[1] ⍝   SIMULATION OF A MARKOV CHAIN.
[2] ⍝   CALLS ∇Z←SS UNIF PAR∇.
[3]   P←+\ 1 0 ↓PAR
[4]   I←2+0×Z←1++/(1 UNIF 0 1)>+\PAR[1;]
[5] LOOP:Z←Z,1++/(1 UNIF 0 1)>P[¯1↑Z;]
[6]   →(SS≥I←I+1)/LOOP
      ∇
```

Function Name: MULTINOM

EXTERNAL DOCUMENTATION

PURPOSE

To generate sample vectors from the multinomial distribution, whose density is

$$f(x) = f(x_1, x_2, \ldots, x_k) = \frac{n!}{x_1! x_2! \cdots x_k!} p_1^{x_1} p_2^{x_2} \cdots p_k^{x_k}$$

where sum of x_i from $i = 1$ to k is equal to n.

In addition to the desired sample size, the user specifies $PAR(1) = n = $ (no. of trials), and $(p_1, p_2, \ldots, p_k) = (PAR(2), PAR(3), \ldots, PAR(k + 1)) = $ probabilities for classes $1, 2, \ldots, k$, respectively, in each trial. The x_i, $i = 1, 2, \ldots, k$ are the number of trials falling into each of the k classes.

ALGORITHM

Each trial can be simulated by calling the general discrete random number generator, DISC, to generate an outcome $1, 2, \ldots,$ or k, with probabilities $p_1, p_2, \ldots, p_k$, respectively. Since each multinomial experiment consists of n trials, this is done n times, in each of the SS experiments. The result is an SS $\times k$ matrix, the ith row of which contains the resulting class counts $x_{i1}, x_{i2}, \ldots, x_{ik}$, for the n trials in the ith experiment. Thus, in the ith row,

$$\sum_{j=1}^{k} x_{ij} = n \quad \text{for each row } i = 1, 2, \ldots, SS$$

INTERNAL DOCUMENTATION

```
SYNTAX: Z←SS MULTINOM PAR
PURPOSE: SIMULATION OF THE MULTINOMIAL
  DISTRIBUTION.
INDEX ORIGIN: 1
AUTHOR: E. SILBERG
LAST UPDATE: 1/79
LEFT ARGUMENT: SAMPLE SIZE (NUMBER OF EXPER-
  IMENTS).
```

RIGHT ARGUMENT: PAR[1]=POSITIVE INTEGER,
REPRESENTING NUMBER OF TRIALS IN EACH
EXPERIMENT, WHERE THE PROBABILITIES FOR THE
K CLASSES ARE PAR[2],PAR[3],...PAR[K+1],
IN EACH TRIAL. (+/1↓PAR)=1.
RESULT: SS×K MATRIX, EACH ROW OF WHICH
REPRESENTS THE RESULTS OF ONE MULTINOMIAL
EXPERIMENT. THE COLUMNS REPRESENT THE
CLASSES. THE ENTRIES ARE THE NO. OF TRIALS
FALLING IN EACH CLASS.

CODE

```
     ∇  Z←SS MULTINOM PAR
[1]  ⍝   MULTINOMIAL RANDOM NUMBER GENERATOR.
[2]  ⍝   CALLS ∇Z←SS DISC PAR∇.
[3]    Z←+/[2]((SS,PAR[1]) DISC 1↓PAR)∘.=
       ⍳⁻1+⍴PAR
     ∇
```

Function Name: NEGBIN

EXTERNAL DOCUMENTATION

PURPOSE

To generate random numbers from the negative binomial distribution, whose density function is as

$$f(x) = \begin{cases} ((r + x - 1)!x)p^r(1 - p)^x, & x = 0, 1, 2, \ldots \\ 0, & \text{otherwise} \end{cases}$$

This is the distribution of the number of failures X before the rth success in a sequence of independent Bernoulli trials from which P(success) $= p$ in each trial. The user specifies $p = PAR(1)$ and $r = PAR(2)$.

ALGORITHM

The object is to generate SS sequences of Bernoulli trials (with parameter p), each ending in the rth success. Let $N(i)$ be the number of trials needed in the ith sequence. Let $N(i) = \Sigma M(ij)$ (Σ from $j = 1$ to r), where $M(ij)$ is the number of trials needed before the jth success. The number of failures, $M(ij) - 1$, is geometric, so $E(M(ij) - 1) = (1 - p)/p$; hence $E(M(ij)) = 1 + (1 - p)/p = 1/p$. Thus, $E(N(i)) = E(\Sigma\, M(ij)) = rE(M(ij)) = r/p$ (Σ from $j = 1$ to r).

Therefore, if $N + \Sigma$ (from $i = 1$ to SS) of $N(i)$ is the total number of trials needed for the entire simulation, then $E(N) = E(\Sigma\, N(i)) = rSS/p$.

Also, $\mathrm{Var}(M(ij) - 1) = \mathrm{Var}(M(ij)) = (1 - p)/p * 2$. Since the $M(ij)$s are independent, $\mathrm{Var}(N(i)) = \Sigma$ (from $j = 1$ to r) $\mathrm{Var}(M(ij)) = (r(1 - p))/p * 2$. Hence, $\mathrm{Var}(N) = \mathrm{Var}(\Sigma\, (N(i)) = (SSr(1 - p))/p * 2$. It is highly probable that $N <$ or $= E(N) + 5\sqrt{(\mathrm{Var}(N))} = (SSr)/p + (5/p)\,\sqrt{(SSr(1 - p))}$. Therefore, generate $N = \max((1/p)\,(SSr + 5\sqrt{(SSr(1 - p))}))$ Bernoulli numbers. To count the number of 0s before each set of r ones:

(1) Use the vector of 0s and 1s to compress the vector (index of N) of trial numbers to a list of the positions of the 1s.

(2) Isolate the rth, $(2r)$th, $(3r)$th, . . . values in this list. (If the length of the list is not divisible by r, then ignore the successes in positions beyond the last multiple of r.)

(3) Find the differences between these position numbers to obtain the total number of trials in each sequence.

(4) Subtract r from each of these values to obtain the number of failures in each sequence. (This subtraction of r is not done until the last step, so that the user may easily switch the program to the other version of the negative binomial, in which the random variable is $Y(i) = X(i) + r =$ no. of trials needed $= r +$ (no. of failures).)

In theory, looping may be necessary to obtain SS sample values; in this case, N is reset to $(r/p)\,(SS$ shape of $Z) =$ (expected number of trials now needed), and the program loops back for additional sample values.

INTERNAL DOCUMENTATION

```
SYNTAX: Z←SS NEGBIN PAR
PURPOSE: SIMULATION OF THE NEGATIVE BINOMIAL
 DISTRIBUTION.
INDEX ORIGIN: 1
AUTHOR: E. SILBERG
LAST UPDATE: 1/79
LEFT ARGUMENT: THE DESIRED SAMPLE SIZE
RIGHT ARGUMENT: PAR[1]= P(SUCCESS); PAR[2]=
NUMBER OF SUCCESSES IN EACH SEQUENCE.
```

RESULT: *RANDOM NUMBERS FROM THE NEGATIVE BINOMIAL DISTRIBUTION; EACH VALUE IS THE NUMBER OF FAILURES IN A SEQUENCE ENDING IN THE (PAR[2])TH SUCCESS.*

CODE

```
     ∇ Z←SS NEGBIN PAR;V;T
[1] ⍝   NEGATIVE BINOMIAL RANDOM NUMBER
    ⍝   GENERATOR.
[2] ⍝   CALLS ∇Z←SS BERN PAR∇.
[3]   Z←0/N←⌈(÷PAR[1])×(SS×PAR[2])+5×(SS×
      PAR[2]×1-PAR[1]*0.5
[4] LOOP:T←((⍴V)⍴((⁻1+PAR[2])⍴0),1)/V←(N
      BERN PAR[1])/⍳N
[5]   Z←Z,-⌿(2,⍴T)⍴T,0
[6]   →(0<N←PAR[2]×(SS-⍴Z)÷PAR[1])/LOOP
[7]   Z←SS↑Z-PAR[2]
     ∇
```

Function Name: PARETO

EXTERNAL DOCUMENTATION

PURPOSE

To generate random numbers from the Pareto distribution, whose density function is

$$f(x) = \begin{cases} \dfrac{(p-1)a^{p-1}}{x^p}, & x > a \\ 0, & x \leq a \end{cases}$$

The user specifies the parameters $a = PAR(1) > 0$ and $p = PAR(2) > 1$.

ALGORITHM

The method used to generate the random sample is a straightforward application of the probability integral transformation. Invert the distribution function:

$$F(x) = 1 - a^{p-1}x^{1-p} = u \text{ implies } a^{p-1}x^{1-p} = 1 - u$$

implies $x^{1-p} = a^{1-p}(1 - u)$ implies $x = a(1 - u)^{1/(1-p)}$.

One could generate u in $R(0, 1)$ and substitute u into the above formula for x: then x would be from the Pareto distribution. However, note that u is $R(0, 1)$ implies $1 - u$ is $R(0, 1)$, so it is more efficient to generate $x(i) = a(u(i)) * (1/(1 - p))$, where $u(i)$ is in $R(0, 1)$. The independent uniform random numbers $u(i), i = 1, 2, \ldots, SS$, are generated by calling the program UNIF.

INTERNAL DOCUMENTATION

```
SYNTAX: Z←SS PARETO PAR
PURPOSE: SIMULATION OF THE PARETO
 DISTRIBUTION
INDEX ORIGIN: 1
AUTHOR: (IMPLEMENTATION): E. SILBERG
LAST UPDATE: 1/79
LEFT ARGUMENT: THE DESIRED SAMPLE SIZE
RIGHT ARGUMENT: THE PARAMETERS A=PAR[1]>0
 AND P=PAR[2]>1 OF THE PARETO DISTRIBUTION
 WITH THE  FOLLOWING DENSITY FUNCTION:
 (P-1)×(A*P-1)×X* ̄P, X>A
RESULT: RANDOM NUMBERS FROM THE PARETO
 DISTRIBUTION SPECIFIED ABOVE.
```

CODE

```
    ∇ Z←SS PARETO PAR
[1] ⍝   PARETO RANDOM NUMBER GENERATOR.
[2] ⍝   CALLS ∇Z←SS UNIF PAR∇.
[3]   Z←PAR[1]×(SS UNIF 0 1)*÷1-PAR[2]
    ∇
```

Function Name: POIS

EXTERNAL DOCUMENTATION

PURPOSE

To generate random numbers from the Poisson distribution, whose density function is

$$f(x) = \begin{cases} \dfrac{e^{-b}b^x}{x!} & \text{for} \quad x = 0, 1, 2, \ldots, \\ 0, & \text{otherwise} \end{cases}$$

This is a Poisson probability distribution with mean b ($b > 0$). The user specifies the parameters lamb $= b$, and SS = the desired sample size.

ALGORITHM

For each Poisson random number desired:

Generate random numbers from the uniform distribution on the interval (0, 1) until the product of these random numbers is less than or equal to exp($-$ lamb). The observation from the Poisson distribution will then be $- 1 +$ number of uniform numbers that have geen generated.

REFERENCE

Schaffer, H. E. (1970). Algorithm 369—Generator of random numbers satisfying the Poisson distribution, *Comm. ACM,* **13**(1), 49.

INTERNAL DOCUMENTATION

```
SYNTAX: Z←SS POIS LAMB;J;K;T;N;H
PURPOSE: SIMULATION OF THE POISSON
 DISTRIBUTION
INDEX ORIGIN: 1
METHOD: THE ALGORITHM OF H.E. SCHAFFER
AUTHOR: G. PODOROWSKY
LAST UPDATE: 7/79
LEFT ARGUMENT: SS, A SCALAR, THE DESIRED
 SAMPLE SIZE
RIGHT ARGUMENT: LAMB=LAMBDA, THE PARAMETER OF
 THE POISSON DISTRIBUTION
RESULT: A VECTOR OF SS RANDOM NUMBERS FROM
 THE POISSON DISTRIBUTION WITH PARAMETER
 LAMB
```

CODE

```
     ∇ Z←SS POIS LAMB;J;K;T;N;H
[1]  ⍝ POISSON RANDOM NUMBER GENERATOR
```

```
[2]    ⍝ CALLS ∇UNIF∇
[3]    ⍝ LAST UPDATE: 7/79
[4]     H←⍳N←0
[5]     →((LAMB≤0)∨(0=Z←*(-LAMB)))/ERR
[6]    LOOP:T←1+K←0
[7]    GEN:J←1 UNIF 0 1
[8]     →(Z≥T←J×T)/OUT
[9]     →((K-1)<K←K+1)/GEN
[10]   OUT:H←H,K
[11]    →(SS>N←N+1)/LOOP
[12]   Z←H
[13]   →0
[14]   ERR;'LAMBDA VALUE INAPPROPRIATE'
        ∇
```

Function Names: POISSONFR, POISSONFR2, POISSONDF, POISSONDF2

EXTERNAL DOCUMENTATION

PURPOSE

To calculate individual and cumulative Poisson probabilities when $x < 30$.

The program POISSONFR computes the Poisson probability

$$f(x) = \frac{b^x e^{-b}}{x!}$$

associated with each positive integer x between the integer limits RANGE, inclusively, for the Poisson distribution with the specified parameter PAR = b. For example, the command 5 10 POISSONFR 7 will yield a (2×6) matrix, the first row of which is $(5, 6, \ldots, 10)$, and the second row of which is

$$\frac{7^x e^{-7}}{x!}, \qquad x = 5, 6, \ldots, 10$$

It is permissible to request the probability for only one point; in this case, RANGE will be a single nonnegative integer, and the result will be a (2×1) array. For example, 5 POISSONFR 7 will yield

$$5$$
$$P(5)$$

If the range is large, it is recommended that the APL utility function PRINTAB be used in conjunction with POISSONFR, in order to produce neat, tabular output. (See discussion of PRINTAB.)

POISSONDF is the cumulative counterpart to POISSONFR. The syntax is identical, but the cumulative probability

$$\sum_{x=0}^{k} \frac{b^x e^{-b}}{x!}$$

will replace the point probability in the output array, for the kth value between the specified limits RANGE. For example, 5 10 POISSONDF 7 will yield a (2×6) matrix, the first row of which is $(5, 6, \ldots, 10)$ and the second row of which is

$$\sum_{x=0}^{k} \frac{7^x e^{-7}}{x!}, \qquad k = 5, 6, \ldots, 10$$

POISSONDF calls POISSONFR in order to calculate point probabilities in the summation.

IMPORTANT

The programs POISSONFR and POISSONDF should not be used to calculate $P(x)$ for $x >$ roughly 30, because the value $x!$ becomes so large beyond some point that a domain error will result.

Instead, for large range values, use the functions POISSONFR2 and POISSONDF2, which will calculate probabilities over an unlimited range, including the low range values handled by POISSONFR. Hence, if the range is broad enough to include small as well as large values of x, then POISSONFR2 and POISSONDF2 are the appropriate functions. For x values ≤ 30, POISSONFR2 uses the usual form of the Poisson frequency function:

$$f(x) = b^x e^{-b}/x!$$

However, for $x > 30$, the approximation

$$x! = (\sqrt{2\pi}) x^{x+0.5} e^{-x}$$

(based on Stirling's formula) is used in the expression for $f(x)$, and then the factors are regrouped to neutralize the large and small components, for numerical balance. For large x,

$$\frac{b^x e^{-b}}{x!} \simeq \frac{1}{\sqrt{2\pi}} x^{-x-0.5} e^x b^x e^{-b} = \frac{1}{\sqrt{2x\pi}} \left(\frac{be}{x}\right)^x e^{-b}$$

Usually, x will be large when b is large, since the distribution is roughly normal for large b, and hence, the user will probably request the probabilities of significant magnitude, which will be for x near b. Therefore, $((b/x)e)^x$ is $\simeq e^x$. The factors $1/\sqrt{2x\pi}$ and e^x will not present any numerical difficulties. Comparison with Poisson tables shows the above approximation to be very good. (For example, see E. C. Molina, "Poisson's Exponential Binomial Limit," New York: Van Nostrand, 1942.)

INTERNAL DOCUMENTATION FOR POISSONDF

```
SYNTAX: Z←RANGE POISSONDF PAR
PURPOSE: CALCULATION OF CUMULATIVE POISSON
  PROBABILITIES (D.F.)
INDEX ORIGIN: 1
AUTHOR: E. SILBERG
LAST UPDATE: 1/79
LEFT ARGUMENT: INTEGER (≥0) OR PAIR OF
  INTEGERS (0≤RANGE[1]<RANGE[2]) INDEXING
  THE BEGINNING AND END VALUES FOR THE LIST
  OF PROBABILITIES TO BE COMPUTED. (ONE
  INTEGER MEANS ONE VALUE IS REQUESTED.)
RIGHT ARGUMENT: (REAL>0) MEAN OF POISSON
  DISTRIBUTION.
RESULT: 2×K MATRIX, THE FIRST ROW OF WHICH
  IS THE SET OF INTEGERS RANGE[1], 1+RANGE[1],
  ...,RANGE[2], AND THE SECOND ROW OF WHICH
  CONTAINS THE CORRESPONDING CUMULATIVE
  POISSON PROBABILITIES(SUMMED FROM P(0)),
  WHERE K=1+RANGE[2]-RANGE[1] (OR K=1).
```


CODE FOR POISSONDF

```
      ∇ Z←RANGE POISSONDF PAR;K
[1] ⍝  CUMULATIVE POISSON PROBABILITIES
    ⍝  (D.F.).
[2] ⍝  CALLS ∇Z←RANGE POISSONFR PAR∇.
[3]    K←RANGE[1]↓0,⍳RANGE[ρRANGE←,RANGE]
[4]    Z←(2,ρK)ρK,(-ρK)↑+\((0,RANGE[ρRANGE])
       POISSONFR PAR)[2;]
      ∇
```

INTERNAL DOCUMENTATION FOR POISSONDF2

SYNTAX: Z←RANGE POISSONDF2 PAR
PURPOSE: CALCULATION OF CUMULATIVE POISSON
  PROBABILITIES (D.F.) OVER AN UNLIMITED
  RANGE.
INDEX ORIGIN: 1
AUTHOR: E. SILBERG
LAST UPDATE: 1/79
LEFT ARGUMENT: INTEGER (≥0) OR PAIR OF
  INTEGERS (0≤RANGE[1]<RANGE[2]) INDEXING THE
  BEGINNING AND END VALUES FOR THE LIST OF
  PROBABILITIES TO BE COMPUTED. (ONE INTEGER
  MEANS ONE VALUE IS REQUESTED.)
RIGHT ARGUMENT: (REAL>0) MEAN OF POISSON
  DISTRIBUTION.
RESULT: 2×K MATRIX, THE FIRST ROW OF WHICH
  IS THE SET OF INTEGERS RANGE[1], 1+RANGE[1],
  ...,RANGE[2], AND THE SECOND ROW OF WHICH
  CONTAINS THE CORRESPONDING CUMULATIVE
  POISSON PROBABILITIES(SUMMED FROM P(0)),
  WHERE K=1+RANGE[2]-RANGE[1] (OR K=1).

CODE FOR POISSONDF2

```
        ∇ Z←RANGE POISSONDF2 PAR;K
[1]   ⍝   CUMULATIVE POISSON PROBABILITIES
      ⍝   (D.F.);UNLIMITED RANGE.
[2]   ⍝   CALLS ∇Z←RANGE POISSONFR2 PAR∇.
[3]     K←RANGE[1]↓0,⍳RANGE[ρRANGE←,RANGE]
[4]     Z←(2,ρK)ρK,(-ρK)↑+\((0,RANGE[ρRANGE])
        POISSONFR2 PAR)[2;]
        ∇
```

INTERNAL DOCUMENTATION FOR POISSONFR

SYNTAX: Z←RANGE POISSONFR PAR
PURPOSE: CALCULATION OF POISSON PROBABILITIES
  (FREQUENCY FUNCTION).
INDEX ORIGIN: 1

AUTHOR: E. SILBERG
LAST UPDATE: 1/79
LEFT ARGUMENT: INTEGER (≥0) OR PAIR OF
 INTEGERS (0≤RANGE[1]<RANGE[2]) INDEXING THE
 BEGINNING AND END VALUES FOR THE LIST OF
 PROBABILITIES TO BE COMPUTED. (ONE INTEGER
 MEANS ONE VALUE IS REQUESTED.)
RIGHT ARGUMENT: (REAL>0) MEAN OF POISSON
 DISTRIBUTION.
RESULT: 2×K MATRIX, THE FIRST ROW OF WHICH
 IS THE SET OF INTEGERS RANGE[1], 1+RANGE[1],
 ...,RANGE[2] AND THE SECOND ROW OF WHICH
 CONTAINS THE CORRESPONDING POISSON PROB-
 ABILITIES, WHERE K=1+RANGE[2]-RANGE[1]
 (OR K=1).

CODE FOR POISSONFR

```
     ∇ Z←RANGE POISSONFR PAR;K
[1] ₳   POISSON PROBABILITIES (FREQUENCY
    ₳   FUNCTION).
[2]   K←RANGE[1]↓0,⍳RANGE[ρRANGE←,RANGE]
[3]   Z←(2,ρK)ρK,(*-PAR)×(PAR*K)÷!K
     ∇
```

INTERNAL DOCUMENTATION FOR POISSONFR2

SYNTAX: Z←RANGE POISSONFR2 PAR
PURPOSE: CALCULATION OF POISSON PROBABILITIES
 (FREQUENCY FUNCTION) OVER AN UNLIMITED RANGE.
INDEX ORIGIN: 1
AUTHOR: E. SILBERG
LAST UPDATE: 1/79
LEFT ARGUMENT: INTEGER(≥0) OR PAIR OF
 INTEGERS (0≤RANGE[1]<RANGE[2]) INDEXING
 THE BEGINNING AND END VALUES FOR THE LIST
 OF PROBABILITIES TO BE COMPUTED. (ONE
 INTEGER MEANS ONE VALUE IS REQUESTED.)
RIGHT ARGUMENT: (REAL>0) MEAN OF POISSON
 DISTRIBUTION.
RESULT: 2×K MATRIX, THE FIRST ROW OF WHICH

*IS THE SET OF INTEGERS RANGE[1], 1+RANGE[1],
...,RANGE[2], AND THE SECOND ROW OF WHICH
CONTAINS THE CORRESPONDING POISSON
PROBABILITIES, WHERE K=1+RANGE[2]-RANGE[1]
(OR K=1).*

CODE FOR POISSONFR2

```
      ∇ Z←RANGE POISSONFR2 PAR;K;K1;T1;T2;
        N;K2
[1] ⍝   POISSON PROBABILITIES (FREQUENCY
    ⍝   FUNCTION);UNLIMITED RANGE
[2]   K1←K←RANGE[1]↓0,⍳RANGE[ρRANGE←,RANGE]
[3]   →(T1←RANGE[1]>30+ρZ←⍳0)/EXTEND
[4]   RANGE←2ρRANGE
[5]   K1←((N←(RANGE[2]×~T2)+30×T2←RANGE[2]
      >30)-RANGE[1]-1)↑K
[6]   →(~T2)/END×~0×ρZ←(*-PAR)×(PAR*K1)÷!K1
[7] EXTEND:K2←((~T1)×ρK1)↓K
[8]   Z←Z,((K2×2×○1)*⁻0.5)×(*K2-PAR)×(PAR÷K2)
      *K2
[9] END:Z←(2,0.5×ρZ)ρZ←K,Z
      ∇
```

Function Name: SAMPLE

EXTERNAL DOCUMENTATION

PURPOSE

To stimulate a random sample, in the range $a \le x \le b$, from a population with given density function F.

ALGORITHM

The program uses the acceptance–rejection method.

NOTE

The uniform density defined on (a, b) will be denoted $U(a, b)$.

$$\text{ss} = \text{size of sample desired}$$
$$\text{need} = \text{ss}$$

Do while (need > 0):

Generate a random sample of size need from $U(a, b)$; Generate a
random sample of size need from $U(0, \max(F \text{ on } (a, b)))$.

Pair the samples pointwise according to order of generation.

For each pair of points:

If

$$F(U(a, b)) > U(0, \max(F \text{ on } (a, b)))$$

then

$$\text{accept } x = U(a, b)$$
$$\text{Else reject } x = U(a, b)$$

need = ss − number already accepted

End.

INTERNAL DOCUMENTATION

```
SYNTAX: Z←F SAMPLE PARM
PURPOSE: SIMULATES RANDOM SAMPLE FROM
 POPULATION WITH GIVEN DENSITY FUNCTION
INDEX ORIGIN: 1
AUTHOR: A. LATTO
LAST UPDATE: 2/80
LEFT ARGUMENT: F: CHARACTER VECTOR
 F←NAME OF DESIRED POPULATION DENSITY
 FUNCTION F MUST ACCEPT VECTOR ARGUMENTS.
RIGHT ARGUMENT: PARM: NUMERIC VECTOR OF
 LENGTH 4
 PARM[1]←LEFT ENDPOINT OF SAMPLE RANGE
 PARM[2]←RIGHT ENDPOINT OF SAMPLE RANGE
 PARM[3]←MAXIMUM VALUE OF FUNCTION F IN
 THE INTERVAL (A,B)
 PARM[4]←SAMPLE SIZE
RESULT: NUMERIC VECTOR OF LENGTH PARM[4]
 CONTAINING SIMULATED RANDOM SAMPLE FROM A
 POPULATION WITH DENSITY FUNCTION IND(A,B)
 ×MIN(F,PARM[3]), WHERE IND(A,B) IS THE IN-
 DICATOR FUNCTION OF THE INTERVAL (A,B).
```

CODE

```
    ∇ Z←F SAMPLE PARM;NEED;TRY
[1] ⍝PROGRAM TO SIMULATE RANDOM SAMPLE FROM
[2] ⍝A POPULATION, GIVEN ITS DENSITY FUNCTION
[3]   Z←⍳0
[4] LOOP:→(0=NEED←PARM[4]-ρZ)/0
[5]   TRY←NEED UNIF 0,PARM[3]
[6]   Z←Z,(≥/[1](2,NEED)ρ(⍕F,' (NEED UNIF
      PARM[1 2])'),TRY)/TRY
[7]   →LOOP
    ∇
```

Function Name: STUDENT

EXTERNAL DOCUMENTATION

PURPOSE

To generate random numbers from the Student's t distribution, whose density function is

$$\frac{G((n + 1)/2)}{\sqrt{n\pi}\ G(n/2)} (1 + x^2/n)^{-(n+1)/2}, \quad -\infty < x < \infty$$

where $G(n/2) = \Gamma(n/2)$.

This is the distribution of a random variable $X = Y/\sqrt{(W/n)}$, where Y is distributed $N(0, 1)$ and W is distributed χ^2 with n degrees of freedom, with W and Y stochastically independent.

ALGORITHM

Let each sample value be $Z(i) = Y(i)/\sqrt{(W(i)/n)}$, where $Y(i)$ distributed $N(0, 1)$ is generated by calling the program GAUSS, and $W(i)$ distributed χ^2 with n degrees of freedom is generated independently by calling the program CHI2 for $i = 1, 2, \ldots, SS$.

INTERNAL DOCUMENTATION

```
SYNTAX: Z←SS STUDENT PAR
PURPOSE: SIMULATION OF THE STUDENT'S T
 DISTRIBUTION
```

```
INDEX ORIGIN: 1
AUTHOR: E. SILBERG
LAST UPDATE: 1/79
RIGHT ARGUMENT: (SCALAR, INTEGER) DEGREES OF
 FREEDOM.
LEFT ARGUMENT: DESIRED SAMPLE SIZE.
RESULT: RANDOM NUMBERS FROM THE STANDARD
 STUDENT'S T DISTRIBUTION WITH THE SPECIFIED
 DEGREES OF FREEDOM.
```

CODE

```
      ∇  Z←SS STUDENT PAR
[1]  ⍝   STUDENT'S T RANDOM NUMBER GENERATOR.
[2]  ⍝   CALLS ∇Z←SS GAUSS PAR∇, ∇Z←SS CHI2
     ⍝   PAR∇.
[3]   Z←(SS GAUSS 0 1)÷((÷PAR)×SS CHI2 PAR)*0.5
      ∇
```

Function Name: UNIF

EXTERNAL DOCUMENTATION

PURPOSE

To generate random numbers from the uniform distribution, whose density is

$$f(x) = \begin{cases} 1/(b - a), & a < x < b \\ 0, & \text{otherwise} \end{cases}$$

The user specifies the parameters a and b in the vector PAR:

$$a = PAR(1) \quad \text{and} \quad b = PAR(2)$$

ALGORITHM

Generate SS random numbers from the uniform distribution on the set of positive integers between 1 and 10 * 10:

$$q(x) = \begin{cases} 10^{-10}, & 1 < x < 10 * 10, x \text{ an integer} \\ 0, & \text{otherwise} \end{cases}$$

Scale these values back to the unit interval (0, 1) by dividing by 1 + 10 *

10. (Division by $1 + 10 * 10$ rather than $10 * 10$ precludes the possibility of obtaining the extreme value $1 = 10 * 10/10 * 10$. Note that 0 can never be obtained, since index origin is 1.) Now multiply these values by the scale factor $(b - a)$, and add the location parameter a, to obtain numbers uniformly distributed between a and b: $0 < u < 1$ implies $0 < (b - a)u < (b - a)$ implies $a < a + (b - a)u < b$.

NOTE

Because of the finiteness of the machine, the values generated are only approximately uniform. Also since $10 * 10 < \infty$, the above method only approximates continuity.

INTERNAL DOCUMENTATION

```
SYNTAX: Z←SS UNIF PAR
PURPOSE: SIMULATION OF THE UNIFORM
 (RECTANGULAR) DISTRIBUTION.
INDEX ORIGIN: 1
AUTHOR: E. SILBERG
LAST UPDATE: 1/79
RIGHT ARGUMENT: (A PAIR OF REAL NUMBERS)
 THE LEFT AND RIGHT ENDPOINTS OF THE INTERVAL
 OVER WHICH THE DENSITY IS UNIFORM: ‾INFINITY
 <PAR[1]<PAR[2]<+INFINITY
LEFT ARGUMENT: (VECTOR OR SCALAR) THE DESIRED
 SAMPLE SIZE.
RESULT: RANDOM NUMBERS FROM THE UNIFORM
 DISTRIBUTION ON THE SPECIFIED INTERVAL.
```

CODE

```
    ∇ Z←SS UNIF PAR
[1] ⍝   UNIFORM RANDOM NUMBER GENERATOR
[2]   Z←PAR[1]+(PAR[2]-PAR[1])×(÷1+10000000000)
    ×?SSρ10000000000
    ∇
```

CHAPTER 23

Special Functions

We give programs for computing some of the classical orthogonal functions as well as the error function.

A reader who wishes to compute other classical orthogonal functions is advised to base the algorithm on the recurrence relations for the system in question.

Function Names: CHEB1 and CHEB2

EXTERNAL DOCUMENTATION

PURPOSE

To calculate numerical values of the orthonormal Chebyshev polynomials of the first and second kind. The function CHEB1 does this for the first kind and CHEB2 for the second kind.

WARNING

Polynomials are normalized.

INTERNAL DOCUMENTATION FOR CHEB1

```
SYNTAX: Z←N CHEB1 X
PURPOSE: TO CALCULATE THE NTH CHEBYSHEV
 POLYNOMIAL OF THE FIRST KIND AT THE
 SPECIFIED VALUES OF X
INDEX ORIGIN: 1
AUTHOR: B. BUKIET
LAST UPDATE: 12/4/79
```

LEFT ARGUMENT: N; A NON-NEGATIVE INTEGER:
 THE NTH POLYNOMIAL IS DESIRED
RIGHT ARGUMENT: X, THE VALUES AT WHICH
 THE POLYNOMIALS ARE CALCULATED
 X IS A NUMBER OR VECTOR
RESULT: Z--A VECTOR OF THE SAME LENGTH AS
 X WITH THE FIRST CHEBYSHEV POLYNOMIALS
WARNING: THE POLYNOMIALS ARE NORMALIZED

CODE FOR CHEB1

```
        ∇ RESULT←N CHEB1 X;SIZEX;P0;P1;I
[1]    ⍝ THIS CALCULATES THE NTH CHEBYSHEV
       ⍝ POLYNOMIAL (OF THE FIRST KIND)
[2]    ⍝ WHERE LEFT ARGUMENT IS NUMBER OF
       ⍝ POLYNOMIAL DESIRED (NTH)
[3]    ⍝ AND THE RIGHT ARGUMENT IS VALUES
       ⍝ X WHERE THEY ARE TO BE CALCULATED
[4]    ⍝ THE POLYNOMIAL IS NORMALIZED
[5]    ⍝ INITIALIZATION
[6]     SIZEX←ρX←,X
[7]     P0←SIZEXρ1
[8]     P1←X
[9]    ⍝ SKIP LOOP IF JUST WANT ZEROETH OR
       ⍝ FIRST POLYNOMIAL
[10]   →(N=0)/ZERO
[11]   →(N=1)/ONE
[12]    I←1
[13]   ⍝ CALCULATE RECURSIVELY
[14]   LOOP:RESULT←(2×X×P1)-P0
[15]    P0←P1
[16]    P1←RESULT
[17]   →(N>I←I+1)/LOOP
[18]   →END
[19]   ZERO:RESULT←P0÷(○1)*0.5
[20]   →0
[21]   ONE:RESULT←P1
[22]   ⍝ NORMALIZATION
[23]   END:RESULT←RESULT÷(○0.5)*0.5
        ∇
```

INTERNAL DOCUMENTATION FOR CHEB2

SYNTAX: Z←N CHEB2 X
PURPOSE: TO CALCULATE THE NTH CHEBYSHEV
 POLYNOMIAL OF THE SECOND KIND AT THE
 SPECIFIED VALUES OF X
INDEX ORIGIN: 1
AUTHOR: B. BUKIET
LAST UPDATE: 12/4/79
LEFT ARGUMENT: N, A NON-NEGATIVE INTEGER--
 THE NTH POLYNOMIAL IS DESIRED
RIGHT ARGUMENT: X, THE VALUES AT WHICH THE
 POLYNOMIAL IS CALCULATED-- X IS A NUMBER OR
 VECTOR
RESULT: Z--A VECTOR OF SAME LENGTH AS X
 WITH THE CHEBYSHEV 2 VALUES AT THE X'S
WARNING: THE POLYNOMIALS ARE NORMALIZED

CODE FOR CHEB2

```
         ∇ RESULT←N CHEB2 X;SIZEX;P0;P1;I
[1]    ⍝ THIS CALCULATES THE VALUE OF THE
       ⍝ NTH CHEBYSHEV POLYNOMIAL OF THE
[2]    ⍝ SECOND KIND FOR THE VALUES OF
       ⍝ X GIVEN
[3]    ⍝ THE LEFT ARGUMENT IS THE POLYNOMIAL
       ⍝ YOU WANT (A NON-NEGATIVE INTEGER) N,
[4]    ⍝ THE RIGHT ARGUMENT IS THE VALUES
       ⍝ WHERE IT IS TO BE CALCULATED (NUMBER
       ⍝ OR VECTOR)
[5]    ⍝ THE POLYNOMIAL IS NORMALIZED
[6]    ⍝ INITIALIZATION
[7]     P0←(SIZEX←⍴(P1←2×X←,X))⍴1
[8]    ⍝ IF WANT ZEROITH OR FIRST POLYNOMIAL
       ⍝ THEN SKIP LOOP
[9]     →(N=0)/ZERO
[10]    →(N=1)/ONE
[11]    I←1
[12] LOOP:RESULT←(2×X×P1)-P0
[13]    P0←P1
```

```
[14]    P1←RESULT
[15]    →(N>I←I+1)/LOOP
[16]    →END
[17]  ZERO:RESULT←P0
[18]    →END
[19]  ONE:RESULT←P1
[20]  ⍝ NORMALIZATION
[21]  END:RESULT←RESULT×(2÷○1)*0.5
        ∇
```

Function Name: ERF

EXTERNAL DOCUMENTATION

PURPOSE

Computes values of error function defined as

$$\mathrm{ERF}(X) = \int_0^x \exp(-t * 2)\, dt$$

ALGORITHM

If all values in the argument are less than 3.16 a truncated Taylor series, accurate to 9 significant digits, will be used to calculate the result. If any value exceeds 3.16, a rational approximation accurate to 6 significant digits will be used.

WARNING

Only nonnegative values allowed.

INTERNAL DOCUMENTATION

```
SYNTAX: Z←ERF X
PURPOSE: CALCULATE THE ERROR FUNCTION
INDEX ORIGIN: 1
AUTHOR: R. RISTOW
LAST UPDATE: 4/72
RESULT: Z IS A VECTOR OF RESULTS CORRESPOND-
  ING TO THE ARGUMENTS.
```

CODE

```
      ∇ Z←ERF X;K
[1]  ⍝ LAST MODIFIED: 5/7/75
[2]    →(,∨/|X>3.16)/S1
[3]    Z←1.128379167095513×X×(((⍴X),1)⍴-X×X)
       ⊥÷(1+2×K)×!K←41-⍳41
[4]    →0
[5]  S1:Z←(×X)×1-(*-X×X)×(((⍴X),1)⍴÷1+0.3275911
       ×|X)⊥ 1.061405429 ‾1.453152027
       1.421413741 ‾0.284496736 0.254829592 0
       ∇
```

Function Name: HAAR

EXTERNAL DOCUMENTATION

PURPOSE

To compute values of the orthonormal Haar function defined in interval $(0, 1)$. The kth function of order m is defined as equal to

$$2 * M \div 2 \quad \text{if } x \text{ is in interval } (K - 1) \times 2^{-M} \text{ to } (K - \tfrac{1}{2}) \times 2^{-M}$$
$$-2 * M \div 2 \quad \text{if } x \text{ is in interval } (K - \tfrac{1}{2}) \times 2^{-M} \text{ to } K \times 2^{-M}$$
$$0 \qquad\qquad \text{else}$$

Here $M = 1, 2, 3, \ldots$ and $K = 1, 2, \ldots, 2 * M$. In addition, for $M = 0$ we have the constant 1 function for $k = 0$ and 1 in $(0, \tfrac{1}{2})$, 1 in $(\tfrac{1}{2}, 1)$ for $k = 1$.

The right argument X can be arbitrary array of numbers in interval $(0, 1)$.

INTERNAL DOCUMENTATION

```
SYNTAX: Z←PAR HAAR X
PURPOSE: TO EVALUATE VALUES OF GIVEN HAAR
 FUNCTION AT ONE OR SEVERAL X-VALUES
INDEX ORIGIN: 1
AUTHOR: U. GRENANDER
LAST UPDATE: 10/80
LEFT ARGUMENT: TWO-VECTOR WITH NUMBER OF
 FUNCTION AS FIRST COMPONENT AND ORDER OF
 FUNCTION AS SECOND COMPONENT
```

RIGHT ARGUMENT: ARRAY OF VALUES IN INTERNAL
(0,1)
RESULT: ARRAY OF VALUES OF THE GIVEN FUNCTION

CODE

```
      ∇ Z←PAR HAAR X;K;M;A
[1]   ⍝COMPUTES VALUES OF KTH HAAR FUNCTION
      ⍝OF ORDER M
[2]   ⍝WHERE K IS FIRST COMPONENT OF 2-VECTOR
      ⍝PAR (LEFT ARGUMENT) AND M IS SECOND
[3]   ⍝COMPONENT
[4]   ⍝RIGHT ARGUMENT X CAN BE ARBITRARY
[5]   ⍝ARRAY OF NUMBERS IN INTERVAL 0 TO 1
[6]   K←PAR[1]
[7]   M←PAR[2]
[8]   →(M=0)/ZERO
[9]   A←2*M
[10]  X←2×K-A×X
[11]  Z←(1<X)∧(X≤2)
[12]  Z←Z-(0<X)∧(X≤1)
[13]  Z←Z×A*0.5
[14]  →0
[15]  ZERO:Z←¯1*K×0.5<X
      ∇
```

Function Name: HERMITE

EXTERNAL DOCUMENTATION

PURPOSE

Computes values of the orthonormal Hermite polynomial of given order at specified points.

WARNING

Polynomials are normalized. Note that program computes polynomials orthogonal w.r.t. weight function $\exp[-(x*2)] \div 2$.

INTERNAL DOCUMENTATION

SYNTAX: $Z \leftarrow N$ *HERMITE X*
PURPOSE: *TO CALCULATE THE NTH HERMITE (HE)*
 POLYNOMIAL AT THE SPECIFIED X VALUES
INDEX ORIGIN: 1
AUTHOR: *B. BUKIET*
LAST UPDATE: 12/4/79
LEFT ARGUMENT: N; *A NON-NEGATIVE INTEGER--*
 THE NTH POLYNOMIAL IS DESIRED
RIGHT ARGUMENT: X, *THE VALUES AT WHICH THE*
 POLYNOMIAL IS TO BE CALCULATED
 X CAN BE A NUMBER OR VECTOR
RESULT: *Z-- A VECTOR OF SAME LENGTH AS Ẋ*
 WITH THE HERMITE (HE) VALUES AT THE X'S
WARNING: *THE POLYNOMIALS ARE NORMALIZED*

CODE

```
        ∇ RESULT←N HERMITE X;P0;P1;I
[1]   ⍝ THIS CALCULATES THE NTH HERMITE
      ⍝ POLYNOMIAL FOR THE X'S GIVEN
[2]   ⍝ LEFT ARGUMENT IS A NON-NEGATIVE
      ⍝ INTEGER N FOR THE POLYNOMIAL
[3]   ⍝ DESIRED (NTH) THE RIGHT ARGUMENT
      ⍝ IS THE VALUES(S) YOU WANT
[4]   ⍝ THE POLYNOMIAL CALCULATED AT.
      ⍝ THE POLYNOMIAL IS NORMALIZED.
[5]   ⍝ INITIALIZATION
[6]    P0←(ρP1←,X)ρ1
[7]    →(N=0)/ZERO
[8]    →(N=1)/ONE
[9]    I←1
[10]  ⍝ IF WANT ZEROITH OR FIRST POLYNOMIAL
      ⍝ THEN SKIP LOOP
[11]  LOOP:RESULT←(X×P1)-I×P0
[12]   P0←P1
[13]   P1←RESULT
[14]   →(N>I←I+1)/LOOP
[15]   →END
[16]  ZERO:RESULT←P0
[17]   →END
```

```
[18]  ONE:RESULT←P1
[19]  ⍝ THE RESULT IS NORMALIZED
[20]  END:RESULT←RESULT÷((!N)×(○2)*0.5)*0.5
      ∇
```

Function Name: LAGUERRE

EXTERNAL DOCUMENTATION

PURPOSE

Computes values of orthonormal Laguerre polynomial of given order at given points.

WARNING

Polynomials are normalized.

INTERNAL DOCUMENTATION

```
SYNTAX: Z←N LAGUERRE X
PURPOSE: TO CALCULATE THE VALUE OF THE
  GENERALIZED LAGUERRE POLYNOMIAL FOR THE
  INPUT VALUES OF ALPHA AND X
INDEX ORIGIN: 1
AUTHOR: B. BUKIET
LAST UPDATE: 12/4/79
LEFT ARGUMENT: NA; A TWO ELEMENT VECTOR
  NA[1]--N A NON-NEGATIVE INTEGER; THE NTH
  POLYNOMIAL IS DESIRED
  NA[2]-- ALPHA; A NUMBER GREATER THAN OR
  EQUAL TO ‾N
RIGHT ARGUMENT: A NUMBER OR VECTOR; VALUES
  WHERE THE POLYNOMIAL IS TO BE EVALUATED
RESULT: Z← A VECTOR OF LENGTH OF X WITH THE
  LAGUERRE POLYNOMIAL VALUES
WARNING: THE VALUES ARE NORMALIZED
```

CODE

```
      ∇ RESULT←NA LAGUERRE X;SIZEX;P0;P1;I
[1]   ⍝ THIS PROGRAM CALCULATES THE VALUE
```

```
            ⍝ OF THE NTH GENERALIZED LAGUERRE
[2]         ⍝ POLYNOMIAL FOR THE VALUES OF X GIVEN
            ⍝ WHERE THE LEFT ARGUMENT IS 2 NUMBERS
[3]         ⍝ N FOLLOWED BY ALPHA AND THE RIGHT
            ⍝ ARGUMENT IS A NUMBER OR VECTOR
[4]         ⍝ THE POLYNOMIAL IS NORMALIZED
[5]         ⍝ INITIALIZATION
[6]          SIZEX←⍴X←,X
[7]          P0←SIZEX⍴1
[8]          P1←1+NA[2]-X
[9]         ⍝ IF ONLY WANT 0ITH OR 1ST POLYNOMIAL.
            ⍝ SKIP LOOP
[10]         →(NA[1]=0)/ZERO
[11]         →(NA[1]=1)/ONE
[12]         I←1
[13]        ⍝ CALCULATION BY RECURSION
[14]        LOOP:RESULT←((P1×(1+NA[2]+2×I)-X)-
             P0×I+NA[2])÷I+1
[15]         P0←P1
[16]         P1←RESULT
[17]         →(NA[1]>I←I+1)/LOOP
[18]         →END
[19]        ZERO:RESULT←P0
[20]         →END
[21]        ONE:RESULT←P1
[22]        ⍝ NORMALIZATION
[23]        END:RESULT←RESULT×((!NA[1])÷!+/NA)*0.5
             ∇
```

Function Name: LEGENDRE

EXTERNAL DOCUMENTATION

PURPOSE

Computes values of orthonormal Legendre polynomials of given order at given set of values.

WARNING

Polynomials are normalized.

INTERNAL DOCUMENTATION

SYNTAX: Z←N LEGENDRE X
PURPOSE: TO CALCULATE THE NTH LEGENDRE
 POLYNOMIAL OF VALUES OF X BY RECURSION
INDEX ORIGIN: 1
AUTHOR: B. BUKIET
LAST UPDATE: 12/4/79
LEFT ARGUMENT: N--NON-NEGATIVE INTEGER1
 INDICATING THAT THE NTH LEGENDRE POLYNOMIAL
 IS DESIRED
RIGHT ARGUMENT: A NUMBER OR VECTOR; THE
 VALUES AT WHICH THE POLYNOMIAL IS CALCULATED
RESULT: Z-- A VECTOR THE SAME LENGTH AS X
 CONTAINING THE LEGENDRE VALUES
WARNING: THE VALUES ARE NORMALIZED

CODE

```
      ∇ RESULT←N LEGENDRE X;SIZEX;P0;P1;I
[1]   ⍝ THIS PROGRAM CALCULATES THE VALUE
[2]   ⍝ OF THE NTH LEGENDRE POLYNOMIAL FOR THE
[3]   ⍝ VALUES OF X GIVEN WHERE N IS THE LEFT
[4]   ⍝ ARGUMENT (A NON-NEGATIVE INTEGER)AND X
      ⍝ (WHICH MAY BE A VECTOR) IS THE RIGHT
[5]   ⍝ ARGUMENT. THE POLYNOMIAL IS NORMALIZED
[6]   ⍝ X IS MADE A VECTOR AND SIZEX IS
      ⍝ ITS LENGTH
[7]    X←,X
[8]    SIZEX←ρX
[9]   ⍝ THE FIRST TWO POLYNOMIALS ARE
      ⍝ INITIALIZED
[10]   P0←SIZEXρ1
[11]   P1←X
[12]  ⍝ IF ONLY WANT ONE OF THE INITIAL
      ⍝ POLYNOMIALS SKIP THE LOOP
[13]   →(N=0)/ZERO
[14]   →(N=1)/ONE
[15]  ⍝  LOOP CALCULATES THE POLYNOMIALS
```

```
        ⍝   UP TO THE ONE DESIRED
[16]    I←2
[17]    LOOP:RESULT←((1+2×(I-1))×X×P1÷(I))-
        (I-1)×P0÷I
[18]    P0←P1
[19]    P1←RESULT
[20]    →(N≥I←I+1)/LOOP
[21]    →END
[22]    ZERO:RESULT←P0
[23]    →END
[24]    ONE:RESULT←P1
[25]    ⍝ THE POLYNOMIAL IS NORMALIZED
[26]    END:RESULT←RESULT×((1+2×N)÷2)*0.5
        ∇
```

CHAPTER 24

Statistics

This chapter contains several programs for descriptive statistics, including the computation of the standard statistics, histogram displays, and frequency tabulations; they are useful for summarizing mathematical experiments involving simulations. It also presents a highly incomplete set of programs for hypothesis testing and estimation.

Function Name: ANALYZE

EXTERNAL DOCUMENTATION

PURPOSE

To compute a list of descriptive statistics on a sample,

$$X = (X_1, X_2, \ldots, X_n)$$

Sample size: N = (shape of X) = number of values.
Sample mean:

$$M = (1/N) \sum_{i=1}^{N} X_i$$

Sample variance:

$$V = (1/N) \sum_{i=1}^{N} (X_i - M)^2$$

Standard deviation: $\sqrt{(V)}$.
Mean deviation:

$$(1/N) \sum_{i=1}^{N} |X_i - M|.$$

451

Median: Let Y_1, Y_2, . . . , Y_n be the ordered sample:

$$\text{MED} = \begin{cases} Y_{(n+1)/2} & \text{if } n \text{ is odd} \\ 0.5(y_{n/2} + y_{(n/2)+1}) & \text{if } n \text{ is even} \end{cases}$$

Mode: Let m = maximum number of times the same number appears in X. Then

$$\text{MODE} = \begin{cases} \text{vector of all values occurring } m \text{ times in } X \text{ if } m > 1, \\ \text{null vector if } m = 1 \text{ (i.e., if all values in } X \text{ are distinct).} \end{cases}$$

Minimum: Y_1 = first-order statistic.
Maximum: Y_n = nth-order statistic.
Range: $Y_n - Y_1$

INTERNAL DOCUMENTATION

```
SYNTAX: ANALYZE X
PURPOSE: STATISTICAL CHARACTERIZATION OF
 A VECTOR OF DATA.
INDEX ORIGIN: 1
AUTHOR: E. SILBERG
LAST UPDATE: 3/79
RIGHT ARGUMENT: VECTOR OF DATA.
RESULT: LIST OF DESCRIPTIVE STATISTICS:
 SAMPLE SIZE, MEAN, VARIANCE, ST. DEV.,
 MEAN DEV., MEDIAN, MODE, MIN., MAX., RANGE.
```

CODE

```
      ∇ ANALYZE X;N;M;V;J;T;MED;MODE;K;U;
        ORDX
[1]   ⍝  DESCRIPTIVE STATISTICS FOR A VECTOR
      ⍝  OF DATA
[2]    'SAMPLE SIZE:          ',⍕N←ρX
[3]    'SAMPLE MEAN:          ',⍕M←(+/X)÷N
[4]    'SAMPLE VARIANCE:      ',⍕V←(+/(X-M)*2)÷N
[5]    'STANDARD DEVIATION:   ',⍕V*0.5
[6]    'MEAN DEVIATION:       ',⍕(+/|X-M)÷N
[7]    ORDX←X[⍋X]
[8]    T←(0.5×N)≠J←⌈0.5×N
[9]    MED←(ORDX[J]×T)+(~T)×0.5×ORDX[J]
        +ORDX[J+1]
```

```
[10]    'MEDIAN:                    ',▼MED
[11]    →(N>ρMODE←((ρU)ρ(ιK)≤1)/U←ORDX[(U=K←
        ⌈/U←+/ORDX∘.=ORDX)/ιρX])/NOMODE+1
[12]    NOMODE:MODE←ι0
[13]    'MODE:                      ',▼MODE
[14]    'MINIMUM:                   ',▼ORDX[1]
[15]    'MAXIMUM:                   ',▼ORDX[ρX]
[16]    'RANGE:                     ',▼ORDX[ρX]-ORDX[1]
        ∇
```

Function Name: FREQTAB1

EXTERNAL DOCUMENTATION

PURPOSE

To provide a one-way frequency table for a given vector of data $X = (x_1, x_2, \ldots, x_n)$.

The user defines the classification scheme by specifying the vector P:

$P(1)$ = left-hand end of first frequency class.
$P(2)$ = class width.
$P(3)$ = number of classes.

Thus, the classes will be as follows:

$P(1), \ldots, P(1) + P(2), \ldots, P(1) + 2P(2), \ldots, P(1) + 3P(2),$
　　((class 1))　　　　　　　(class 2))　　　　　　　(class 3))
　　　　　$\ldots, P(1) + P(3) \times P(2)$
　　　　　　(class $P(3)$)))

where ((indicates closed interval and (indicates open interval. (Note that the first interval is closed at both ends and the last interval is closed at the right-hand end.)

The output of the program is the frequency of occurrence of x values in the kth interval, for each $k = 1, 2, \ldots, P(3)$; i.e., the explicit result is a vector z in which

$$Z(k) = (\text{no. of } x_i\text{s in the } k\text{th class})/n$$

Since the first and last intervals are closed at the extreme ends, x values falling at the left and right end points will fall into the first and last classes, respectively. If the smallest x value is $\geq P(1)$ and the largest x value is $\leq P(1) + P(3) \times P(2)$, then $(+/Z) = 1$.

INTERNAL DOCUMENTATION

SYNTAX: Z←P FREQTAB1 X
PURPOSE: ONE-WAY FREQUENCY TABLE
INDEX ORIGIN: 1
AUTHOR: STATPACK2, MODIFIED BY E. SILBERG
LAST UPDATE: 3/79
LEFT ARGUMENT: P IS A THREE-ELEMENT VECTOR
 DEFINING THE CLASSIFICATION INTERVALS:
 P[1]=LEFT-HAND END OF FIRST FREQ. CLASS,
 P[2]=CLASS WIDTH, P[3]=NO. OF CLASSES.
RIGHT ARGUMENT: VECTOR OF DATA.
RESULT: Z IS A VECTOR OF FREQUENCIES, GIVING
 THE FRACTION OF X VALUES FALLING IN EACH
 CLASS. ((ρZ)= NO. OF CLASSES)

CODE

```
     ∇  Z←P FREQTAB1 X
[1]  ₳ ONE-WAY FREQUENCY TABLE FOR A VECTOR
     ₳ OF DATA.
[2]    Z←(÷ρX)×+/(ιP[3])∘.=(X=P[1])+⌈(X-P[1])
     ÷P[2]
     ∇
```

Function Name: FREQTAB2

EXTERNAL DOCUMENTATION

PURPOSE

To provide a two-way frequency table for a given set of ordered pairs

$$M = \begin{cases} M11 & M21 & \cdots & Mn1 \\ M12 & M22 & \cdots & Mn2 \end{cases}$$

The user defines the classification scheme by specifying the matrix P:

For row classification (i.e., for the abscissas),

$P(1; 1)$ = left-hand end of first frequency class.
$P(1; 2)$ = class width.
$P(1; 3)$ = number of classes.

For column classification (i.e., for the ordinates),

$P(2; 1)$ = left-hand end of first frequency class.
$P(2; 2)$ = class width.
$P(2; 3)$ = number of classes.

Thus the two-dimensional classification table Z will be as follows:

The ordered pair $(Mk1, Mk2)$, $1 \leq k <$ or $= n$ will be classified into the row into which $Mk1$ falls, and column into which $Mk2$ falls. Since the (i, j)th entry of the table is $Z(i; j)$ = (no. of $(Mk1, Mk2)$s in row i, column $j)/n$, it follows that $(+/+/Z) = 1$, provided that the following conditions hold:

(1) $P(1; 1) \leq \min(Mk1)$.
(2) $P(2; 1) \leq \min(Mk2)$.
(3) $\max(Mk1) \leq P(1; 1) + P(1; 3) \times P(1; 2)$.
(4) $\max(Mk2) \leq P(2; 1) + P(2; 3) \times P(2; 2)$.

(I.e., both axes of the frequency table are broad enough to include all of the ordered pairs.)

INTERNAL DOCUMENTATION

```
SYNTAX: Z←P FREQTAB2 M
PURPOSE: TWO-WAY FREQUENCY TABLE
INDEX ORIGIN: 1
AUTHOR: STATPACK2, MODIFIED BY E. SILBERG
LAST UPDATE: 3/79
LEFT ARGUMENT: P IS A 2×3 MATRIX DEFINING
  THE TWO-DIMENSIONAL CLASSIFICATION SCHEME:
  FOR ROW CLASSIFICATION,P[1;1]=LEFT-HAND END
  OF FIRST FREQUENCY CLASS,P[1;2]=CLASS WIDTH,
  P[1;3]= NO. OF CLASSES. P[2;] GIVES THE
  CORRESPONDING INFO. FOR COL. CLASSIFICATION.
RIGHT ARGUMENT: M IS A 2×N MATRIX, EACH COL.
  OF WHICH IS A PAIR OF VALUES TO BE
  CLASSIFIED INTO THE TWO-DIM. STRUCTURE
  DEFINED BY P: M[1;K] WILL BE USED TO LOCATE
  THE CORRECT ROW, AND M[2;K], TO LOCATE THE
  CORRECT COL., K=1,...,N.
```

RESULT: Z IS A FREQUENCY MATRIX, IN WHICH
THE IJTH ENTRY IS THE FRACTION OF PAIRS
FALLING INTO ROW I, COL. J OF THE
CLASSIFICATION TABLE.

CODE

```
     ∇ Z←P FREQTAB2 M;R;C
[1] ⍝ TWO-WAY FREQUENCY TABLE FOR A SET OF
    ⍝ ORDERED PAIRS.
[2]   R←(M[1;]=P[1;1])+⌈(M[1;]-P[1;1])÷P[1;2]
[3]   C←(M[2;]=P[2;1])+⌈(M[2;]-P[2;1])÷P[2;2]
[4]   Z←(÷1↓⍴M)×P[;3]⍴+/(⍳×/P[;3])∘.=C+P
      [2;3]×R-1
     ∇
```

Functions Names: HISTOGRAM1 and HISTOGRAM2

EXTERNAL DOCUMENTATION

PURPOSE

To provide a simple histogram, printed at the terminal, for a vector of data $X = (x_1, x_2, \ldots, x_n)$.

The user specifies the number of intervals (N). The intervals will be defined as follows. Let $x(1) = \min x_i$, and $x(n) = \max x_i$, and let $a = (x(n) - x(1))/N$. Then the intervals are as follows:

```
   ((      ))              (    ))            (    ))       · · ·    (    ))
x(1), . . . , x(1) + a,   . . . , x(1) + 2a,   . . . , x(1) + 3a,         . . . , x(n)
```

Each program classifies the x_i into these intervals and calculates the relative frequencies of the intervals; i.e., Rel. freq. of interval $j = ($ no. of x_i in interval $j)/n, j = 1, 2, \ldots, N$.

For each of the programs, output is a list of the midpoints of the intervals, the corresponding relative frequencies, and a plot of the histogram. The only difference between HISTOGRAM1 and HISTOGRAM2 is that HISTOGRAM2 calls the function PLOT, while HISTOGRAM1 constructs a simpler histogram directly. The histograms will differ as follows:

1. HISTOGRAM2 prints a labeled histogram; HISTOGRAM1's plot is only partially labeled.

2.   HISTOGRAM2 prints a vertical frequency graph (i.e., largest frequency = longest vertical line of 'quad's), with *x* scale from left to right; HISTOGRAM1 prints a horizontal frequency graph (i.e., largest frequency = longest horizontal line of 'quad's), with *x* scale from top to bottom.

Thus, the advantage of using HISTOGRAM2 is that the plot is more refined, but since the program PLOT is long and complex, HISTOGRAM2 is much more expensive than HISTOGRAM1.

NOTE

To execute HISTOGRAM2 the workspace must contain plotting function; see introduction to Chapter 20.

INTERNAL DOCUMENTATION FOR HISTOGRAM1

```
SYNTAX: H←N HISTOGRAM1 X
PURPOSE: TO CREATE A HISTOGRAM FROM A VECTOR
  OF DATA
INDEX ORIGIN: 1
AUTHORS: E. SILBERG, G. PODOROWSKY; FROM
  ORIGINAL BY W. PRAGER
LAST UPDATE: 7/79
LEFT ARGUMENT: (INTEGER≥2) NUMBER OF
  INTERVALS INTO WHICH RANGE IS TO BE DIVIDED
RIGHT ARGUMENT: VECTOR OF DATA
RESULT: A HISTOGRAM, WITH THE MIDPOINTS OF
  THE INTERVALS ON THE LEFT AND THE
  CORRESPONDING RELATIVE FREQUENCIES TO THE
  RIGHT OF THE GRAPH.
```

CODE FOR HISTOGRAM1

```
      ∇ H←N HISTOGRAM1 X;MIN;RIGHT;MIDS;F;
        R;M;K;L
[1]  ⍝RELATIVE FREQUENCIES AND QUICK HISTOGRAM.
[2]    L←⍉((ρF),1)ρF←(÷ρX)×+/(⍳N)∘.=(X=MIN)
       +⌈(X-MIN)÷R←((⌈/X)-MIN←⌊/X)÷N
[3]    K←⍉((ρMIDS),1)ρMIDS←0.5×(MIN,¯1↓RIGHT)
       +RIGHT←MIN+R×⍳N
[4]    G←' ⎕'[1+F∘.≥⍳⌈/F←⌊0.5+F×30]
```

```
[5]   'MIDPOINTS  ',((10+(1↓ρG))ρ' '),
      'RELATIVE'
[6]   'OF INTERVALS',((10+(1↓ρG))ρ' '),
      'FREQUENCIES'
[7]   ' '
[8]   H←K,((N,18-1↓ρK)ρ' '),G,((N,3)ρ' '),L
      ∇
```

INTERNAL DOCUMENTATION FOR HISTOGRAM2

SYNTAX: N HISTOGRAM2 X
PURPOSE: TO CREATE A HISTOGRAM FROM A
VECTOR OF DATA.
INDEX ORIGIN: 1
AUTHOR: E. SILBERG
LAST UPDATE: 3/79
LEFT ARGUMENT: (INTEGER≥2) NUMBER OF
INTERVALS INTO WHICH RANGE IS TO BE DIVIDED.
RIGHT ARGUMENT: VECTOR OF DATA
RESULT: (1) LIST OF RELATIVE FREQUENCIES AND
MIDPOINTS OF THE RESPECTIVE INTERVALS.
(2) HISTOGRAM. GLOBAL VARIABLES CREATED: XY
IS A (2×N) MATRIX: XY[1;] = RELATIVE
FREQUENCIES; XY[2;] = MIDPOINTS.

CODE FOR HISTOGRAM2

```
      ∇ N HISTOGRAM2 X;MIN;RIGHT;MIDS;F;R
[1]  ⍝   RELATIVE FREQUENCIES AND HISTOGRAM.
     ⍝   CALLS PLOT.
[2]   F←(÷ρX)×+/(ιN)∘.=(X=MIN)+⌈(X-MIN)÷
      R←((⌈/X)-MIN←⌊/X)÷N
[3]   MIDS←0.5×(MIN,¯1↓RIGHT)+RIGHT←MIN+R×ιN
[4]   'RELATIVE FREQUENCIES'
[5]   □←XY←(2,N)ρF,MIDS
[6]   'MIDPOINTS OF INTERVALS'
[7]   ' '
[8]   10 PLOT XY[1;] VS XY[2;]
      ∇
```

Function Name: HISTOSPLINE

EXTERNAL DOCUMENTATION

PURPOSE

To compute the quadratic spline transformation of a matrix, and thus create the histospline that is associated with a given histogram. This spline function $S(x)$ defined on the entire real line has the following properties:

(1) In each interval $(x(j), x(j+1))$ for $j = 0, 1, . . . , n$ ($x(0) = -$ inf, $x(n+1) =$ inf), $S(x)$ is given by a polynomial of degree 2 or less.

(2) $S(x)$ and its first derivative are continuous everywhere.

(3) The transformation $S(x)$ between the histogram and the histospline is a homeomorphism; i.e., small changes in the histogram h produce only small changes in the image function S, and vice versa.

ALGORITHM

The histospline is produced by first creating a unit histospline (called a deltaspline) and then using the data to transform the spline into the appropriate shape. The deltaspline is created as follows.

The deltaspline is determined in each cell of the grid by a triplet of numbers (the coefficients of 1, x, $x * 2$ in the quadratic function, where local coordinates are used such that the cell is always represented by the segment $0 \le x \le 1$). In all, three linear relations are imposed upon the triplets. Two linear relations connect the triplet for one cell with the triplet for the next; these restrictions result from the continuity of both the deltaspline and its first derivative. A third linear relation connects the triplet for any cell with the histogram content of that cell (which is always 0 or 1 since a unit histogram is used in constructing the deltaspline).

Once the deltaspline has been constructed, the data are used to form the desired spline by adding together appropriately translated versions of the deltaspline. If the data is not in the correct form (i.e., if the data does not have (1) less than 100 ordinates or (2) evenly spaced abscissas), then HISTER will be called to change the data into a more suitable form for HISTOSPLINE. *For description of Hister, see below*.

NOTE

After HISTOSPLINE has been executed with a command such as XY gets HISTOSPLINE H, the spline and histogram can be drawn on a plotter if plotting software is available.

REFERENCE

Boneva, L., Kendall, D., and Stefanov, I. (1971). Spline transformations: Three new diagnostic aids for the statistical data analyst, *J. Roy. Statist. Soc. Ser. B,* **33,** 1–70.

INTERNAL DOCUMENTATION

```
SYNTAX: XY←HISTOSPLINE H
PURPOSE: DETERMINES QUADRATIC SPLINE TRANS-
 FORMATION OF A GIVEN MATRIX
INDEX ORIGIN: 1
AUTHOR: D. PETERS
LAST UPDATE: 7/79
RIGHT ARGUMENT: H IS A 2×N MATRIX WITH
 H[1;] A LIST OF HISTOGRAM ORDINATES
 (HEIGHTS)
 H[2;] A LIST OF CORRESPONDING ABSCISSAS
 (MIDPOINTS)
RESULT: XY IS A VECTOR OF CARTESIAN POINTS
 OF THE FORM X,Y,X,Y... IN FOUR STRINGS TO
 BE USED TO PLOT A SPLINE, A HISTOGRAM AND
 THE X-AXIS.
GLOBAL VARIABLES SET:
 ABS IS A VECTOR OF SPLINE ABSCISSAS, ρABS
 ABOUT 500
 ORD IS A VECTOR OF SPLINE ORDINATES, ρORD
 ABOUT 500
TO USE: DESIGNED FOR USE WITH TSP/HP
 PLOTTERS AND THE PLOT FUNCTIONS IN THIS
 WORKSPACE. TO PLOT HISTOSPLINE, TYPE DRAW1
 XY. THIS ASSUMES DEFINITION OF PLOTTING
 FUNCTION DRAW1 FOR USE AT PLOTTER.
NOTE: IF INPUT HISTOGRAM DOES NOT 1) HAVE
 LESS THAN 100 ORDINATES OR 2) HAVE EVENLY
 SPACED ABSCISSAS (MIDPOINTS) THEN HISTER IS
 CALLED TO TRANSFORM H. A WARNING IS GIVEN.
```

CODE

```
     ∇ XY←HISTOSPLINE H;FDIFF;DELT;DELTA;
       N;PTS;N;INTV;STEPS;SPLINE
[1]  ⍝ CHECK FOR CONSECUTIVE STANDARD
```

```
        ⍝ ABSCISSA INTERVALS
[2]     →(∧/FDIFF[1]=FDIFF←1↓H[2;]-¯1⌽(H←H
        [;⍋H[2;]])[2;])/SKIP
[3]     ⍝ NOT FOUND, CALL HISTER TO WORK
        ⍝ ON H
[4]     H←HISTER H
[5]     ⍝ SET UP A UNIT HISTOSPLINE IN DELTA
[6]     SKIP:DELT←(13,3)⍴0
[7]     DELT[⍳7;2]← 0.00081279 ¯0.0030334
        0.011321 ¯0.042249 0.15768 ¯0.58845
        2.1962
[8]     DELT[⍳7;3]← 0.00023463 ¯0.00087566
        0.003268 ¯0.012196 0.045517 ¯0.16987
        0.63397
[9]     DELT[7+⍳6; 2 3]←⊖⌽(2,6)⍴(-DELT
        [1+⍳6;2]),DELT[1+⍳6;3]
[10]    DELT[;1]←(7⌽3,12⍴0)-3×DELT[;3]+0.5
        ×DELT[;2]
[11]    DELTA←((3,N)⍴-¯1+⍳N)⌽[2](3 1 2)⍉
        ((N,3,13)⍴,⍉DELT),(N,3,(N←(⍴H)[2])-
        1)⍴0
[12]    ⍝ CREATE THE COEFFICIENTS OF THE
        ⍝ HISTOSPLINE
[13]    SPLINE←DELTA+.×H[1;]
[14]    ⍝ CREATE THE ORDINATES, AIMING FOR 500
        ⍝ POINTS
[15]    ORD←,(⍉SPLINE)+.×(3,STEPS)⍴(PTS*2),
        (PTS←(¯1+⍳STEPS)÷STEPS),(STEPS←⌊0.5+
        500÷(⍴SPLINE)[2])⍴1
[16]    ⍝ CREATE THE ABSCISSAS, SHIFTING
        ⍝ ACCORDING TO H[2;]
[17]    ABS←(H[2;1]-6×INTV)+(INTV←H[2;2]-
        H[2;1])×(¯1+⍳STEPS×N+12)÷STEPS
[18]    ⍝ 'HISTODRAW' GIVES COORDINATES FOR
        ⍝ THE HISTOGRAM AND X-AXIS
[19]    ⍝ ARRANGE POINTS IN XYXY...SEQUENCE
        ⍝ FOR PLOTTING
[20]    XY←(ABS XYXY ORD) AND HISTODRAW H
[21]    ' '
[22]    ' ABSCISSA (''ABS'') VALUES ARE ',
        (⍕ABS[1]),'(',(⍕4⍕INTV÷STEPS),')',
        ⍕ABS[STEPS×N+12]
```

```
[23]    '      ',(⍕STEPS×N+12),' VALUES.'
[24]    ' MAXIMUM ORDINATE (''ORD'') VALUE IS
        ',⍕⌈/ORD
        ∇
```

INTERNAL DOCUMENTATION FOR HISTER

```
SYNTAX: ∇H̲← HISTER H∇
PURPOSE: CONVERTS A GENERAL HISTOGRAM TO
 ONE SUCH THAT
 1) THE NUMBER OF ORDINATES ≤ 100
 2) THE ABSCISSAS (MIDPOINTS) ARE EVENLY
 SPACED
 ∇HISTER∇ IS CALLED BY ∇HISTOSPLINE∇.
INDEX ORIGIN: 1
AUTHOR: D. PETERS
LAST UPDATE: 3/79
RIGHT ARGUMENT: H IS A 2×N MATRIX SUCH THAT
 H[1;] IS A LIST OF HISTOGRAM ORDINATES
 (HEIGHTS)
 H[2;] IS A LIST OF HISTOGRAM ABSCISSAS
 (MIDPOINTS)
RESULT: H̲ IS A 2×M HISTOGRAM MATRIX SATIS-
 FYING 1) AND 2) ABOVE AND PRESERVING THE
 BASIC CHARACTER OF H. THE ABSCISSA INTERVAL
 IS CHOSEN TO BE THE MAXIMUM OF
 1) THE INTERVAL MOST COMMON IN H
 2) THE GREATEST POWER OF TEN WHICH KEEPS THE
 NUMBER OF ABSCISSAS ≤ 100
GLOBAL VARIABLES SET:
 HISTOGRAM ← H̲ FOR USERS OF ∇HISTOSPLINE∇
NOTE: THE MESSAGE PRINTED IS FOR USERS OF
 ∇HISTOSPLINE∇ DHISTOGRAM1
```

CODE FOR HISTER

```
       ∇ H̲←HISTER H;RANGE;INTV;MAI;ZDIFF;
         FDIFF;FILL;FULL;PDIFF;NABS;POS;SCAN
[1]    ⍝ FIND MINIMUM ALLOWABLE INTERVAL (MAI)
[2]    MAI←10*⁻2+⌈10⊛RANGE←H[2;(⍴H)[2]]-
       (H←H[;⍋H[2;]])[2;1]
```

```
[3]   ⍝ CHOOSE INTV TO BE LARGER OF MAI AND
      ⍝ MODE OF FDIFF
[4]   INTV←MAI⌈PDIFF[1↑⍒+/PDIFF∘.=PDIFF
      ←(FDIFF>0)/FDIFF←1↓H[2;]-¯1⌽H[2;]]
[5]   ⍝ ROUND ALL ABSCISSAS TO MULTIPLES
      ⍝ OF INTV
[6]   H[2;]←INTV×⌊0.5+H[2;]÷INTV
[7]   ⍝ COMBINE ORDINATES OF COLUMNS WHOSE
      ⍝ ABSCISSAS WERE ROUNDED EQUAL...
[8]   ⍝ ...AND KNOCK OUT THE DUPLICATES
[9]   NABS←H[2;1],H[2;ZDIFF←((ZDIFF>0)/
      ZDIFF←(FDIFF≠0)×⍳FDIFF←(1↓H[2;])
      -¯1↓H[2;])+1]
[10]  H←(2,⍴POS)⍴SCAN[POS[1]],(SCAN[⌽1↓+
      \POS]-(SCAN←+\H[1;])[⌽¯1↓+\POS])
      H[2;+\POS←+/NABS∘.=H[2;]]
[11]  ⍝ PAD H WITH (0,J) COLUMNS TO GIVE
      ⍝ CONTINUOUS EQUAL INTERVALS
[12]  H←H,(2,⍴FILL)⍴((⍴FILL)⍴0),FILL←(~FULL
      ∈H[2;])/FULL←H[2;1]+INTV×¯1+⍳1+
      (H[2;(⍴H)[2]]-H[2;1])÷INTV
[13]  ⍝ RESEQUENCE H TO SHUFFLE IN THE PADDING
[14]  H←HISTOGRAM←H[;⍋H[2;]]
[15]  ⍝ ALERT USER OF HISTOSPLINE
[16]  ' '
[17]  ' INPUT HISTOGRAM NOT USED --
      ''HISTOGRAM'' CREATED AND USED'
      ∇
```

Function Name: KOLMOG

EXTERNAL DOCUMENTATION

PURPOSE

Computes confidence band for distribution function using Kolmogorov's asymptotic distribution.

WARNING

Sample size must be large.

INTERNAL DOCUMENTATION

SYNTAX: V←CC KOLMOG X
PURPOSE: TO COMPUTE KOLMOGOROV CONFIDENCE
 BANDS AND TO SET UP COORDINATES FOR PLOTTING
 THESE BANDS
INDEX ORIGIN: 1
AUTHOR: T. RICHMAN
LAST UPDATE: 7/79
LEFT ARGUMENT: CC IS THE CONFIDENCE
 COEFFICIENT, A SCALAR. IT MUST BE ONE OF THE
 FOLLOWING: .9 .95 .98 .99 .995 .998 .999
RIGHT ARGUMENT: X IS A VECTOR CONTAINING
 THE N SAMPLE OBSERVATIONS.
RESULT: V IS A VECTOR CONTAINING THE
 COORDINATES FOR PLOTTING THE CONFIDENCE
 BANDS.
NOTES: 1) V IS NOT PRINTED. THE 3×N MATRIX
 W IS PRINTED:
W[1;] LISTS THE OBSERVATIONS
W[2;] LISTS CORRESPONDING LOWER LIMITS OF
 CONFIDENCE BAND
W[3;] LISTS CORRESPONDING UPPER LIMITS OF
 CONFIDENCE BAND
 THE BOUNDS OF W ARE SUCH THAT F(X) FALLS
 WITHIN THESE LIMITS WITH CONFIDENCE
 COEFFICIENT DESIRED.
 2) TO PLOT THE CONFIDENCE BANDS ON A
 PLOTTER, COPY W INTO WORKSPACE WITH
 PLOTTING FUNCTIONS. GLOBAL VARIABLE E
 FOR BREAKS IN PLOTTING.

CODE

```
       ∇ V←CC KOLMOG X;M
[1]   ⍝LAST UPDATE: 5-16-79
[2]   ⍝COMPUTES KOLMOGOROV CONFIDENCE BANDS
[3]   ⍝NEED CONTINUOUS DISTRIBUTION AND
      ⍝LARGE SAMPLE SIZE N
[4]    M← 2 7 ρ 0.9 0.95 0.98 0.99 0.995
       0.998 0.999 1.22 1.36. 1.52 1.63 1.73
       1.86 1.95
[5]    →L0×⍳CC∊M[1;]
```

```
[6]    'CONFIDENCE COEFFICIENT MUST BE ONE
       OF SEVEN SPECIFIED'
[7]    →0
[8]    L0:DN←((N←ρX)*⁻0.5)×DN←(CC=M[1;])/M[2;]
[9]    LL←(⁻2×DN)+UL←DN+FN←(÷N)×♠X←X[♠X]
[10]   W←(3,N)ρX,(UL←1⌊UL),(LL←0⌈LL)
[11]   V←,⍉V←(4,(2×N))ρX,X,(⁻1⌽UL),(⁻1⌽LL),
       X,X,UL,LL
[12]   V[(1+4×N),(2+4×N)]←E,E
[13]   V←2↓V
[14]   W
       ∇
```

Function Name: MODE

EXTERNAL DOCUMENTATION

PURPOSE

To calculate the sample mode (or modes) of a vector of continuous data $X = (X_1, X_2, \ldots, X_n)$, where the mode is defined as follows: Let L be the lag; i.e., order the data and form the x sequences $(X(k), X(k + 1), \ldots, X(x + L))$ for all integers k: $1 \le k \le n - L$, where L is a fixed positive integer. The mode is the midpoint of the shortest interval $(X(k), X(k+L))$. In the program, L is arbitrarily set to $\sqrt{(n)}$.

ALGORITHM

Order the data and compare the lengths of all intervals containing $L + 1$ ordered values. Find the minimum length, and take M = midpoint of the corresponding interval, as the mode:

$$M = (X(k) + X(k+L))/2$$

If there is more than one interval of minimal length, then M will be a vector of the modes. If a value is repeated in X, then this repeated value may be a mode; if so, then it will occur only once in the vector M.

INTERNAL DOCUMENTATION

```
SYNTAX:  Z←MODE X
PURPOSE: MODE(S) OF A VECTOR OF DATA.
```

```
INDEX ORIGIN: 1
AUTHOR: E. SILBERG
LAST UPDATE: 3/79
RIGHT ARGUMENT: VECTOR OF DATA.
RESULT: THE MODE(S) OF X, WHERE THE MODE
 IS DEFINED AS THE MIDPOINT OF THE SHORTEST
 INTERVAL CONTAINING 1+((ρX)*.5) ORDERED
 X[I]'S.
```

CODE

```
      ∇ M←MODE X;L;A;D;I
[1]   ⍝ MODE OF A VECTOR OF DATA.
[2]   L←⌊(ρX)*0.5
[3]   I←(D=⌊/D←-/A←(2,(ρX)-L)ρ(L↓X),(-L)
      ↓X←X[⍋X])/ι(ρX)-L
[4]   M←(1,≠/(2,‾1+ρM)ρ(‾1↓M),(1↓M))/M←0.5
      ×+/A[;I]
      ∇
```

Function Name: QUANT

EXTERNAL DOCUMENTATION

PURPOSE

To compute a list of sample quantiles of the given vector of data $X = (x_1, x_2, \ldots, x_n)$.

Input consists of X and a vector P of decimal fractions (not necessarily ordered) representing the requested quantiles. (For example, to request the quartiles, $P = (0.25 \quad 0.5 \quad 0.75)$). The only restriction on the entries of P is that $0 < p_j \leq 1$ for all j. Let $F * (y)$ be the empirical d.f. of $(x_1, x_2, \ldots, x_n)$; i.e., $F * (y) = $ (no. of $x_i \leq y$)/n. Let $x(1), \ldots, x(n)$ be the ordered sample. If $F * (x(m)) < p_j < F * (x(m + 1))$, then the value computed for the p_j quantile will be $gp = (\frac{1}{2}) (x(m) + x(m + 1))$.

ALGORITHM

Order the sample: $x(1), \ldots, x(n)$.

There are $(n - 1)$ intervals between x values. If p is the fraction representing the quantile desired, and if $j = \min(p(n - 1))$ and $k = $

max($p(n - 1)$), then the pth sample quantile is

$$qp = (X(1 + j) + X(1 + k))/2 = \begin{cases} X(1 + j) \text{ if } p = j/(n - 1), \\ \quad \text{where } j \text{ is an integer,} \\ \quad 1 \le j \le n - 1 \\ (X(1 + j) + X(2 + j))/2 \\ \quad \text{if } j/(n - 1) < p < (j + 1)/(n - 1) \\ \quad \text{where } j \text{ is an integer,} \\ \quad 1 \le j \le n - 2 \end{cases}$$

INTERNAL DOCUMENTATION

```
SYNTAX: Q←P QUANT X
PURPOSE: QUANTILES OF A VECTOR OF DATA.
INDEX ORIGIN: 1
AUTHOR: E. SILBERG
LAST UPDATE: 3/79
LEFT ARGUMENT: A VECTOR OF DECIMAL FRACTIONS
  SPECIFYING THE QUANTILES REQUESTED. (E.G.,
  .25 MEANS FIRST QUARTILE.)
RIGHT ARGUMENT: THE VECTOR OF DATA.
RESULT: A VECTOR OF LENGTH(ρP), GIVING THE
  REQUESTED QUANTILES OF X.
```

CODE

```
     ∇ Q←P QUANT X;I
[1] ⍝   QUANTILES OF A VECTOR OF DATA.
[2]    Q←0.5×+/(2,ρP)ρX[(⌊I),⌈I←1+P×⁻1+ρX
       ←X[⍋X]]
     ∇
```

Function Name: RUNS1

EXTERNAL DOCUMENTATION

PURPOSE

To perform a nonparametric one-sample runs test for randomness.

It is used to test the hypothesis that observations have been drawn at random from a single population. Too many or too few observed runs implies a nonrandom arrangement of sample values, and leads to the rejection of the hypothesis.

EXAMPLE

Let $X = X_1, X_2, \ldots, X_N$ denote the sample and let $M =$ sample median. Compute $X - M$ and assign a value of 1 to those $X_I - M > 0$ (observations above median), and a value of 0 to those $X_I - M < 0$ (observations below median) for all I. Let $N_1 =$ total number of 0s and $N_2 =$ total number of 1s. Count the number of runs (consecutive strings of 0s or 1s). Reject H_0 when there are too many or too few observed runs. Note that if the sample size N is odd, $M =$ one of the sample values $(X_I - M = 0)$. Such an observation should be omitted from the test procedure.

Compute the probability

$$P(X \leq R) = \sum_{k=2}^{R} P(X = k) = \sum_{k=2}^{R} \sum_{y_1+y_2=k} P(y_1, y_2)$$

$$= \sum_{k=2}^{R} \sum_{y_1+y_2=k} \frac{((N1 - 1)!(y_1 - 1))((N2 - 1)!(y_2 - 1))}{((N1 + N2)!N1)}$$

$$P(X \geq R) = 1 - P(X < R) = 1 - P(X \leq R - 1)$$

Note that the minimum number of runs $= 2$. The maximum number of runs $= N1 + N2$. For $N1$ and $N2$ both >10, a normal approximation is used:

$$Z = \frac{R + 0.5 - E(R)}{\sqrt{(\text{Var } (R))}}, \qquad E(R) = \frac{2(N1)(N2)}{N1 + N2} + 1$$

$$\text{Var}(R) = \frac{2(N1)(N2)(2(N1)(N2) - N1 - N2)}{((N1 + N2) * 2) - (N1 + N2 - 1)}$$

For two-sample runs test, see RUNS2.

REFERENCES

Dixon, W. J., and Massey, F. J. (1957). "Introduction to Statistical Analysis," 2nd ed. McGraw-Hill, New York.
Lehmann, E. L. (1975). "Nonparametrics." Holden-Day, San Francisco, California.

NOTE

Calls functions *GAUSSDF* and *ERF*.

INTERNAL DOCUMENTATION

```
SYNTAX: ∇E←RUNS1 X∇
PURPOSE: PERFORMS NONPARAMETRIC ONE-SAMPLE
 RUNS TEST FOR RANDOMNESS.
AUTHOR: T. RICHMAN
LAST UPDATE: 5-9-79
INDEX ORIGIN: 1
RIGHT ARGUMENT: X IS AN N-VECTOR OF SAMPLE
 OBSERVATIONS.
RESULT: OBSERVATIONS ABOVE AND BELOW THE
 SAMPLE MEDIAN ARE ASSIGNED VALUES 1 AND 0
 RESPECTIVELY.
HO: OBSERVATIONS ARE DRAWN AT RANDOM
 FROM A SINGLE POPULATION. THUS TOO MANY
 OR TOO FEW RUNS WILL LEAD TO THE REJECTION
 OF HO.
IF AT LEAST ONE SAMPLE SIZE (OF 1'S AND
 0'S) IS ≤10 THEN THE PROBABILITY OF
 OBSERVING ≤R RUNS OR ≥R RUNS UNDER HO IS
 COMPUTED DIRECTLY.
IF SAMPLE SIZES ARE BOTH >10 A NORMAL
 APPROXIMATION IS USED.
E[1]= R= NUMBER OF RUNS
E[2]= P(X≤R)
E[3]= P(X≥R)
GLOBAL VARIABLES: M IS A GLOBAL VARIABLE
 EQUAL TO THE SAMPLE MEDIAN.
```

CODE

```
       ∇ E←RUNS1  X;X1;D;X2;N1;N2;R;K;Y1;
         Y2;V;W;PR;PR1;PR2;Z
[1]    ⍝LAST UPDATE: 5-9-79; CALLS GAUSSDF
       ⍝WHICH CALLS ERF
[2]    ⍝ONE SAMPLE RUNS TEST FOR RANDOMNESS
[3]    M←(÷2)×X1[⌊D]+(X1←X[⍋X])[⌊0.5+D←(÷2)
       ×1+⍴X]
[4]    N1←(⍴X2)-N2←+/X2←0<X2←(0≠X-M)/X-M
[5]    K←1+⍳¯1+R←1++/0≠¯1↓X2-1⌽X2
[6]    →L0×⍳(N1>10)∧N2>10
[7]    Y2←K-Y1←(0.5×1=2|K)+K÷2
```

```
[8]     W←(ρK)⌽V←Y1,Y2
[9]     PR1←+/PR←(((V-1)!N1-1)×((W-1)!N2-1)
        ÷N1!N1+N2
[10]    PR2←1-PR1-+/PR[(ρK),ρV]
[11]    E←R,PR1,PR2
[12]    →0
[13]   ⍝NORMAL APPROXIMATION FOR N1 AND N2 >10
[14]   LO:R←R,R-1
[15]    Z←(R+0.5-1+2×N1×N2÷N1+N2)÷((2×N1×N2×
        (2×N1×N2)-N1+N2)÷(N1+N2-1)×(N1+N2)
        *2)*0.5
[16]    PR←(Z GAUSSDF 0 1)[2;]
[17]    E←R[1],PR[1],PR[2]←1-PR[2]
[18]    'NORMAL APPROXIMATION USED'
        ∇
```

Function Name: RUNS2

EXTERNAL DOCUMENTATION

PURPOSE

To perform a nonparametric two-sample run test for comparing probability distributions of two random samples.

It is used to test the hypothesis that the two random samples come from populations having the same distribution. Too many or too few observed runs lead to the rejection of the hypothesis.

EXAMPLE

Let $X = X_1, X_2, \ldots , X_{N_1}$ and let $Y = Y_1, Y_2, \ldots , Y_{N_2}$. Rank the two samples together in one series according to size. Count the number of runs (consecutive strings of items from each sample). Reject H_0 when there are too many or too few observed runs. Note that sample sizes N_1 and N_2 should be fairly close to each other. Note also that ties between samples are not permitted, because ranking the values would then be arbitrary. Duplicate values within one sample are, however, allowed.

Compute the probability

$$P(X \leq R) = \sum_{k=2}^{R} P(X = k) = \sum_{k=2}^{R} \sum_{y_1+y_2=k} P(y_1, y_2)$$

$$= \sum_{k=2}^{R} \sum_{y_1+y_2=k} \frac{((N1 - 1)!(y_1 - 1))((N2 - 1)!(y_2 - 1))}{((N1 + N2)!N1)}$$

$$P(X \geq R) = 1 - P(X < R) = 1 - P(X \leq R - 1)$$

Note that the minimum number of runs = 2. The maximum number of runs = $N1 + N2$. For $N1$ and $N2$ both >10, a normal approximation is used:

$$Z = \frac{R + 0.5 - E(R)}{\sqrt{(\mathrm{Var}(R))}}, \qquad E(R) = \frac{2(N1)(N2)}{N1 + N2} + 1$$

$$\mathrm{Var}(R) = \frac{2(N1)(N2)(2(N1)(N2) - N1 - N2)}{((N1 + N2) * 2) - (N1 + N2 - 1)}$$

For one-sample runs test, see RUNS1.

NOTE

Calls functions GAUSSDF and ERF.

REFERENCES

Dixon, W. J., and Massey, F. J. (1957). "Introduction to Statistical Analysis," 2nd ed. McGraw-Hill, New York.
Lehmann, E. L. (1975). "Nonparametrics." Holden-Day, San Francisco, California.

INTERNAL DOCUMENTATION

```
SYNTAX: ∇E←A RUNS2 B∇
PURPOSE: PERFORMS NONPARAMETRIC TWO-SAMPLE
 RUNS TEST FOR COMPARING PROBABILITY
 DISTRIBUTIONS OF TWO UNPAIRED RANDOM SAMPLES.
AUTHOR: T. RICHMAN
LAST UPDATE: 5-9-79
INDEX ORIGIN: 1
LEFT ARGUMENT: A IS A VECTOR OF LENGTH N
 CONTAINING A RANDOM SAMPLE OF OBSERVED
 VALUES.
RIGHT ARGUMENT: B IS A VECTOR OF LENGTH M
 CONTAINING ANOTHER RANDOM SAMPLE. NOTE THAT
 N AND M SHOULD BE FAIRLY CLOSE TO EACH OTHER.
RESULT: TIES BETWEEN THE TWO SAMPLES ARE NOT
 PERMITTED. IF THEY DO OCCUR, A MESSAGE IS
 PRINTED AND E← EMPTY VECTOR.
H0: OBSERVATIONS ARE DRAWN AT RANDOM FROM
 A COMMON DISTRIBUTION. TOO MANY OR TOO FEW
 RUNS WILL LEAD TO THE REJECTION OF H0.
```

IF AT LEAST ONE SAMPLE SIZE IS ≤10 THEN THE
PROBABILITY OF OBSERVING ≤R RUNS OR ≥R RUNS
UNDER H0 IS COMPUTED DIRECTLY.
IF BOTH SAMPLE SIZES ARE >10 A NORMAL
APPROXIMATION IS USED.
E[1]= R= NUMBER OF RUNS
E[2]= P(X≤R)
E[3]= P(X≥R)

CODE

```
        ∇ E←A RUNS2 B;X;N1;N2;R;K;Y1;Y2;V;
          W;PR;PR1;PR2;Z
[1]     ⍝LAST UPDATE: 5-9-79; CALLS GAUSSDF
        ⍝WHICH CALLS ERF
[2]     ⍝TWO SAMPLE RUNS TEST FOR COMPARING
        ⍝PROB DISTRIBUTIONS
[3]     ⍝TIES BETWEEN SAMPLES ARE NOT PERMITTED-
        ⍝SEE EXTERNAL DOCUMENTATION
[4]     ⍝SAMPLE SIZES N1 AND N2 SHOULD BE FAIRLY
        ⍝CLOSE TOGETHER, ESPECIALLY
[5]     ⍝FOR LARGE VALUES.
[6]        →L0×⍳0=+/A∊B
[7]        'NO TIES ALLOWED BETWEEN SAMPLES'
[8]        E←⍳0
[9]        →0
[10] L0:N2←(⍴X)-N1←+/0=X←X[⍋X←A,B]∊B
[11]    K←1+⍳¯1+R←1++/0≠¯1↓X-1⌽X
[12]    →L1×⍳(N1>10)∧N2>10
[13]    Y2←K-Y1←(0.5×1=2|K)+K÷2
[14]    W←(⍴K)⌽V←Y1,Y2
[15]    PR1←+/PR←(((V-1)!N1-1)×(W-1)!N2-1)
        ÷N1!N1+N2
[16]    PR2←1-PR1-+/PR[(⍴K),⍴V]
[17]    E←R,PR1,PR2
[18]    →0
[19]    ⍝NORMAL APPROXIMATION FOR N1 AND N2 >10
[20] L1:R←R,R-1
[21]    Z←(R+0.5-1+2×N1×N2÷N1+N2)÷((2×N1×N2
        ×(2×N1×N2)-N1+N2)÷(N1+N2-1)×(N1+N2)
        *2)*0.5
[22]    PR←(Z GAUSSDF 0 1)[2;]
```

```
[23]    E←R[1],PR[1],PR[2]←1-PR[2]
[24]    'NORMAL APPROXIMATION USED'.
        ∇
```

Function Name: SPECTR1

EXTERNAL DOCUMENTATION

PURPOSE

Performs statistical spectral analysis of time series from real stationary stochastic process. It estimates the spectral density function of the process by averaging the periodogram with triangular window.

Program calls the auxiliary programs PDGRAM, NINVFFT, and TRISMOOTH, as well as the program FACTOR (see Chapter 18) and MRFFT (see Chapter 17).

INTERNAL DOCUMENTATION FOR SPECTR1

```
SYNTAX: ∇Z←M SPECTR1 X∇
PURPOSE: ESTIMATION OF THE SPECTRAL DENSITY
   OF A TIME SERIES BY SMOOTHING THE PERIODO-
   GRAM OF THE STRING X BY CONVOLUTION WITH A
   TRIANGULAR WINDOW HAVING M POSITIVE
   ELEMENTS.
INDEX ORIGIN: 0 OR 1.
AUTHOR: G. PODOROWSKY, D. E. MCCLURE
LAST UPDATE: 02/08/81
RIGHT ARGUMENT: X: A REAL-VALUED VECTOR WITH
   N ELEMENTS OR A COMPLEX-VALUED VECTOR WITH
   N ELEMENTS (2×N MATRIX WHOSE ROWS CORRE-
   SPOND TO THE REAL AND IMAGINARY PARTS OF
   THE VECTOR).
LEFT ARGUMENT: M: AN ODD INTEGER LESS THAN
   OR EQUAL TO N. M IS THE NUMBER OF POSITIVE
   ELEMENTS OF THE TRIANGULAR WEIGHT-FUNCTION
   WHICH IS USED TO SMOOTH THE PERIODOGRAM OF
   X.
RESULT: Z: A VECTOR WITH N ELEMENTS GIVING
   THE VALUES OF AN ESTIMATE OF THE SPECTRAL
   DENSITY FUNCTION OF X AT THE N EQUALLY-
   SPACED POINTS 2×K×PI÷N, FOR K=0,1,...,N-1.
```

REMARKS ABOUT USE: SPECTR1 CALLS THE
FUNCTIONS PDGRAM AND TRISMOOTH. IF THE USER
WISHES TO EXPERIMENT WITH DIFFERENT DEGREES
OF SMOOTHING, THEN IT WILL BE MORE EFFICIENT
TO COMPUTE THE PERIODOGRAM ONCE (USING
PDGRAM) AND THEN TO EXAMINE THE EFFECTS
OF SMOOTHING BY USING TRISMOOTH ON THE
PERIODOGRAM.

CODE FOR SPECTR1

```
      ∇  Z←M SPECTR1 X
[1]   Z←M TRISMOOTH PDGRAM X
      ∇
```

INTERNAL DOCUMENTATION FOR PDGRAM

SYNTAX: ∇I←PDGRAM X∇
PURPOSE: COMPUTES THE PERIODOGRAM OF THE
FINITE RECORD X OF A DISCRETE PARAMETER
TIME SERIES.
INDEX ORIGIN: 0 OR 1.
AUTHOR: D. E. MCCLURE
LAST UPDATE: 02/08/81
ARGUMENT: X: A REAL-VALUED VECTOR WITH N
ELEMENTS OR A COMPLEX-VALUED VECTOR WITH
N ELEMENTS (2×N MATRIX WHOSE ROWS CORRESPOND
TO THE REAL AND IMAGINARY PARTS OF THE
VECTOR).
RESULT: I: A VECTOR WITH N ELEMENTS GIVING
THE VALUES OF THE PERIODOGRAM OF X AT THE
N EQUALLY SPACED POINTS 2×K×PI÷N, FOR
K=0,1,...,N-1.
REMARKS ABOUT USE: PDGRAM CALLS THE FUNCTIONS
FACTOR AND NINVFFT. THE RESULT I MAY BE
USED AS INPUT FOR A SMOOTHING FUNCTION
SUCH AS TRISMOOTH TO FORM SPECTRAL ESTIMATES
ON [0,2×PI).
IF THE PRIME FACTORIZATION OF N CONTAINS
LARGE FACTORS, THEN 'WS FULL' MAY RESULT.

CODE FOR PDGRAM

```
      ∇ I←PDGRAM X;N;PR;□IO
[1]   □IO←1
[2]   PR←FACTOR N←¯1↑ρX
[3]   PR←PR[2;1++/(ι¯1↑I)∘.>I←+\PR[1;]]
[4]   I←(÷○2×N)×+/(PR NINVFFT X)*2
      ∇
```

INTERNAL DOCUMENTATION FOR NINVFFT

SYNTAX: ∇Z←PR NINVFFT X∇
PURPOSE: COMPUTES N TIMES THE INVERSE
 FAST-FOURIER-TRANSFORM OF A REAL OR
 COMPLEX-VALUED STRING WITH N ELEMENTS.
 USED IN PERIODOGRAM ANALYSIS.
INDEX ORIGIN: 0 OR 1.
AUTHOR: D. E. MCCLURE
LAST UPDATE: 02/08/81
RIGHT ARGUMENT: X: A REAL-VALUED VECTOR WITH
 N ELEMENTS OR A COMPLEX-VALUED VECTOR WITH
 N ELEMENTS (2×N MATRIX WHOSE ROWS CORRESPOND
 TO THE REAL AND IMAGINARY PARTS OF THE
 VECTOR).
LEFT ARGUMENT: PR: EITHER THE SCALAR N (IF
 REASONABLY SMALL) OR A VECTOR OF POSITIVE
 INTEGERS WHOSE PRODUCT IS N.
RESULT: Z: COMPLEX VECTOR (2×N MATRIX)
 WHICH IS N TIMES THE INVERSE DISCRETE
 FOURIER TRANSFORM OF X.
REMARKS ABOUT USE: NINVFFT CALLS THE FAST-
 FOURIER-TRANSFORM PROGRAM MRFFT.
 IF AN ELEMENT OF PR IS TOO LARGE, THEN
 'WS FULL' WILL RESULT.

CODE FOR NINVFFT

```
      ∇ Z←PR NINVFFT X;N
[1]   X←(2,N)ρ(,X),(N←¯1↑ρX)ρ0
[2]   Z←PR MRFFT(2 1 ↑X),⏀ 0 1 ↓X
      ∇
```

INTERNAL DOCUMENTATION FOR TRISMOOTH

SYNTAX: ∇Z←M *TRISMOOTH VEC*∇
PURPOSE: TO SMOOTH THE ELEMENTS OF A VECTOR
 BY CONVOLVING THE VECTOR WITH A TRIANGULAR
 WINDOW HAVING M POSITIVE ELEMENTS.
INDEX ORIGIN: 0 *OR ·1.*
AUTHOR: D. E. MCCLURE
LAST UPDATE: 02/08/81
RIGHT ARGUMENT: VEC: A VECTOR WITH N ELEMENTS
 TO BE SMOOTHED.
LEFT ARGUMENT: M: AN ODD INTEGER LESS THAN
 OR EQUAL TO N. M IS THE NUMBER OF POSITIVE
 ELEMENTS OF THE TRIANGULAR WEIGHT-FUNCTION
 WHICH IS CONVOLVED WITH VEC.
RESULT: Z: A VECTOR OF N ELEMENTS FORMED BY
 AVERAGING, WITH TRIANGULAR WEIGHTS, THE
 ELEMENTS OF VEC.
REMARKS: VEC IS EXTENDED PERIODICALLY,
 FORWARD AND BACKWARD, BEFORE SMOOTHING SO
 THAT THE CONVOLUTION OPERATION IS DONE ON
 THE DISCRETE CIRCLE WITH N POINTS.
 TRISMOOTH MAY BE APPLIED TO THE OUTPUT OF
 PDGRAM TO COMPUTE SPECTRAL ESTIMATES (E.G.,
 FHAT←5 TRISMOOTH PDGRAM X).

CODE FOR TRISMOOTH

```
      ∇  Z←M TRISMOOTH VEC;N;□IO;WIN;J;I
[1]   □IO←1
[2]   N←ρVEC
[3]   WIN←J↓WIN÷+/WIN←(⌽WIN),1,WIN←1-(ιJ)
      ÷(1+J←(M-1)÷2)
[4]   Z←WIN[1]×VEC
[5]   →(M≤I←1)/0
[6]   SM:Z←Z+WIN[I+1]×(ιφVEC)+(-I)φVEC
[7]   →(J≥I←I+1)/SM
      ∇
```

Function Name: STEMLEAF

EXTERNAL DOCUMENTATION

PURPOSE

To produce a stem and leaf representation of a batch of numbers.

DEVELOPMENT

A stem and leaf representation is similar to displaying numbers in histogram form. Each stem corresponds to a particular interval of specified length on the real line into which elements from the batch are placed (stems will be denoted by the left end point of the interval they represent). The length of an interval corresponding to a stem is called the step-size and must be a power of ten. Associated with each stem are leaves, which show the positions of elements in the stem. Each stem usually has 10 positions labeled 0, 1, 2, 3, . . . , 9 and each position is itself an interval of length step-size/10. If for some reason less than 10 positions are desired per stem, the step-size can be multiplied by a scaling factor (usually 1). Thus, if the step-size = 10 and the scaling factor = 0.5, then each stem is an interval of length 5 with 5 positions of length 1.

To classify a batch element, you specify the stem and the leaf into which the element falls.

EXAMPLE

Display 30 47 48.7 54 60 61 65.2 71 with step-size = 10 and scale factor = 1.

RESULT

(1)	03|0
(3)	04|78
(4)	05|4
(4)	06|015
(1)	07|1

The stems are 03,04,05,06,07, corresponding to the intervals (30,40), (40,50) (50,60), (60,70), (70,80). The leaves are shown to the right of the stems. The number 47 was placed in the 2nd stem and the 7th leaf in that stem. The number 65.2 was placed in the 4th stem and the 5th leaf.

The stem and leaf form used in this program also provides cumulative counts of elements from both bottom and top of the display, converging at the interval containing the median (these are the numbers in parentheses at the left of each stem). Recall that the position of the median in a batch of numbers is equal to $\frac{1}{2}$ (1 + batch size). In the above example, the median is in the 4.5 position of the batch and lies between the 3rd and 4th stems. Thus cumulative counts are done from the 1st to the 3rd stem and from the 5th to the 4th stem.

REFERENCE

Mosteller, J., and Tukey, J. (1977). "Data Analysis and Regression," Addison-Wesley, Reading, Massachusetts.

INTERNAL DOCUMENTATION

```
SYNTAX: Z←S STEMLEAF X
PURPOSE: TO PRODUCE A STEM AND LEAF REPRE-
  SENTATION OF A BATCH OF NUMBERS.
INDEX ORIGIN: 1
AUTHOR: T. FERGUSON, ADAPTED FROM D. MCNEIL,
  INTERACTIVE DATA ANALYSIS
LEFT ARGUMENT: A VECTOR OF TWO VALUES. THE
  FIRST VALUE IS THE SCALING FACTOR, TYPICALLY
  0.5,1, OR 2. THE SECOND VALUE IS THE STEP-
  SIZE OF THE PLOT, AND MUST BE AN INTEGER
  POWER OF TEN.
RIGHT ARGUMENT: THE SCALAR, VECTOR OR MATRIX
  OF DATA TO BE USED IN THE STEM=AND=LEAF
  DISPLAY.
RESULT: Z, A MATRIX CONTAINING THE PRINTOUT
  OF THE STEM-AND-LEAF PLOT.
EXAMPLE:   1 10 STEMLEAF 30 47 48.7 54 60
  61 65.2 71

  (1)      03|0
  (3)      04|78
  (4)      05|4
  (4)      06|015
  (1)      07|1

THE STEMS, THE NUMBERS LEFT OF THE BAR, ARE
```

IN UNITS OF THE STEP-SIZE. THE LEAVES ARE THE CORRESPONDING INTEGERS RIGHT OF THE BAR, REPRESENTING THE DECIMAL PART OF THE STEP-SIZE OF EACH DATA POINT. THE LEFT-MOST COLUMN CONTAINS THE CUMULATIVE COUNTS FROM BOTH THE TOP AND BOTTOM OF THE DISPLAY, CONVERGING AT THE INTERVAL CONTAINING THE MEDIAN.

CODE

```
     ∇ Z←S STEMLEAF X;A;G;M;N;Q;K;SS
[1]    M←SS×⌊(X←X[⍋X←,X])[1]÷SS←S[2]
[2]    MD←(1+ρX)÷2+K←0
[3]    A←'0123456789|- ',Z←''
[4]  L:K←K+Q←ρW←1+⌊10⊤|(10÷SS)×(X<M←M+G←S[1]
       ×SS)/X
[5]    T←10ρ'(',(⍕(K,Q)[1+MD≤K←K⌊ρX]),')'
[6]    Z←Z,100ρT,A[(N+12),(1+ 10 10 ⊤⌊|(M-G×N←
       X[1]≥0)÷SS),11,W],90ρ' '
[7]    →(0<ρX←(ρW)↓X)/L
[8]    Z←((0.01×ρZ),100)ρZ
     ∇
```

Function Name: SIGN

EXTERNAL DOCUMENTATION

PURPOSE

To perform a nonparametric sign test on two random samples to determine if they are drawn from a common distribution.

It is useful to compare two materials or treatments under various sets of conditions.

ASSUMPTIONS

(1) There are pairs of observations on the two things being compared (thus the two sample sizes must be equal).

(2) Each of the two observations of a given pair was made under similar conditions.

(3) The different pairs were observed under different conditions.

(4) Differences between paired observations are independent.

In theory, the test is based on sampling from continuous distributions. Thus there should be no ties in paired comparisons. However, in practice ties do occur with $p > 0$ due to limits on precision or integer-valued random variables. Such ties should be omitted, and the test should proceed with reduced sample size.

EXAMPLE

Let X and Y represent measurements made on treatments A and B. Let N = the number of pairs of observations. Pairs of observations and their differences are denoted by

$$(X_1, Y_1), (X_2, Y_2), \ldots, (X_N, Y_N)$$
$$X_1 - Y_1, \quad X_2 - Y_2, \quad \ldots, \quad X_N - Y_N$$

If any of the $X_I - Y_I = 0$, they are excluded and N is reduced.

H_0: U = median of $(X_I - Y_I) = 0$ for all I. That is, each difference has a probability distribution (which need not be the same for all differences) with median equal to zero.

H_1: U is not = 0 (two-sided) or $U > 0$ (one-sided).

Reject H_0 when there are too many or too few positive differences.

ALGORITHM

Compute probability as binomial with $p = \frac{1}{2}$. $PR(X \le S) = \Sigma$ from $k = 0$ to S of $PR(X = K)$, where

$$PR(X = k) = (N!k)((\tfrac{1}{2}) * k)((\tfrac{1}{2}) * (N - k))$$
$$= (\tfrac{1}{2} * N)(N!/k!(N - k)!)$$

For two-sided tests, double $\min(p, 1 - p)$. For $N > 25$, a normal approximation is used to compute the probability:

$$Z = (S + \tfrac{1}{2} - N/2)/\sqrt{\tfrac{1}{2} \tfrac{1}{2} N}) = (S + \tfrac{1}{2} - N/2)/0.5\sqrt{(n)}$$

For extensions of the sign test, see Dixon and Massey.

REFERENCES

Dixon, W. J., and Massey, F. J. (1957). "Introduction to Statistical Analysis," 2nd ed. McGraw-Hill, New York.
Putter, J. (1955). Treatment of Ties in Some Nonparametric Tests, *Ann. of Math. Statist.* **26,** 368–386.

INTERNAL DOCUMENTATION

```
SYNTAX: ∇E←X SIGN Y∇
PURPOSE: PERFORMS NONPARAMETRIC SIGN TEST
 ON TWO SETS OF PAIRED DATA TO DECIDE IF
 THEY ARE DRAWN FROM A COMMON DISTRIBUTION.
AUTHOR: T. RICHMAN
LAST UPDATE: 5-9-79
INDEX ORIGIN: 1
LEFT ARGUMENT: X IS AN N-VECTOR WHOSE
 ELEMENTS ARE OBSERVATIONS SAMPLED
 FROM DIFFERENT CONDITIONS.
RIGHT ARGUMENT: Y IS ALSO AN N-VECTOR WHOSE
 ELEMENTS ARE INDIVIDUALLY PAIRED WITH THOSE
 OF X. IT IS ASSUMED THAT DIFFERENCES BETWEEN
 PAIRS ARE INDEPENDENT.
RESULT: FOR N≤25, THE PROBABILITY OF OB-
 SERVING ≤S POSITIVE DIFFERENCES (X-Y) IS
 COMPUTED DIRECTLY. TIES (DIFFERENCES = 0)
 ARE EXCLUDED FROM THE TEST.
 FOR N>25 THE DESIRED PROBABILITY IS COM-
 PUTED VIA A NORMAL APPROXIMATION.
 H0: DISTRIBUTION OF PAIRED DIFFERENCES
 HAVE MEDIAN EQUAL TO ZERO. AN EXCESS OF
 POSITIVE (OR NEGATIVE) DIFFERENCES WILL
 LEAD TO THE REJECTION OF H0.
 FOR TWO-SIDED TESTS, DOUBLE MIN(P, 1-P).
 E[1]= S =NUMBER OF + SIGNS
 E[2]= SAMPLE SIZE N (EXCLUDING TIES)
 E[3]= P(X≤S;N)
```

CODE

```
      ∇ E←K SIGN Y;P;N;D;S;Z
[1]   ⍝LAST UPDATE: 5-9-79; CALLS GAUSSDF
```

```
        ⍝WHICH CALLS ERF
[2]     ⍝NOTE THAT TIES ARE OMITTED
[3]     N←⍴D←((X-Y)≠0)/X-Y
[4]     S←+/D>0
[5]     →L0×⍳N>25
[6]     E←S,N,+/(÷2*N)×(0,⍳S)!N
[7]     →0
[8]     ⍝NORMAL APPROXIMATION FOR N>25
[9]     L0:Z←(S+0.5-N÷2)÷0.5×N*0.5
[10]    E←S,N,(Z GAUSSDF 0 1)[2;]
[11]    'NORMAL APPROXIMATION USED'
        ∇
```

Function Name: TRENDS

EXTERNAL DOCUMENTATION

PURPOSE

To apply the two-way frequency table to the case of two-dimensional classification of paired consecutive values in a vector of data.

For $X = (x(1), x(2), \ldots, x(n))$, the resulting frequency table z classifies the ordered pairs $(x(k), x(k + 1))$, $1 \le k \le n - 1$ in two dimensions. (See the description of FREQTAB2 for more information about the table.) The user specifies the matrix P (defining the classification scheme), which will be used when FREQTAB2 is called.

INTERNAL DOCUMENTATION

```
SYNTAX: Z←P TRENDS X
PURPOSE: TWO-DIMENSIONAL SPREAD OF PAIRED
  CONSECUTIVE VALUES OF A VECTOR OF DATA.
INDEX ORIGIN: 1
AUTHOR: E. SILBERG
LAST UPDATE: 3/79
LEFT ARGUMENT: P IS A 2×3 MATRIX DEFINING
  THE TWO-DIMENSIONAL CLASSIFICATION SCHEME:
  FOR ROW CLASSIFICATION,P[1;1]=LEFT-HAND
  END OF FIRST FREQUENCY CLASS,P[1;2]=CLASS
  WIDTH, P[1;3]= NO. OF CLASSES. P[2;] GIVES
```

THE CORRESPONDING INFO. FOR THE COL.
CLASSIFICATION.
RIGHT ARGUMENT: VECTOR OF DATA.
RESULT: Z IS A FREQUENCY MATRIX, IN WHICH
THE IJTH ENTRY IS THE FRACTION OF PAIRS
(X[K],X[K+1]) FALLING INTO ROW I, COL. J,
OF THE CLASSIFICATION TABLE, WHERE K=1,2,
...,N-1.

CODE

```
      ∇ T←P TRENDS X
[1] ⍝   TWO-DIMENSIONAL SPREAD OF PAIRED
    ⍝   CONSECUTIVE VALUES
[2] ⍝   IN A VECTOR OF DATA. CALLS ∇Z←P
    ⍝   FREQTAB2 M∇.
[3]   T←P FREQTAB2(2,⁻1+⍴X)⍴(⁻1↓X),1↓X
      ∇
```

CHAPTER 25

Utilities

The following programs are useful when editing APL programs and comparing several by timing them. Code is also included for organizing and listing the content of workspaces. There is also a program that organizes the printing of large $2 \times N$ arrays in legible form.

Function Names: COMPOSE, EDIT

EXTERNAL DOCUMENTATION

PURPOSE

To create and edit describe variables.

The program COMPOSE can be used to set up describe variables or any other character-string variables. There are two features of COMPOSE that make it both easy to use and efficient. First, COMPOSE allows the user to enter the text exactly as it should appear when the finished variable is called—there is no guessing involved. Secondly, COMPOSE stores the text in vector form, so that no space is wasted.

This second feature, however, makes it difficult to edit the variable manually. Therefore, the program EDIT is available to simplify the process. EDIT interrogates the user to find out what changes should be made and it will make as many changes as the user desires.

INTERNAL DOCUMENTATION FOR COMPOSE

```
SYNTAX: COMPOSE NAME
PURPOSE: ASSISTS IN CREATING DESCRIBE
  VARIABLE FOR APL DOCUMENTATION
INDEX ORIGIN: 1
AUTHORS: B. BENNETT, P. KRETZMER
```

```
LAST UPDATE: 11-19-78
RIGHT ARGUMENT: NAME: CHARACTER STRING
 NAME←NAME OF GLOBAL VARIABLE TO BE
 CREATED
TO USE: CALL FUNCTION; ENTER TEXT EXACTLY
 AS IT SHOULD APPEAR WHEN THE FINISHED
 VARIABLE IS CALLED. WHEN FINISHED, ENTER
 QUIT ON A NEW LINE.
```

CODE FOR COMPOSE

```
      ∇ COMPOSE NAME;CR;TEXT;LINE;LNTX
[1]   ⍝ASSISTS IN CREATING DESCRIBE VARIABLE
      ⍝FOR APL DOCUMENTATION
[2]   ⍝LAST UPDATE: 11-19-78
[3]   CR←⎕AV[203]
[4]   TEXT←⍞
[5]   RET:LINE←⍞
[6]   LNTX←LINE,'   '
[7]   →((LNTX[1]='Q')∧(LNTX[3]='I')∧
      (LNTX[4]='T'))/QUIT
[8]   TEXT←TEXT,CR,LINE
[9]   →RET
[10] QUIT:⍎NAME,'←TEXT'
      ∇
```

INTERNAL DOCUMENTATION FOR EDIT

```
SYNTAX: AV←EDIT V
PURPOSE: TO MAKE CHANGES IN A DESCRIBE
 VARIABLE
INDEX ORIGIN: 1
AUTHOR: B. BENNETT
LAST UPDATE: 7/2/79
RIGHT ARGUMENT: NAME OF THE DESCRIBE VARIABLE
 BEING ALTERED
RESULT: USER WILL BE PROMPTED FOR DESIRED
 CHANGES
NOTE: THIS FUNCTION SHOULD BE CALLED IN THE
 FOLLOWING MANNER:
 DESCRIBE←EDIT DESCRIBE (TO EDIT THE
 VARIABLE 'DESCRIBE')
```

CODE FOR EDIT

```
      ∇  AV←EDIT V;W
[1]    'OLD STRING:'
[2]    L←1↑,(∧≠(⁻1+ιρW)⌽(W←⎕)∘.=V)/ιρV
[3]    →OUT×ι(L=0)
[4]    'NEW STRING:'
[5]    AV←((L-1)↑V),⎕,(-(ρV)-L+(ρW)-1)↑V
[6]    V←AV
[7]    'MORE CORRECTIONS? [Y/N]'
[8]    →(1↑⎕='Y')/1
[9]    →0
[10]  OUT:'NOT FOUND'
[11]   AV←V
      ∇
```

Function Name: LISTALL

EXTERNAL DOCUMENTATION

PURPOSE

To list in alphanumeric order the code of all functions in a workspace.

ALGORITHM

Using the APL system function, quad NL 3 creates a character matrix of all function names in the workspace. Since GLIST has been copied, LISTALL, SORT1, and DLB will be included in this matrix. To drop these unwanted names, first create a matrix whose columns are the character strings "LISTALL", "LIST", "DLB", and "SORT1". By comparing the original matrix with this new one, find the positions in the original matrix of LISTALL, LIST, DLB, and SORT1 and then drop these names from the original. Call SORT1 to arrange the function names in alphanumeric order. Call LIST to print the code of the functions.

NOTE

For a detailed description of alphanumeric sorting see the external documentation to the program SORT1 in this chapter.

INTERNAL DOCUMENTATION

SYNTAX: LISTALL
PURPOSE: TO LIST IN ALPHA-NUMERIC ORDER,
 THE CODE OF ALL FUNCTIONS IN A WORKSPACE.
INDEX ORIGIN: 1
AUTHOR: D. MCCLURE·
LAST UPDATE: 7/22/80
TO USE: INTO A CLEAR WORKSPACE, LOAD THE
 WORKSPACE FOR WHICH A LISTING OF CODES
 IS DESIRED. COPY WORKSPACE LIST1. TYPE
 LISTALL.
RESULT: THE CODES OF ALL FUNCTIONS IN THE
 WORKSPACE ARRANGED IN ALPHA-NUMERIC ORDER.

CODE

```
       ∇ LISTALL
[1]    N←¯1↑ρNA←□NL 3
[2]    M←1↑ρNA
[3]    MAT←⍉(5 7 ρ'LISTALLLISTFNSDLB      SORT1
       '),(5,(N-7))ρ' '
[4]    NA←NA[((~∨/NA∧.=MAT)/⍳M);]
[5]    NA←N SORT1 NA
[6]    LISTFNS,NA,(M-4)ρ' '
       ∇
```

Function Name: LISTDV

EXTERNAL DOCUMENTATION

PURPOSE

To display all or some of the describe variables in a workspace.

ALGORITHM

Create two character matrices *Get* and *Fns* using the APL system functions quad NL 2 and 3.

Get = quad NL 2 = character matrix whose rows are the names of the
                    global variables in the active workspace.
Fns = quad NL 3 = character matrix whose rows are the names of the
                    functions in the active workspace.

Call SORT1 to sort the rows of *Get* alphanumerically.

Because describe variables are named according to the rule 'D name
of function described,' the following procedure will produce a matrix
whose rows are the names of the describe variables:

(1) Drop from *Get* all rows whose first character is not the letter D.
(2) Temporarily remove the first column of *Get* and then drop from
Get all rows that are not equivalent to a row of *Fns*.

Print out the vector representation of *Get* (*Get* is now the matrix of
names of describe variables):

$$
let\ names\ = \begin{cases} \text{vector representation of } Get \text{ if user desires all describe} \\ \text{variables in the active workspace.} \\ \text{vector of describe variables} \qquad\qquad \text{else} \\ \text{desired by user.} \end{cases}
$$

Do for each name in names;

   Display the describe variable associated with the current name.
   End.

INTERNAL DOCUMENTATION

```
SYNTAX: LISTDV
PURPOSE: TO DISPLAY ALL OR SOME OF THE
 DESCRIBE VARIABLES IN A WORKSPACE.
INDEX ORIGIN: 1
AUTHOR: R. COHEN
LAST UPDATE: 9/28/80
TO USE: TYPE LISTDV. THE COMPUTER WILL GIVE
 YOU A LIST OF ALL DESCRIBE VARIABLES IN
 YOUR ACTIVE WORKSPACE AND ASK WHETHER
 YOU WANT ALL OR SOME OF THEM DISPLAYED.
 IF YOU REQUEST A PARTIAL LIST, THE COMPUTER
 WILL ASK YOU TO TYPE THE LIST YOU DESIRE.
 MAKE SURE THAT WHEN YOU ENTER THIS LIST
```

ALL OF THE NAMES ARE SEPARATED FROM ONE
ANOTHER BY AT LEAST ONE BLANK.
RESULT: THE DESIRED PARTIAL OR FULL DISPLAY
OF DESCRIBE VARIABLES IN THE ACTIVE WORK-
SPACE.
WARNING: THE LIST OF DESCRIBE VARIABLES
THE COMPUTER PROVIDES YOU CONTAINS ONLY
THOSE DESCRIBE VARIABLES WHOSE RESPECTIVE
FUNCTIONS ARE IN THE ACTIVE WORKSPACE. IF
YOU WANT TO˘SEE A DESCRIBE VARIABLE WHOSE
RESPECTIVE FUNCTION ISN'T IN THE ACTIVE
WORKSPACE SPECIFY THE OPTION FOR A PARTIAL
DISPLAY AND INCLUDE THAT NAME IN THE LIST
GIVEN TO THE COMPUTER.

CODE

```
        ∇ LISTDV;V;NAMES;CURNAM;GET;BL;I;
          ALPH;FNS;S;SELECT
[1]     □IO←1
[2]     →((ρGET←□NL 2)= 0 0)/MSG
[3]     GET←(⁻1↑ρGET) SORT1 GET
[4]     GET←GET[('D'=GET[;1])/ι(ρGET)[1];]
[5]     →((S←(ρGET)[2]-1)>(ρFNS←□NL 3)[2])
        /FIX
[6]     GET←GET,((ρGET)[1],(ρFNS)[2]-S)ρ' '
[7]     →PICK
[8]     FIX:FNS←FNS,((ρFNS)[1],S-(ρFNS)[2])ρ''
[9]     PICK:GET←GET[(+/GET[;1↓ι(ρGET)[2]]
        ×.=⍉FNS)/ι(ρGET)[1];]
[10]    →(0=ρNAMES←,GET,((ρGET)[1],2)ρ' ')/MSG
[11]    'THE DESCRIBE VARIABLES IN THIS WORK-
        SPACE ARE:'
[12]    □←NAMES
[13]    □←' '
[14]    'ENTER 1 IF YOU WANT ALL OF THE
        DESCRIBE VARIABLES PRINTED OUT.'
[15]    'ENTER 2 IF YOU ARE GOING TO WANT
        ONLY A PARTIAL LIST PRINTED.'
[16]    □IO←0
```

```
[17]    →(□=1)/PRINT
[18]    'ENTER THE LIST OF DESCRIBE VARIABLES
        YOU WANT PRINTED OUT.'
[19]    NAMES←□
[20]   PRINT:NAMES←(φDLBφDLB NAMES),' '
[21]   AGN:BL←((' '=NAMES)/ιρNAMES)[0]
[22]    □←' '
[23]    □←' '
[24]    □←CURNAM←BL↑NAMES
[25]    ⍙CURNAM
[26]   NAMES←DLB(1+BL)↓NAMES
[27]    →(0≠ρNAMES)/AGN
[28]    □IO←1
[29]    →0
[30]  MSG:'THERE ARE NO DESCRIBE VARIABLES
        IN THIS WORKSPACE WHICH'
[31]    'CORRESPOND TO ANY OF THE FUNCTIONS
        IN THIS WORKSPACE.'
        ∇
```

Function Name: PRINTAB

EXTERNAL DOCUMENTATION

PURPOSE

To help the user work with $2 \times N$ arrays, where N is so large that each row of the array takes several lines to print.

In this case it would be difficult to visually match up the entries of the first and second rows of the array, since many lines may intervene. PRINTAB transforms such an array to a readable form:

If $a_1, a_2, \ldots, a_N$ are the entries of row 1, and $b_1, b_2, \ldots b_N$ are the entries of row 2 in the table T specified by the user, and if the user has requested $C = c$ columns per row of output, then the following will be printed:

$$
\begin{array}{cccc}
a_1 & a_2 & \cdots & a_c \\
b_1 & b_2 & \cdots & b_c \\
\\
a_{c+1} & a_{c+2} & \cdots & a_{2c} \\
b_{c+1} & b_{c+2} & \cdots & b_{2c} \\
& & \vdots & \\
a_{(m-1)c+1} & a_{(m-1)c+2} & \cdots & a_{mc} \\
b_{(m-1)c+1} & b_{(m-1)c+2} & \cdots & b_{mc} \\
\\
a_{mc+1} & \cdots & & a_{mc+r} \\
b_{mc+1} & \cdots & & b_{mc+r}
\end{array}
$$

where m is the largest integer such that $N = mc + r,\ 0 \leqq r \leqq m$.

INTERNAL DOCUMENTATION

```
SYNTAX: C PRINTAB T
PURPOSE: TO PRINT A SET OF ORDERED PAIRS
 IN NEAT, TABULAR FORM
INDEX ORIGIN: 1
AUTHOR: E. SILBERG
LAST UPDATE: 1/79
LEFT ARGUMENT: (INTEGER>0) NO. OF TABLE
 ENTRIES PER ROW
RIGHT ARGUMENT: 2×N MATRIX, EACH COLUMN
 OF WHICH IS AN ORDERED PAIR
RESULT: OUTPUT IS A TABLE CONTAINING ROWS
 AS FOLLOWS:
C ENTRIES FROM T[1;] IMMEDIATELY FOLLOWED
BELOW BY THE CORRESPONDING C ENTRIES
FROM T[2;], CONTINUED UP TO THE MAXIMUM
RECTANGULAR DIMENSIONS ((⌊(ρT)[2]÷C),C).
THEN A SHORTER PAIR OF ROWS OF THE
REMAINING (C|(ρT)[2]) PAIRS FROM T
WILL BE PRINTED.
```

CODE

```
      ∇ C PRINTAB T;R;S;D;N
[1]  ⍝ CREATES A NEAT TABLE FROM A SET OF
     ⍝ ORDERED PAIRS.
[2]   →(0=N←1↓ρS←(0,-R←C|1↓ρT)↓T)/PRINT2
```

```
[3]    (DρS[1;]),[1.5](D←(N÷C),C)ρS[2;]
[4]    →(R=0)/0
[5]    '    '
[6]    PRINT2:(2,-R)↑T
       ∇
```

Function Name: SORT1

EXTERNAL DOCUMENTATION

PURPOSE

To sort into alphanumeric order the rows of a character matrix. Alphanumeric order is similar to alphabetical order except that in alphanumeric order character strings containing one or more integers can be ordered. The well known rule for alphabetical order is A before B and B before C, etc. In alphanumeric order the rule is still A before B and B before C but it is extended to all letters before 0, 0 before 1, etc. Three additional characters are considered in this version of the alphanumeric order: the blank character, the comma and the period. The rule can now fully be stated as: the blank character before the comma, the comma before the period, the period before all letters and all letters before all integers.

ALGORITHM

All characters in the above categorization are given a numerical value between 1 and 39. A blank is 1, an A is 4, and a 9 is 39.

For each row of the character matrix, create a vector of numerical representations of characters in a row. (Note: this vector can be considered to be coded in base 40. Just as 4 5 6 in base 10 reads 4 100s, 5 10s, and 6 1s, 4 5 6 in base 40 reads 4 1600s, 5 40s and 6 1s.)

Decode the vector into its specific numeric value (4 5 6 becomes 6606)

End.

Find the permutation that would numerically order (ascending) the vector of integers calculated above and order the rows of the inputted character matrix according to this permutation.

INTERNAL DOCUMENTATION

```
SYNTAX: M←K SORT1 L
PURPOSE: TO SORT INTO ALPHA-NUMERIC ORDER
  THE ROWS OF A CHARACTER MATRIX.
```

INDEX ORIGIN: 1
AUTHOR: D. MCCLURE
LAST UPDATE: 7/22/80
RIGHT ARGUMENT: THE CHARACTER MATRIX TO
 BE SORTED ROWWISE.
LEFT ARGUMENT: THE NUMBER OF COLUMNS IN THE
 CHARACTER MATRIX ABOVE.
RESULT: THE NOW ALPHA-NUMERICALLY SORTED
 CHARACTER MATRIX.(FOR AN EXPLANATION OF
 ALPHA-NUMERIC SORTING, CONSULT THE EXTERNAL
 DOCUMENTATION TO THIS PROGRAM.)

CODE

```
      ∇ M←K SORT1 L;ALPH;B;N;I;PERM
[1]   ⍝ SORTS INTO ALPHABETICAL ORDER CHAR-
      ⍝ ACTER MATRIX CREATED BY NEWTABLE
[2]   ⍝ LAST MODIFIED: 5/7/75
[3]   ⍝ AUTHOR: D. MCCLURE
[4]    ALPH←' ,.ABCDEFGHIJKLMNOPQRSTUVWXYZ
       0123456789'
[5]    M←⍳0
[6]    B←1+⍴ALPH
[7]    N←(⍴L)[1]
[8]    I←1
[9]  S1:M←M,B⊥ALPH⍳K↑L[I;]
[10]   →S1×⍳N≥I←I+1
[11]   M←L[PERM←⍋M;]
      ∇
```

Function Name: SPLICE

EXTERNAL DOCUMENTATION

PURPOSE

To splice together two APL functions, and thus save the user the trouble of copying code from one function to another when there is enough code to make this task burdensome.

Suppose F_1 is a function containing a piece of code, on lines n_1 through n_2, that one would like to place between lines n_3 and n_4 of a second function F_2. For convenience (and in accordance with the names used in the APL function SPLICE), let the desired piece of code in F_1 be called INSR (for insert) and let the code in F_2 be called MAIN. The object is to splice INSR together with MAIN below line n_3 of MAIN, and above line n_4 of MAIN. If n_3 and n_4 are consecutive integers, then no code is deleted from F_2 when INSR is inserted; however, if $n_4 = n_3 + k$, where $k > 1$, then the intervening $k - 1$ lines of MAIN (i.e., lines $n_3 + 1, n_3 + 2, \ldots, n_3 + k - 1$) will be deleted. n_1, n_2, n_3, n_4 may be any integers such that $1 \leq n_1 < n_2$ and $0 \leq n_3 < n_4$. Thus INSR may be any block of code two or more lines in length, starting with line 1 of F_1. INSR may be inserted into F_2 anywhere below line 0 and above the last line.

To accomplish the task described above, the command is

$$n_1 \quad n_2 \quad n_3 \quad n_4 \quad \text{SPLICE} \quad `F_1 \quad F_2\text{'}$$

The result will be the character string "F_2," naming the function into which the piece has been inserted. Thus SPLICE changes the function F_2; however, F_1 will remain unchanged. Note that all lines will be properly renumbered in F_2 when the change is made.

INTERNAL DOCUMENTATION

```
SYNTAX: Z←N SPLICE NAMES
PURPOSE: TO SPLICE TOGETHER PARTS OF APL
 FUNCTIONS.
INDEX ORIGIN: 1
AUTHOR: E. SILBERG
LAST UPDATE: 2/79
RIGHT ARGUMENT: A CHARACTER STRING CONSISTING
 OF TWO FUNCTION NAMES, SEPARATED BY A
 BLANK: E.G., 'F1 F2'.
 THE FIRST NAME IS THE FUNCTION FROM WHICH
 A PIECE IS TO BE TAKEN; THE SECOND NAME IS
 THE FUNCTION INTO WHICH THIS PIECE IS TO
 BE INSERTED.
LEFT ARGUMENT: A VECTOR OF FOUR INTEGERS:
 N[1],N[2] = LINES ON WHICH THE PIECE OF F1
 BEGINS AND ENDS.
 N[3],N[4]=LINES OF F2 WHICH ARE TO PRECEDE
 AND FOLLOW THE PIECE OF F1.
RESULT: THE NAME OF THE FUNCTION INTO WHICH
 THE PIECE HAS BEEN INSERTED.
```

CODE

```
      ∇ Z←N SPLICE NAMES;INSR;MAIN;A;B;D;S
[1] ⍝   PROGRAM TO SPLICE TOGETHER TWO APL
    ⍝   FUNCTIONS.
[2]   INSR←((1+N[2]-N[1]),1↓⍴A)↑(N[1],0)
      ↓A←⎕CR(¯1+B←NAMESι' ')↑NAMES
[3]   D←(⍴INSR),⍴MAIN←⎕CR B↓NAMES
[4]   INSR←(D[1],S←D[2]⌈D[4])↑INSR
[5]   MAIN←(D[3],S)↑MAIN
[6]   Z←⎕FX(((1+N[3]),S)↑MAIN),[1] INSR,
      [1]((-D[3]-N[4]),S)↑MAIN
      ∇
```

Function Name: TIMER

EXTERNAL DOCUMENTATION

PURPOSE

To time execution of a program.

This program can be used to time one program or to time two different programs to determine their relative efficiencies.

ALGORITHM

Timer simply indexes the time indicator of the Accounting Information (AI) before and after executing the specified program, and finds the difference to determine the amount of computer time elapsed.

INTERNAL DOCUMENTATION

```
SYNTAX: TIMER F
PURPOSE: EXECUTES A USER-DEFINED FUNCTION
 AND COMPUTES AMOUNT OF CPU TIME USED
INDEX ORIGIN: 1
AUTHOR: E. SILBERG
LAST UPDATE: 7/2/79
RIGHT ARGUMENT: A CHARACTER STRING--INSIDE
 QUOTES SHOULD BE THE FUNCTION NAME AND THE
```

INPUT TO THAT FUNCTION, AS IT WOULD NORMALLY
APPEAR WHEN CALLING THE FUNCTION
RESULT: 1) USUAL OUTPUT OF THE USER-DEFINED
FUNCTION
2) CPU TIME USED IN FUNCTION EXECUTION

CODE

```
     ∇ TIMER F;T
[1]   T←□AI[2]
[2]   ⍎F
[3]   ' '
[4]   'TIME:   ',⍕(-T)+□AI[2]
     ∇
```

Function Name: TITLEC

EXTERNAL DOCUMENTATION

PURPOSE

To make changes in a program without having to either retype the program, lose the original, or change workspaces.

This service can be quite convenient if one wishes to experiment with several versions of one basic program. In order to use TITLEC in this way, the user must first define the new function (the copied version) by writing a header line under the new title. Once this has been done, the command of TITLEC '*Fn*1 *Fn*2' can be entered (where *Fn*1 = name of original function, *Fn*2 = proposed name of copy). The result will be the character string '*Fn*2,' naming the copied version which has just been created.

TITLEC can also serve a second purpose. It can be used to add the code (except header line) of *Fn*1 onto the end of *Fn*2, without altering *Fn*1. The same command of TITLEC '*Fn*1 *Fn*2' should be used and the result will again be the character string '*Fn*2,' naming the function to which lines have been added.

NOTE

If the result is a number rather than '*Fn*2,' then one of the following two errors has probably occurred:

(1) *Fn*2 had not been defined by a header prior to calling TITLEC or

(2) One of the functions was in suspension prior to issuing the TITLEC command.

INTERNAL DOCUMENTATION

```
SYNTAX: TITLEC NAMES
PURPOSE: TO CHANGE A FUNCTION NAME, OR TO
 ADD THE LINES OF FN1 ONTO THE END OF FN2
INDEX ORIGIN: 1
AUTHOR: G. PODOROWSKY
LAST UPDATE: 6/14/79
RIGHT ARGUMENT: A CHARACTER STRING CONSISTING
 OF TWO FUNCTION NAMES, SEPARATED BY A BLANK;
 E.G. 'F1 F2'
 THE FIRST NAME IS THE OLD FUNCTION NAME
 (OR THE ONE WHOSE LINES ARE BEING ADDED
 ONTO FN2
 THE SECOND NAME IS THE NEW FUNCTION NAME
 (OR THE ONE WHICH IS BEING LENGTHENED)
RESULT: THE NAME OF THE NEW (OR LENGTHENED)
 FUNCTION (FN2)
WARNING: THE NEW FUNCTION MUST ALREADY
 HAVE BEEN DEFINED, AT LEAST IN A HEADER.
```


CODE

```
      ∇ Z←TITLEC NAMES;B;A;ADD;NEWP;S;D
[1]  ⍝ PROGRAM TO CHANGE THE NAME OF A
     ⍝ FUNCTION, OR TO ADD THE LINES
[2]  ⍝ OF FN1 ONTO THE END OF FN2.
[3]  ⍝ FN2 MUST ALREADY BE DEFINED BY AT
     ⍝ LEAST A HEADER LINE.
[4]   ADD←(1,0)↓A←□CR(¯1+B←NAMESι' ')↑NAMES
[5]   D←(ρADD),ρNEWP←□CR B↓NAMES
[6]   ADD←(D[1],S←D[2]⌈D[4])↑ADD
[7]   NEWP←(D[3],S)↑NEWP
[8]   Z←□FX NEWP,[1] ADD
      ∇
```

References

Abramowitz M., and Stegun, I. A. (eds.) (1964). "Handbook of Mathematical Functions," National Bureau of Standards Applied Mathematics Series, Vol. 55, pp. 890–891. Washington, D.C.

Andersen, C., et al. (1962). In "Conformal Mapping, Selected Numerical Methods," (Christian Gram, ed.). 114–261. National Bureau of Standards, Government Printing Office, Washington, D.C.

"APL Language" (1976). IBM publication GC26-3847-2, File No. S370-22.

Artstein, Z., and Vitale, R. A. (1974). A strong law of large numbers for random compact sets, Pattern Analysis Rept. #30, Div. Appl. Math. Brown University, Providence, Rhode Island.

Atkins, A. O. L., and Hall, M. (eds.) (1971). "Computers in Algebra and Number Theory," Amer. Math. Soc., Providence, Rhode Island.

Auslander, L., Dunham, B., and North, J. H. (1976). An example of the use of the computer as an experimental tool in mathematical research, Adv. in Math. 22, 52–63.

Banchoff, T. F. (1970). Computer animation and the geometry of surfaces in 3- and 4-space, invited address at the International Congress of Mathematics, Finnish Academy of Sciences, Helsinki.

Banchoff, T. F., and Strauss, C. (1974). Real time computer graphics techniques in geometry, in "Proceedings of Symposia in Applied Mathematics," Vol. 20, pp. 105–111. Amer. Math. Soc., Providence, Rhode Island.

Banks, H. T. (1975). "Modeling and Control in the Biomedical Sciences," Lecture Notes in Biomathematics, No. 6. Springer-Verlag, Berlin and New York.

Beck, R. E., and Kolman, B. (1977). "Computers in Nonassociative Rings and Algebras." Academic Press, New York.

Blum, E. K. (1972). "Numerical Analysis and Computation: Theory and Practice." Addison-Wesley, Reading, Massachusetts.

Boneva, L., Kendall, D., and Stefanov, F. (1971). Spline transformations: Three new diagnostic aids for the statistical data analyst, J. Roy. Statist. Soc. Ser. B 33, 1–70.

Boyce, W., and DiPrima, R. (1965). "Elementary Differential Equations and Boundary Value Problems," 2nd ed. Wiley, New York.

Brahic, A. (1971). Numerical study of a simple dynamical system. I. The associated plane area-preserving mapping. Astronom. Astrophys. 12, 98–110.

Braun, M. (1977). Invariant curves, homoclinic points, and ergodicity in area preserving mappings (to appear).

Childs, L. (1979). "A Concrete Introduction to Higher Algebra." Springer-Verlag, Berlin and New York.

Churchhouse, R. F., and Herz, J. C. (eds.) (1968). "Computers in Mathematical Research." North-Holland Publ., Amsterdam.

Cooley, J. W., and Tukey, J. W. (1965). An algorithm for the machine calculation of complex Fourier series, Math. Comp. 19, 297–301.

Deboor, C. (1978). "A Practical Guide to Splines," Springer-Verlag, Berlin and New York.

Dixon, W. J., and Massey, F. J. (1957). "Introduction to Statistical Analysis," 2nd ed. McGraw-Hill, New York.

Feller W. (1971). "An Introduction to Probability Theory and Its Applications," Vol. 2, p. 47. Wiley, New York.

Fermi, E., Pasta, J., and Ulam, S. (1955). Studies of non-linear problems, U.S. Dept. of Commerce, Washington, D.C.

Freiberger, W. F., and Grenander, U. (1971). "A Short Course in Computational Probability and Statistics." Springer-Verlag, Berlin and New York.

Frolow, I. (1978). Abstract growth patterns, Pattern Analysis Rept. #58. Brown University, Providence, Rhode Island.

Geman, S. (1980). A limit theorem for the norm of random matrices, *Ann. Probab.* **8** (2), 252–261.

Gerald, C. F. (1970). "Applied Numerical Analysis," 2nd ed., pp. 474–482. Addison-Wesley, Reading, Massachusetts.

Gilman, L., and Rose, A. J. (1976). "APL—An Interactive Approach." Wiley, New York.

Graustein, W. (1951). "Differential Geometry." Dover, New York.

Grenander, U. (1963). "Probabilities on Algebraic Structures." Almqvist and Wiksell, Stockholm; Wiley, New York.

Grenander, U. (1976). Disproving a conjecture by heuristic search, Pattern Analysis Rept. #43. Brown University, Providence, Rhode Island.

Grenander, U. (1978). "Pattern Analysis, Lectures in Pattern Theory," Vol. I. Springer-Verlag, Berlin and New York.

Grenander, U. (1981). "Regular Structures, Lectures in Pattern Theory," Vol. III. Springer-Verlag, Berlin and New York.

Grenander, U., and Silverstein, J. W. (1977). Spectral analysis of networks with random topologies, *SIAM J. Appl. Math.* **32**, 499–519.

Hellerman, H., and Smith, I. A. (1976): *APL/360. Programming and Applications.* McGraw-Hill, New York.

Heyer, H. (1977). "Probability Measures on Locally Compact Groups," Springer-Verlag, Berlin and New York.

Hill, I. D., and Pike, M. C. (1978). Collected Algorithms from ACM, Algorithm No. 299, P1-R1-P3-0. Association for Computing Machinery, New York.

Iverson, K. E. (1962). "A Programming Language." Wiley, New York.

Klein, F. (1979). "Development of Mathematics in the 19th Century," (English ed.) transl. by M. Ackerman, Math. Sci. Press, Brookline, Massachusetts.

Kronmal, R. A., and Peterson, A. V. (1978). On the alias method for generating random variables from a discrete distribution, Tech. Rept. #17. Dept. of Biostatistics, School of Public Health and Community Medicine, Seattle, Washington.

LaSalle, J. P. (ed.) (1974). The influence of computing on mathematical research and education, *in* "Proceedings of Symposia in Applied Mathematics," Vol. 20. Amer. Math. Soc., Providence, Rhode Island.

Leech, J., and Howlett, J. (eds.) (1970). "Computational Problems in Abstract Algebra." Pergamon Press, Oxford.

Lehmann, E. L. (1975). "Nonparametrics." Holden-Day, San Francisco, California.

Lehmer, D. H. (1954). Number theory on the SWAC, *in* "Proceedings of Symposia in Applied Mathematics," Vol. 6, pp. 103–108. McGraw-Hill, New York.

Lehmer, D. H. (1968). Machines and pure mathematics, *in* "Computers in Mathematical Research," pp. 1–7. North-Holland Publ., Amsterdam.

Lehmer, D. H. (1974). The influence of computing on research in number theory, "Proceedings of Symposia in Applied Mathematics," Vol. 20, pp. 3–12. Amer. Math. Soc., Providence, Rhode Island.

McClure, D. E. (1977). Strategies for plotting a large number of isolated points (personal communication).

McNeil, D. (1977). "Interactive Data Analysis." Wiley, New York.

Molina, E. C. (1942). "Poisson's Exponential Binomial Limit." Van Nostrand-Reinhold, Princeton, New Jersey.

Moser, J. (1973). "Stable and Random Motion in Dynamical Systems." Princeton Univ. Press, Princeton, New Jersey.

Mosteller, F., and Tukey, J. W. (1968). Data analysis, including statistics. In "The Handbook of Social Psychology" (G. Lindzey and E. Aronson, eds.), Vol. II, pp. 80–203. Addison-Wesley, Reading, Massachusetts.

Mosteller, F., and Tukey, J. (1977). In "Data Analysis and Regression," pp. 43–47. Addison-Wesley, Reading, Massachusetts.

Muller, D. E. (1965). A method for solving algebraic equations using an automatic computer, Math. Comp. 10, 208–215.

Nijenhuis, A., and Wilf, H. S. (1978). "Combinatorial Algorithms." Academic Press, New York.

Nitecki, Z. (1971). "Differentiable Dynamics. An Introduction to the Orbit Structure of Diffeomorphisms." M.I.T. Press, Cambridge, Massachusetts.

Olver, F. W. J. (1974). "Introduction to Asymptotics and Special Functions," p. 45. Academic Press, New York.

Pólya, G. (1954). "Induction and Analogy in Mathematics." Princeton Univ. Press, Princeton, New Jersey.

Putter, J. (1955). Treatment of trees in some nonparametric tests, Ann. of Math. Statist. 26, 368–386.

Reingold, E. M., Nievergelt, J., and Deo, N. (1977). "Combinatorial Algorithms: Theory and Practice." Prentice-Hall, Englewood Cliffs, New Jersey.

Schaffer, H. E. (1970). Algorithm 369—Generator of random numbers satisfying the poisson distribution, Comm. ACM. 13 (1), 49.

Shamos, M. I. (1976). Problems in computational geometry, Ph.D. Thesis. Dept. of Computer Science, Yale University, New Haven, Connecticut.

Silverstein, J. W. (1976). Asymptotics applied to a neural network, Biol. Cybernet. 22, 73–84.

Silverstein, J. W. (1977a). Stability analysis and correction of pathological behavior of neural networks, Pattern Analysis Rept. #48. Brown University, Providence, Rhode Island.

Silverstein, J. W. (1977b). On the randomness of eigen-vectors generated from networks with random topologies, Pattern Analysis Rept. #47. Brown University, Providence, Rhode Island.

Silverstein, J. W. (1980). Describing the behavior of eigen-vectors of random matrices using sequences of measures on orthogonal groups, North Carolina State Univ. Report. Raleigh, North Carolina.

Singleton, R. C. (1969). An algorithm for computing the mixed radix fast Fourier transform, IEEE Trans. Audio Electroacoust. AU-17, 93–104.

Sokolnikoff, I. S. (1951). "Tensor Analysis." Wiley, New York.

Stewart, G. W. (1973). "Introduction to Matrix Computation," pp. 139–142. Academic Press, New York.

Tadikamalla, P. (1978a). Computer generation of gamma random variables, Comm. ACM, 21, 419–422.

Tadikamalla, P. (1978b). Computer generation of gamma random variables II, Comm. ACM. 21, 925–928.

Tihonov, A. N. (1963). Solutions of incorrectly formulated problems, *Soviet Math.* **4**, 1035–1038.

Town, D. E. (1978). Restoration analysis of noise disrupted density functions, Pattern Analysis Rept. #63. Brown University, Providence, Rhode Island.

Trantor, C. J. (1971). "Integral Transforms in Mathematical Physics," pp. 61–71. Chapman and Hall, London.

Uspensky, J. V. (1948). "Theory of Equations." McGraw-Hill, New York.

Vitale, R. A. (1973). An asymptotically efficient estimate in time series analysis, *Quart. Appl. Math.* **30**, 421–440.

Warschawski, S. E. (1956). Recent results in numerical methods of conformal mapping, *in* "Proceedings of Symposia in Applied Mathematics," (J. H. Curtiss, ed.), Vol. 6, pp. 219–250. McGraw-Hill, New York.

Wielandt, H. (1956). Error bounds for eigenvalues of symmetric integral equations, *in* "Proceedings of Symposia in Appl. Mathematics," (John H. Curtiss, ed.), Vol. 6, pp. 261–282. McGraw-Hill, New York.

Wilkinson, J. H. (1965). "The Algebraic Eigenvalue Problem." Oxford Univ. Press (Clarendon), London and New York.

Wilkinson, J. H., and Reinsch, C. (1971). "Handbook for Automatic Computation," Vol. II, pp. 339–358. Springer-Verlag, Berlin and New York.

Young, D., and Gregory, R. (1973). "A Survey of Numerical Mathematics," Vol. II, pp. 924–930. Addison-Wesley, Reading, Massachusetts.

Solutions to Exercises

In general there are many solutions to the exercises and we usually give only one, which the reader may be able to improve substantially.

Chapter 9

Section 2

3. $X \div Y + Y = 0$
4. $(!X+1) \times (!Y+1) \div !X+Y+1$
5. $M \geq ?M+N$

Section 3

1. $^{-}1+A+\iota 1+B-A$
2. $(A-K)+K \times \iota N+1$
3. $V \leftarrow M < \iota N$

Another way is to use dyadic "ρ" as in the statement

$V \leftarrow (M\rho 0),(N-M)\rho 1$

4. $(T\rho {}'A{}'),M\rho {}'B{}'$
5. $(3 \times N)\rho {}'ABC{}'$

Section 4

1. $\sim C \in W$
2. $(V[1]\in W)+(V[2]\in W)+(V[3]\in W)+V[4]\in W$

This is clumsy and better ways can be found using tools to be introduced in later chapters.

3. $(1\in V)\vee (2\in V)\vee (3\in V)$

Section 5

2. $('BW')[1+2|8\ 8\ \rho\ (\iota8),9-\iota8]$
4. $5\times M+\lozenge M$

Section 6

2. $A[1;;]\lceil A[2;;]\lceil A[3;;]\lceil A[4;;]$

We shall be able to do this better later on.

3. $A \leftarrow 2\ 2\ 25\ \rho\ 0$
   $A[1;;] \leftarrow ?\ 2\ 25\ \rho\ 10$
   $A[2;;] \leftarrow 10+?2\ 25\ \rho\ 10$

Chapter 10

Section 2

1. $X \perp \div\ !\ N-0,\iota N$
2. $X \leftarrow {}^{-}1+?N\rho 2$

and then

   $.5\ \perp\ X$
3. $X \leftarrow 2\times{}^{-}1+?N\rho 2$

and then

   $(\div 3)\ \perp\ X$

Section 3

1. $1\ 1\ \lozenge\ M$

as in the text, or

   $(,M)[1+0,(N+1)\times\iota N-1]$

2. $(2,N)\rho(\iota N),\phi\iota N$
3. $X \leftarrow \iota N$

and then

   $LOG \leftarrow \circledast X$

and finally

   $\lozenge(2,N)\rho X,LOG$

4. $X \leftarrow \bigcirc(2 \div N) \times \iota N$

and

$COS \leftarrow 2 \bigcirc X$

and

$SIN \leftarrow 1 \bigcirc X$

Now put

$TABLE \leftarrow \lozenge(3,N)\rho X,COS,SIN$

5. $(1\phi V)+(\bar{\ }1\phi V)-2\times V$
7. $(1\phi M)+(\bar{\ }1\phi M)$
     $+(1\phi[1]M)+(\bar{\ }1\phi[1]M)-4\times M$

Section 4

1. $(0 \neq K \mid \iota N) / \iota N$
2. $(0 = 1 \mid X) / X$
3. $((V \geq A) \wedge V \leq B) / V$
4. $F[\iota N],(L\rho 0),F[N+\iota M]$

Can also be done by using expand as

$((N\rho 1),(L\rho 0),M\rho 1)\backslash F$

Section 5

1.   First compute

$ORDER \leftarrow V[\blacktriangle V]$

and then

$ORDER[N-1] \neq ORDER[N]$

This can also be done by dyadic "iota" twice.

2. $(\phi(L\rho 2)\top X)\iota 1$
3. $(\rho CHAR) \geq CHAR \iota \;'T'$

Can also be done using "ϵ."

Section 6

2. $((\iota\rho V) = V\iota V) / V \leftarrow A,B$

Here we have used multiple assignments in a single statement, which of course can be avoided by using two statements instead of one.

3. $(\sim(\iota N)\in A)/\iota N$
4. $A,(\sim B\in A)/B$

   for the union, and

   $(A\in B)/A$

   for the intersection.

6. $V \leftarrow \iota N,$

   then

   $V[I] \leftarrow J$

   and

   $V[J] \leftarrow I$

7. $V1[\spadesuit V2]$

8. $10\ ^-4\ \bar{\varphi}\ M$

   Also print out M as it is to see the difference in format.

9. $'THE\ FIRST\ ROOT\ HAS\ THE\ VALUE\ ',\ \bar{\varphi}\ R1,$
   $'AND\ THE\ SECOND\ ROOT\ HAS\ THE\ VALUE\ ',$
   $\bar{\varphi}\ R2$

Chapter 11

Section 2

1. $(+/\div\iota N)-\circledast N$
2. Calculate first

   $X \leftarrow .05\times\iota 100$

   and

   $V \leftarrow X\times(\star X)\div X+\star 2\times X$

   Then do for example

   $V[V\iota\lceil/V]$

3. The "alternating product"

$$\frac{v_1 v_3 \cdots}{v_2 v_4 \cdots}$$

4. $(\wedge / \wedge / M \geq 0) \wedge \wedge / 1 = + / M$
5. (a) $\wedge / \wedge / M = \lozenge M$
    (b) $\wedge / \wedge / M = - \lozenge M$
    (c) $\wedge / \vee / M = \lceil M$

The last statement could be done with residue too.

6. $GEOMETRIC \leftarrow * (\div \rho V) \times + / \circledast V$

and

$HARMONIC \leftarrow \div (\div \rho V) \times + / \div V$
7. $\wedge / A \in B$
8. $\wedge / (1 \downarrow V) > {}^{-}1 \downarrow V$
9. $\wedge / A [\spadesuit A] = B [\spadesuit B]$

Section 3

1. $1 + + / (+ \backslash TRANS [I ;]) < 1E^{-}6 \times ?1E6$
2. $\times \backslash 1 - (X \div \iota N) * 2$
3. ${}^{-}1 \downarrow V$

Section 4

1. $\phi [1] M$
2. $+ / [2] A$
3. First find the Boolean vector of the nontrivial "ones"

    $NONTRIVIAL \leftarrow \qquad \vee / MV \neq 0$

    Then compress

    $NONTRIVIAL / [1] MV$

Section 5

1. First define a matrix with the v's as columns

    $SUBSPACE \leftarrow S \circ . * RS$

    where

    $S \leftarrow \iota N$

    and with the exponents in the $(R + 1)$-vector

    $RS \leftarrow 0 , \iota R$

    Then "divide quad" gives the projection

$SUBSPACE+.×F⊟SUBSPACE$

(see APL Language (1976), p. 41).

2. $DISTANCE⌊.+DISTANCE$

for one step. Same idea for two steps, and so on.

3.   If X is vector and Q the matrix

$X+.×Q+.×X$

Section 6

1. $+/1\ 1\ ⍉\ M,$

See Exercise 1 in Chapter 3, Section 3, or

$+/+/M×(⍳N)∘.=⍳N$

where we multiply M *elementwise* by the identity matrix and then sum all elements. Here N is the dimension of M.

2.   First find the dimension

$DIM←ρDIAG$

and then form the identity matrix

$I\ ←\ DIM∘.=DIM$

Finally the desired diagonal matrix is obtained as

$(DIAG∘.×DIAGρ1)×I$

3. $(⍳N)∘.>⍳N$

4.   If N is a dimension of M,

$⊟((⍳N)∘.=⍳N)-X×M$

5.   Introduce the matrix

$M←(Nρ1)∘.×C,(N-R)ρ0$

and the circulant

$CIRC\ ←\ (0,-⍳N-1)⌽M$

Then form

$CIRC+.×F$

6. Form the matrix with entries $v_k v_l$ and divided by norm of V. We then get

$$(V \circ . \times V) \div (+ / V \star 2) \star . 5$$

7. To test for commutativity we do

$$\wedge / \wedge / M = \lozenge M$$

(see Exercise 5 in Chapter 4, Section 2). For associativity we need that $(x_i x_j)x_k$ always equals $x_i(x_j x_k)$. But $x_i x_j$ means looking up the (i, j) element in the matrix M. Hence we get the truth value

$$\wedge / \wedge / \wedge / M[M;] = M[;M]$$

8. Form the matrix COS with entries $\cos kx_l$, with the x_k arranged in the vector

$$X \leftarrow \circ (2 \div L) \times 0, \iota L - 1$$

and

$$K \leftarrow 0, \iota N$$

with the matrix

$$COS \leftarrow 2 \circ X \circ . \times K$$

Then the Fourier transform is, with C as the vector of coefficients,

$$COS + . \times C$$

This is all right for small L and N; otherwise the fast Fourier transform must be used.

9. Recall that Vandermonde's determinant can be evaluated as $\Pi_{i>j}(x_i - x_j)$. To compute this product we form the outer product

$$M \leftarrow X \circ . - X$$

where X, of course, is the vector $(x_1, x_2, \ldots, x_n)$. We now add the identity matrix elementwise

$$M + (\iota N) \circ . = \iota N$$

and do $\times / \times /$ over all elements. This gives us the square of the determinant except for the sign, which is easy. This is a bit wasteful and could be improved.

10. To get $\{\min(i, j); i, j = 1, 2, \ldots, n\}$ we write

$$(\iota N) \circ . \lfloor (\iota N) \quad \text{and} \quad (\iota N) \circ . \lceil \iota N$$

for $\{\max(i, j),\ i, j = 1, 2, \ldots, n\}$.

11. First make

$$M00 \leftarrow (N,N)\rho 0$$

then

$$M00[N;N]\leftarrow 1$$

and so on for the other matrices. Then the Laplacian is the sum of four such matrices minus four times the fifth one.

12. Simply do

$$V\circ.=\iota N$$

13. One way of doing this is to use dyadic transposition (see APL Language (1976), pp. 51–52):

$$M \leftarrow L+.\times\ 1\ 2\ 1\ 4\ \lozenge\ O\circ.\times O.$$

Chapter 12

Section 1

1. Something like the following

```
      ∇ RESULT ← A UNION B
[1] ⍝ COMPUTES UNION OF SETS A AND B
[2]   RESULT←((ιρV) = VιV)/V←A,B
      ∇
```

2.
```
      ∇ RESULT ← N POWERSUM P
[1] ⍝ COMPUTES THE SUM OF PTH POWER OF N
    ⍝ FIRST NATURAL NUMBERS
[2]   RESULT ← +/[1](ιN)∘.*ιP
      ∇
```

Section 2

1.
```
      ∇ Z←A F MAX
[1]   I←1
[2]   Z←ι0
[3]   Z←Z,(A×*I)+I*2
[4]   I←I+1
[5]   →((¯1↑Z)<MAX)/3
      ∇
```

This is not very good. A label should have been used in [3] instead of a specific reference number. In this small program it does not matter much, but it is good practice to use labels.

2.
```
      ∇Z←INIT FIBONACCI N
[1]   SERIES←INIT
[2]   LOOP:→(N>ρSERIES←SERIES,+/¯2↑SERIES)
      /LOOP
      ∇
```

3.
```
      ∇Z ← P BERNOULLI SUM
[1]   ⍝COMPUTES SUMS OF BERNOULLI VARIABLES
      ⍝UNTIL 'SUM' IS REACHED
[2]   ⍝PROBABILITY IN BERNOULLI IS LEFT
      ⍝ARGUMENT
[3]   Z ← ⍳0
[4]   Z ← Z, P ≥ 1E¯6×?1E6
[5]   →(SUM >+/Z)/4
      ∇
```

4.
```
      ∇ KSI←EPS FIXPOINT INITIAL; NAME; KSI;
        NEW
[1]   ⍝FINDS THE FIXPOINT KSI OF INCREASING
      ⍝FUNCTION F
[2]   ⍝ON INTERVAL [0,1] WITH F(0)=0,F(1)=1
[3]   ⍝LEFT ARGUMENT IS TOLERANCE,RIGHT
      ⍝ARGUMENT IS INITIAL VALUE
[4]   ⍝F SHOULD HAVE ONE EXPLICIT ARGUMENT
      ⍝AND EXPLICIT RESULT
[5]   KSI←INITIAL
[6]   'WHAT IS NAME OF FUNCTION?'
[7]   NAME ← ⍞
[8]   LOOP:KSINEW ← ⍎ NAME,' ',⍕ KSI
[9]   →(EPS > |KSI-KSINEW)/0
[10]  KSI←KSINEW
[11]  →LOOP
      ∇
```

The statement in [7] uses "quote quad," the "quad" overstruck with a "quote." This prepares for accepting character data.

5.
```
      ∇ MATRIX1 ← MATRIXNORM MATRIX
[1]   MATRIX1 ← MATRIX
[2]   LOOP: MATRIX1 ← MATRIX+.×MATRIX1
[3]   →(TEST MATRIX1)/0
```

```
[4]    →LOOP
       ∇
```

This function calls another function for testing whether the sum is smaller or equal to ϵ:

```
       ∇ ANSWER ← TEST M
[1]  ANSWER ← EPS ≥ +/+/|M
       ∇
```

Section 4

1.
```
       ∇ U←B POISSON PAR;POINTS;R;FI;COS
[1]  ⍝COMPUTES POISSON FORMULA FOR CIRCLE
[2]  ⍝B IS N-VECTOR OF BOUNDARY VALUES
[3]  ⍝PAR IS TWO VECTOR (R,FI) IN POLAR
     ⍝COORDINATES
[4]  N←ρB
[5]  POINTS ← ○(2÷N)×⍳N
[6]  R←PAR[1]
[7]  FI←PAR[2]
[8]  COS←2○ FI-POINTS
[9]  U←(÷N)×+/(1-R*2)×B÷1+(-2×R×COS)+R*2
       ∇
```

2. (a)   To test whether MAP is surjective we can write, for example

```
       ∇ ANSWER ← MAP SURJECTIVE M
[1]  ⍝COMPUTES TRUTHVALUE OF THE MAP-
     ⍝FUNCTION (VECTOR) TO BE SURJECTIVE
[2]  ⍝FROM SET (1,2,...N) TO (1,2,...M)
[3]   ANSWER←∧/(⍳M)∈MAP
       ∇
```

(b)   For injective we can do

```
       ∇ ANSWER←MAP INJECTIVE M
[1]  ⍝COMPUTES TRUTHVALUE OF THE MAP-
     ⍝FUNCTION (VECTOR) TO BE INJECTIVE
[2]  ⍝FROM SET (1,2,...N) TO (1,2,...M)
[3]  N←ρMAP
[4]   ANSWER←∧/(N-1)=+/MAP∘.≠MAP
       ∇
```

If the amusing outer product in [4] takes too much storage use a looping program.

(c)   To test for bijectivity we just combine the two given functions.

(d)   To get the inverse we define

```
     ∇ RESULT←MAP INVERSE ARGUMENT
[1]  ⍝COMPUTES INVERSE OF SURJECTIVE MAP-
[2]  ⍝FUNCTION(VECTOR) FROM SET (1,2,...N)
     ⍝TO (1,2,...M)
[3]  ⍝EXPLICIT RESULT IS SET, POSSIBLY EMPTY
[4]  RESULT ← (MAP=ARGUMENT)/⍳⍴MAP
     ∇
```

3.   See program library in Part IV: function GRAMS employs idea that can be used.

4.   See Example 1 in Chapter 4, Section 5.

5.   (a)   ∇ ANSWER ← SUPERHARMONIC F;F1
```
[1]  N←(⍴F)[1]
[2]  F1←(1⌽F)+(⁻1⌽F)+(1⌽[1]F)+(⁻1⌽[1]F)-4×F
[3]  F1←F1[1+⍳N-2;1+⍳N-2]
[4]  ANSWER ← ∧/∧/F1≥0
     ∇
```

(b)   Same but with equality in statement [4]. Here we have not treated the "function" F (really a matrix) as periodic, which motivates the truncation in statement [3].

6.   ∇ RESULT ← PERMUTE1 N;MAT
```
[1]  ⍝COMPUTES ALL PERMUTATIONS OF ORDER N
[2]  ⍝RESULT IS STORED AS MATRIX
[3]  RESULT ← 1 1⍴1
[4]  →(N=1)/0
[5]  MAT←(PERMUTE1 N-1),N
[6]  RESULT←MAT
[7]  LOOP:MAT[;N,N-1]→MAT[;(N-1),N]
[8]  RESULT←RESULT,[1]MAT
[9]  N←N-1
[10] →LOOP×N≠1
     ∇
```

Another more sophisticated way of generating all the permutations is given by

```
     ∇ RESULT←PERMUTE2 N;PERM;M
[1]  ⍝COMPUTES ALL PERMUTATIONS OVER N
     ⍝OBJECTS
[2]  ⍝RESULT IS IN MATRIX FORM
```

```
[3]      →(ιN=RESULT←1 1ρ1)/0
[4]      PERM←PERMUTE2 N-1
[5]      M←0
[6]      RESULT←ιM
[7]      LOOP:→(ιN<M←M+1)/0
[8]      L←(~(ιN)∈M)\PERM
[9]      L[;M]←N
[10]     RESULT←((M×!N-1),N)ρ(,RESULT),,L
[11]     →LOOP
         ∇
```

Of course, these functions can be used only for small values of N.

7. Use previous function on $N-\rho V$ objects, then rearrange if needed.
8. We have

$$c_k = \sum_{l=0}^{k} a_l b_{k-l}, \qquad k = 0, 1, \ldots, m + n$$

so that

$$c_0 = a_0 b_0$$
$$c_1 = a_0 b_1 + a_1 b_0$$
$$c_2 = a_0 b_2 + a_1 b_1 + a_2 b_0$$
$$\vdots$$
$$c_{m+n} = a_n b_m$$

This leads to the function definition

```
         ∇  C ← A MULTIPLY B;NA;NB;A1;B1
[1]      NA ← ρA
[2]      NB ← ρB
[3]      A1 ← (NBρ0),A
[4]      B1 ← (ϕB),NAρ0
[5]      C ← (¯1+NA+NB)ρ0
[6]      I ← 1
[7]      LOOP:C[I] ← +/(IϕA1)×B1
[8]      I ← I+1
[9]      →(I ≤ NA+NB-1)/LOOP
         ∇
```

This program is not very efficient.

9. If the two complex vectors are given by the two matrices

$$Z1 = \begin{pmatrix} a_1 & a_2 & \cdots & a_r \\ b_1 & b_2 & \cdots & b_r \end{pmatrix} \qquad \text{and} \qquad Z2 = \begin{pmatrix} u_1 & u_2 & \cdots & u_r \\ v_1 & v_2 & \cdots & v_r \end{pmatrix}$$

we should compute

$$Z = \begin{pmatrix} a_1 u_1 - b_1 v_1 & a_2 u_2 - b_2 v_2 & \cdots & a_r u_r - b_r v_r \\ a_1 v_1 + b_1 u_1 & a_2 v_2 + b_2 u_2 & \cdots & a_r v_r + b_r u_r \end{pmatrix}$$

This is easy:

```
        ∇ Z ← Z1 MULT Z2
[1]  ⍝COMPUTES PRODUCT ELEMENTWISE OF TWO
[2]  ⍝COMPLEX MATRICES GIVEN AS TWO-ROW
     ⍝MATRICES
[3]  ⍝FIRST ROW CONTAINS REAL PARTS, SECOND
     ⍝IMAGINARY PARTS
[4]  R ← (ρZ1)[2]
[5]  Z ← (2,R)ρ0
[6]  Z[1;]←(Z1[1;]×Z2[1;])-Z1[2;]×Z2[2;]
[7]  Z[2;]←(Z1[1;]×Z2[2;])+Z1[2;]×Z2[1;]
        ∇
```

WARNING

Arguments must be matrices with two rows also when $r = 1$!

To get the powers we loop

```
        ∇ RESULT ← K POWER Z;R;I
[1]   ⍝RESULT IS INTEGRAL K'TH POWER OF Z
[2]   ⍝RESULT AND Z ARE GIVEN AS 2×R
[3]   ⍝MATRICES, REAL PARTS IN FIRST ROW
[4]    R←(ρZ)[2]
[5]    RESULT←(2,R)ρ(Rρ1),Rρ0
[6]    I←0
[7]    →(K=0)/0
[8]  LOOP:RESULT← Z MULT RESULT
[9]    I←I+1
[10]   →(I<K)/LOOP
        ∇
```

WARNING

Argument Z must be a matrix with two rows also when $r = 1$!

10.
```
        ∇ RESULT←C TAYLOR Z;R;I;N
[1]   ⍝COMPUTES FINITE POWER SERIES WITH
[2]   ⍝COEFFICIENTS IN 2×(N+1) MATRIX C
```

```
[3]    ⍝ARGUMENT Z IS COMPLEX VECTOR GIVEN
       ⍝AS 2×R MATRIX
[4]    N←¯1+(⍴C)[2]
[5]    R ← (⍴Z)[2]
[6]    I←2
[7]    RESULT ← C[;1]∘.×R⍴1
[8]    →(N=0)/0
[9]    LOOP: RESULT←RESULT+(C[;I]∘.×R⍴1)
       MULT (I-1) POWER Z
[10]   I←I+1
[11]   →(I≤N+1)/LOOP
       ∇
```

11.
```
       ∇ INTEGRAL ← ENDPTS TRAPEZE N;A;B;D
[1]    ⍝COMPUTES TRAPEZOIDAL RULE OVER(A,B)
       ⍝WITH N INTERVALS
[2]    ⍝ENDPTS IS TWO-VECTOR (A,B)
[3]    A ← ENDPTS[1]
[4]    B ← ENDPTS[2]
[5]    D ← (B-A)÷N
[6]    INTEGRAL←D×+/(.5,((N-1)⍴1),.5)×F A,
       A+D×⍳N
       ∇
```

This would be easy to extend to Simpson's rule and similar quadrature formulas. More challenging would be to write a program for
Gaussian quadrature.

12.
```
       ∇ K ← IJ CODE SHAPE ;V
[1]    ⍝COMPUTES NATURAL NUMBERING IN MATRIX
       ⍝OF SIZE SHAPE
[2]    ⍝IJ IS TWO-VECTOR = (I,J)
[3]    V ← ⍳ SHAPE[1]×SHAPE[2]
[4]    K ← (SHAPE ⍴V)[IJ[1];IJ[2]]
       ∇
```

In [3] we have introduced the vector V of numbers used for K; they
are rearranged in matrix form in [4].

```
       ∇ IJ ← K DECODE SHAPE;X
[1]    ⍝COMPUTES INVERSE FUNCTION FOR CODE
[2]    IJ ← 2⍴0
[3]    IJ[1] ← ⌈ K ÷ SHAPE[2]
```

```
[4]    IJ[2]←(SHAPE[2]×X=0)+X←SHAPE[2]|K
       ∇
```

13. The following is clumsy but works. It tests that each row and column is a permutation of ιN.

```
       ∇ ANSWER GROUP MTABLE ;M
[1]    M←(ρ MTABLE)[1]
[2]    I←1
[3]    ANSWER←1
[4]    LOOP:ANSWER←ANSWER ∧∧/(ιN) = MTABLE
       [I;⍋MTABLE[I;]]
[5]    ANSWER←ANSWER∧∧/(ιN) = MTABLE[⍋MTABLE
       [;I];I]
[6]    I←I+1
[7]    →(I≤M)/LOOP
       ∇
```

This can also be used to test whether a subset is a subgroup. One could use the same idea as in GROUP to test for a subgroup being normal.

14. We shall call PERMUTE1 and then fill in the $l = (n!)^2$ entries in MTABLE using a double loop with the row counter I and column counter J.

```
       ∇ MTABLEPERM N;L;PERM;I;J
[1]    L←!N
[2]    PERM←PERMUTE1 N
[3]    MTABLE←(L,L)ρ0
[4]    I←J←1
[5]    LOOP:MTABLE[I;J]←(∧/PERM=(Lρ1)∘.×PERM
       [I;PERM[J;]])/ιL
[6]    I←I+1
[7]    →(I≤L)/LOOP
[8]    I←1
[9]    J←J+1
[10]   →(J≤L)/LOOP
       ∇
```

15. Assuming that the order m of the group is already defined as a global variable we write

```
       ∇ Z←DIVIDE X
[1]    Z←(MTABLE[X;]=1)+.×ιM
       ∇
```

16.   The P and Q vectors representing the two measures must of course be of equal length. We get the program simply as

```
     ∇ R←P CONVOLVE Q;G;M;GINV
[1]  M←ρP
[2]  G←ιM
[3]  GINV←DIVIDE G
[4]  R←Q[MTABLE[GINV;]]+.×P
     ∇
```

More efficient algorithms would apply the FFT idea.

17.   Let the subgroup be named N in the APL code. It is a set containing the element 1. We now just generate one coset after the other, looking for remaining elements not yet accounted for by a compression statement

```
     ∇ COSETS ← M COSET N;R;S;SET
[1]  R←ρN
[2]  S←M÷R
[3]  COSETS←(S,R)ρ0
[4]  COSETS[1;]←N
[5]  I←2
[6]  REMAIN←(~(ιM)∈N)/ιM
[7]  LOOP:SET←MTABLE[REMAIN[1];N]
[8]  COSETS[I;]←SET
[9]  REMAIN←(~SET∈REMAIN)/REMAIN
[10] I←I+1
[11] →(I≤S)/LOOP
     ∇
```

18.   Just find, for given group element x, the coset number = row number in COSETS for the row that contains x. Use compression.

```
     ∇ NUMBER←ISOMORPHISM X
[1]  NUMBER←(∨/X∈COSETS)/ιS
     ∇
```

The multiplication table for F can be computed similarly. With the same ordering of cosets as above:

```
∇ RESULT M FACTORGROUP F
X←COSETS[1;]
RESULT←XιMTABLE[X;X]
∇
```

The reader may be interested in finding what happens with convolution power

$$P^{\nu*} = P \text{ CONVOLVE } P \text{ CONVOLVE } P \cdots P$$

(ν occurrences of P) as ν tends to infinity. The most convenient way would be to write a looping program doing

$$R \leftarrow R \text{ CONVOLVE } P$$

repeatedly or (a faster way) using successive "convolution squaring"

$$R \leftarrow R \text{ CONVOLVE } R$$

Index

Pure and Applied Mathematics

A Series of Monographs and Textbooks

Editors **Samuel Eilenberg and Hyman Bass**

Columbia University, New York

RECENT TITLES

CARL L. DeVITO. Functional Analysis

MICHIEL HAZEWINKEL. Formal Groups and Applications

SIGURDUR HELGASON. Differential Geometry, Lie Groups, and Symmetric Spaces

ROBERT B. BURCKEL. An Introduction to Classical Complex Analysis: Volume 1

JOSEPH J. ROTMAN. An Introduction to Homological Algebra

C. TRUESDELL AND R. G. MUNCASTER. Fundamentals of Maxwell's Kinetic Theory of a Simple Monatomic Gas: Treated as a Branch of Rational Mechanics

BARRY SIMON. Functional Integration and Quantum Physics

GRZEGORZ ROZENBERG AND ARTO SALOMAA. The Mathematical Theory of L Systems.

DAVID KINDERLEHRER and GUIDO STAMPACCHIA. An Introduction to Variational Inequalities and Their Applications.

H. SEIFERT AND W. THRELFALL. A Textbook of Topology; H. SEIFERT. Topology of 3-Dimensional Fibered Spaces

LOUIS HALLE ROWEN. Polynominal Identities in Ring Theory

DONALD W. KAHN. Introduction to Global Analysis

DRAGOS M. CVETKOVIC, MICHAEL DOOB, AND HORST SACHS. Spectra of Graphs

ROBERT M. YOUNG. An Introduction to Nonharmonic Fourier Series

MICHAEL C. IRWIN. Smooth Dynamical Systems

JOHN B. GARNETT. Bounded Analytic Functions

EDUARD PRUGOVEČKI. Quantum Mechanics in Hilbert Space, Second Edition

M. SCOTT OSBORNE AND GARTH WARNER. The Theory of Eisenstein Systems

K. A. ZHEVLAKOV, A. M. SLIN'KO, I. P. SHESTAKOV, AND A. I. SHIRSHOV. Translated by HARRY SMITH. Rings That Are Nearly Associative

JEAN DIEUDONNÉ. A Panorama of Pure Mathematics; Translated by I. Macdonald

JOSEPH G. ROSENSTEIN. Linear Orderings

AVRAHAM FEINTUCH AND RICHARD SAEKS. System Theory: A Hilbert Space Approach

ULF GRENANDER. Mathematical Experiments on the Computer

IN PREPARATION

ROBERT B. BURCKEL. An Introduction to Classical Complex Analysis: Volume 2

HOWARD OSBORN. Vector Bundles: Volume 1, Foundations and Stiefel-Whitney Classes

RICHARD V. KADISON AND JOHN R. RINGROSE. Fundamentals of the Theory of Operator Algebras

BARRETT O'NEILL. Semi-Riemannian Geometry: With Applications to Relativity

EDWARD B. MANOUKIAN. Renormalization

E. J. McSHANE. Unified Integration

A. P. MORSE. A Theory of Sets, Revised and Enlarged Edition